TWO-PHASE EMISSION DETECTORS

TWO-PHASE EMISSION DETECTORS

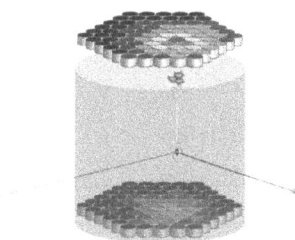

Dmitry Yu Akimov
National Research Nuclear University MEPhI (Moscow Engineering Physics Institute), Russia

Alexander I Bolozdynya
National Research Nuclear University MEPhI (Moscow Engineering Physics Institute), Russia

Alexey F Buzulutskov
Budker Institute of Nuclear Physics, Russia
Novosibirsk State University, Russia

Vitaly Chepel
University of Coimbra, Portugal
Laboratory of Instrumentation and Experimental Particle Physics (LIP), Portugal

World Scientific

NEW JERSEY · LONDON · SINGAPORE · BEIJING · SHANGHAI · HONG KONG · TAIPEI · CHENNAI · TOKYO

Published by

World Scientific Publishing Co. Pte. Ltd.

5 Toh Tuck Link, Singapore 596224

USA office: 27 Warren Street, Suite 401-402, Hackensack, NJ 07601

UK office: 57 Shelton Street, Covent Garden, London WC2H 9HE

Library of Congress Control Number: 2021942570

British Library Cataloguing-in-Publication Data
A catalogue record for this book is available from the British Library.

ISBN 978-981-123-108-7 (hardcover)
ISBN 978-981-123-109-4 (ebook for institutions)
ISBN 978-981-123-110-0 (ebook for individuals)

For any available supplementary material, please visit
https://www.worldscientific.com/worldscibooks/10.1142/12126#t=suppl

Typeset by Stallion Press
Email: enquiries@stallionpress.com

This monograph is dedicated to the memory of

Boris Anatolyevich Dolgoshein

who made a significant contribution to the development of experimental methods for detecting elementary particles, including the development of track spark and streamer chambers, solid state silicon photomultipliers, and, which is especially important for the topic of this book, two-phase emission detectors.

Preface

This book is devoted to the technology of two-phase emission detectors and the features of their application for solving problems of fundamental importance for constructing a physical picture of the world in which we live. Among all the possible working media for emission detectors, special attention is paid to condensed noble gases, in which quasi-free electrons resulting from ionization of the medium by charged particles can exist for quite a long time (tens of milliseconds), and can also be pulled by fairly moderate electric fields into equilibrium gas phase. Those working media are excellent scintillators, and in their equilibrium gas phases during the drift of electrons heated by an electric field, intense electroluminescence is possible as well as the development of avalanche electron signal multiplication. It is also important to note that noble gases are available in large quantities as a by-product of the production of oxygen from the atmosphere for the metallurgical industry. It should be noted that argon is the third most abundant component of the atmosphere, followed by nitrogen and oxygen. The global production of xenon, the most promising material for detectors, is currently about 70 tons per year, which is significantly more than the global production of the most popular detector materials based on crystalline sodium iodide and cesium iodide. It is also very important that noble gases, due to their chemical inertness, can be relatively easily purified from impurities, which is an indispensable condition for creating large mass detectors for recording processes with extremely small cross-sections. The most suitable for detecting penetrating radiation are heavy noble gases such as argon, krypton and xenon. As shown in this monograph, all of those make such media the most attractive for supporting research in modern experimental nuclear physics and elementary particle physics.

The monograph summarizes the authors' many years of experience on the development of technology for two-phase emission detectors and their applications. This work was carried out in the framework of cooperation with many national and international organizations and collectives. The authors would like to express their special gratitude to the International Atomic Energy Agency (IAEA) for the indicated interest to the topic covered by this book, to the State Corporation of the Russian Federation ROSATOM, the Russian Foundation for Basic Research, the administration of National Research Nuclear University MEPhI, Institute for Theoretical and Experimental Physics of National Research Center "Kurchatov Institute", Institute of Nuclear Physics named after G.I. Budker SB RAS and University of Coimbra for support in development of technology of emission two-phase detectors and all collaborators, with whom we had a privilege to work together in XENON10, ZEPLIN-III, LUX, LZ, DarkSide, CDMS, RED-100 collaborations. Also the authors appreciate granting agencies, supporting their research in the field of R&D of two-phase emission detectors and their applications to experimental physics in particularly the Russian Science Foundation (grant #18-12-00135).

Contents

Introduction

One of the rapidly developing areas of modern experimental nuclear physics is non-accelerator experiments using low-background detectors. Such experiments, as a rule, are aimed at solving problems that are of fundamental importance for understanding the structure of the Universe, checking the Standard Model of elementary particles, and looking for new physics behind the observable world. The most interesting tasks include the search for dark matter in the form of new weakly interacting particles, the search for double neutrinoless beta decay, the determination of the magnetic moment of the neutrino, the study of neutrino oscillation and new types of interaction of elementary particles, such as coherent neutrino scattering off heavy nuclei.

All these processes, occurring with extremely low cross-sections, require the development of efficient large-mass detectors capable of detecting small energy releases down to individual ionization electrons. An effective method to do this is the emission method of detecting ionizing particles in two-phase media, which has been proposed at MEPhI 50 years ago. The beginning of this kind of work was initiated by the research in the laboratory headed by Boris A. Dolgoshein of the properties of condensed noble gases in order to develop a tracking streamer chamber with a high-density working medium. In the course of these works, the emission of ionization electrons from liquid argon in fairly moderate electric fields was discovered, and it was shown that ionizing particles can be detected

by pulling the ionization electrons out of the condensed noble gases
and saturated hydrocarbons into the equilibrium gas phase. As
a result of a series of experimental works that followed, it was
demonstrated that emission detectors can combine the high detection
efficiency of ionizing particles inherent for liquid or solid working
media with the possibility of physically amplifying the ionization
signal, as successfully can be done in gas detectors (Rodionov,
1969; Dolgoshein *et al.*, 1970). Soon after that an emission streamer
chamber for visualization of high-energy particle tracks in condensed
media was constructed (Bolozdynya *et al.*, 1977a). However, after
successful testing of the emission streamer chamber at the ITEP
proton synchrotron, it was realized that emission streamer chambers
due to long drift time of electrons in condensed working medium are
too slow devices to be used at modern high-luminosity accelerators.
After that, the emission method has been quite successfully studied
for visualization of gamma radiation fields in nuclear medicine
(Egorov *et al.*, 1983).

With the development of non-accelerator physics and the successful completion of a series of low-background experiments to search for
rare nuclear decays and exotic particles, it became clear that emission
detectors can bring significant advantages to these studies due to the
possibility of constructing large mass "wall-less" detectors with a low
registration threshold (Bolozdynya *et al.*, 1995; Bolozdynya, 1999). In
a relatively short time that has elapsed since then, emission detectors
have occupied a unique niche in the arsenal of advanced methods of
modern experimental and applied physics as detectors for cold dark
matter in the form of massive weakly ionizing particles (WIMPs). It
is significant that during the life of one generation of researchers, the
development of technology of emission detectors has gone away from
miniature devices for methodological studies with a working volume
of about 1 cm^3 to physical installations with a working volume of
several cubic meters.

Liquid xenon, the most commonly used in emission detectors
until recently, has not only excellent protective properties against
natural gamma background, but also against neutrons. Neutron
background can be suppressed by identifying such events by the

multiple elastic scattering of neutrons in liquid xenon. However, a noticeable efficiency is possible only for detectors with a mass of about one ton or more. In particular, for this reason, in recent years, projects are underway to create liquid-xenon two-phase emission detectors for underground low-background detectors weighing up to 40 tons (Aalberts *et al.*, 2016). For registration of high-energy particles, liquid argon is more preferable, and now there are projects for development of liquid argon two-phase emission detectors with a working medium mass of up to 10 kilotons (Cuesta *et al.*, 2019).

Two-phase emission detectors based on pure noble gases as a working medium have noticeable advantages:

(1) Detection of the signal from one event in two channels — ionization and scintillation;
(2) High sensitivity to weak ionization signals down to single electrons;
(3) Practically unlimited size and mass of the working medium.

The use of all these advantages allows creating a "wall-less" detector that is record-breaking in sensitivity to rare and slightly ionizing particles. The principle of operation of the emission: "wall-less" detector is as follows:

(1) The registered radiation interacts with the condensed working medium of the detector, excites and ionizes the atoms, as a result of which the first signal *S1* appears (Fig. 1.1), manifested in the form of scintillation of condensed noble gases. This signal can be used as a trigger for retrieving coordinate information in the TPC mode.
(2) Due to the applied electric field, the ionization electrons drift to the free surface of the condensed detector working medium and exit (emit) into the equilibrium gas phase or vacuum, where they generate a second, amplified signal proportional to the number of ionization electrons *S2* (Fig. 1.1). The amplification of this second signal can be achieved in different ways: electroluminescence of a noble gas, avalanche multiplication of electrons in a rarefied gas phase, acceleration of electrons in a vacuum, etc.

The sensor matrix is used to register the two-dimensional distribution of secondary particles and determine the coordinates of the primary interaction over the plane made up by the input windows of the top sensor matrix (for example, an array of photomultipliers). Since the second signal *S2* is delayed relative to the first *S1* at the time of the ionization electron drift, the third coordinate of the initial point of interaction is determined by the delay time between the first and the second signals.

(3) Knowledge of the position of the interaction point in the three-dimensional space is used to select events occurring in sensitive volume *A* (Fig. 1). For sufficiently large and massive detectors constructed of materials with high absorptivity for nuclear radiation (liquid xenon, for example), the volume *B*, surrounding sensitive volume *A*, plays the role of a shielding from radioactive radiation of surrounding materials. In active protection mode, layer *B* is used for sifting out events in volume*A*, which are correlated in time with events in volume *B*. This makes it possible to exclude events that occurred as a result of multiple scatterings of background particles such as neutrons. An additional powerful method for eliminating background events is the identification of the nature of the detected particle by the ration of the energy values spent for ionization and for excitation of the detector working medium (respectively, signals *S1* and *S2* in Fig. 1).

Electron background discrimination against *S2/S1* signals was studied by several groups of *XMASS, XENON, LUX* collaborations, and was successfully used in data analysis in dark matter searches by the *ZEPLIN, XENON, LUX* collaborations. Dense nuclear recoil tracks produce a smaller *S2/S1* value than less dense electron tracks.

Today two-phase emission detectors found the best application in the most sensitive at the moment experiments searching for cold dark matter in the form of weakly interacting massive particles (WIMPs). A number of successful experiments arranged by ZEPLIN, XENON, LUX and PandaX collaborations with LXe emission detectors during 10 years period reduced allowed region of existence for WIMPs with mass of 40–50 GeV/c^2 from $8.8 \cdot 10^{-44}$cm^2 (reported by XENON-10 collaboration in 2006) down to $1.1 \cdot 10^{-46}$cm^2 (reported by LUX collaboration at the end of 2016). Detector LZ of the second generation (G2) is installed at Davis' cage of the Homestake mine by joint collaboration of former LUX and ZEPLIN collaborations and will use 6 ton LXe active mass emission detector in order to reach sensitivity below 10^{-47} cm^2 for spin-independent WIMP-nucleon interactions. With the increasing detector mass and their sensitivity, solar neutrino interactions become an irreducible source of background for WIMP search experiments. Multi-ton active mass WIMP detectors of the upcoming G3 generation shall become, even with naturally occurring isotope abundances, sensitive to double-beta decay at the modern level of sensitivity and solar neutrino interactions via elastic coherent scattering off of xenon nuclei. Detectors of G3 generation such as DarkSide-20k can achieve spin-independent cross-sections for WIMPs as low as $\sim 7.4 \cdot 10^{-48}$ cm^2 ($6.9 \cdot 10^{-47}$ cm^2) for WIMPs of 1 TeV/c^2 (10 TeV/c^2) mass (Aalseth *et al.*, 2018).

The RED-100 detector recently constructed at NRNU MEPhI will be used for investigation of recently discovered elastic coherent electron neutrino scattering off heavy nuclei. The detector can be installed practically on the Earth's surface in vicinity to low energy neutrino sources such as NPP nuclear reactors or accelerators such as the Spallation Neutron Source (Akimov *et al.*, 2017b).

The Deep Underground Neutrino Experiment (DUNE) is a dual-site experiment for long baseline neutrino oscillation studies, and for neutrino astrophysics and nucleon decay searches. DUNE will comprise four 10 kiloton liquid argon time-projection-chamber (LAr TPC) modules placed at the Sanford Underground Research Facility (South Dakota, USA). One of these modules will profit from the emission detector technology where the charge is extracted, amplified, and detected in gaseous argon above the liquid surface allowing a fine readout pitch, a low energy threshold, and good pattern reconstruction of the events. To gain experience in building and operating such a large-scale emission LAr detector, a dual-phase prototype ProtoDUNE-DP of a $6 \times 6 \times 6$ m^3 working volume is currently being assembled at the CERN Neutrino Platform (Cuesta *et al.*, 2019).

Thus, the detector technology invented at MEPhI 50 years has demonstrated a great potential to be used in a variety of fundamental research programs. This monograph considers the technology's basic features taking into account new developments introduced into experimental practice in the last ten years after the publication of the first monograph exclusively devoted to two-phase emission detectors (Bolozdynya, 2010).

Chapter 1

Historical Review of Development of Two-Phase Emission Detectors

1.1. Introduction

An electron emission from dense media has been used to detect radiation since Heinrich Rudolf Hertz discovered the external photoelectric effect in 1887. In 1888–1890, the photoelectric effect was systematically studied by the Russian physicist Alexander G. Stoletov, who published six works on this subject. In 1898, Joseph John Thomson experimentally established that the flow of electric charge emerging from a metal as a result of an external photoelectric effect is a stream of electrons recently discovered by him. Therefore, an increase in the photocurrent with an increase in illumination should be understood as an increase in the number of knocked out electrons. At present, by the external photoelectric effect we mean the emission of hot electrons from photocathodes, the thickness of which is less than the electron path to thermalization, that is about 20–30 nm for solid photocathode materials.

The external photoelectric effect is widely used in one of the most popular tools of experimental physics — vacuum photo-electron multipliers (PMTs) — for detecting photons. It should be noted that at the dawn of the development of PMT technology, Radio Corporation of America (RCA), the first company to master the industrial production of PMTs, also considered gas filled photomultipliers with gas gain for detecting photons in the visible and infrared regions (Zvorykin and Ramberg, 1949).

The emission of quasi-free electrons from massive samples of condensed noble gas (solid argon) was first observed by a young English physicist Hutchinson (1948). However, neither he nor his contemporaries at that time realized the new possibilities that opened up for particle registration when using detectors with a two-phase working medium.

1.2. The Birth of the Idea of Two-phase Emission Detectors

The idea of using electron emission from condensed dielectrics to detect particles is the brainchild of track detectors for high-energy physics. With the rapid development of high-energy physics in the 50–60s of the last century, the task of developing a controllable by external counters detector technology for recording tracks of high-energy particles has become particularly urgent. Improving the technology of projection spark chambers has led to the development of a streamer chamber in which a spark discharge caused by ionization electrons under the influence of a nanosecond high-voltage pulse breaks off at the streamer stage of its development. Streamers are formed in the direction of the electric field, starting from the initial ionization electrons along the track of the ionizing particle and breaking off at a length of several mm. The streamer chain is photographed and after processing the film, the coordinates of the track of the ionizing particle are determined from it.

The first operable streamer cameras were created in the USSR in 1963 by George E. Chikovani with colleagues at the Institute of Physics of the Academy of Sciences of the Georgian SSR and, independently, by Boris A. Dolgoshein with colleagues at the Moscow Engineering Physics Institute. In terms of image contrast and resolution, streamer cameras are inferior to bubble cameras, however, controllability by an external trigger allows them to be used to study processes that have low probabilities. Gases He, H_2, mixtures of $Ne + He$, $He + CH_4$, $D_2 + CH_4$ at a pressure of 1 bar are normally used as the working gas of streamer chambers. The coordinate resolution of the streamer camera is determined by the dimensions of the streamers, which, as a rule, have diameters of $\sim 1\,\mathrm{mm}$ and lengths of $\sim 5\,\mathrm{mm}$, with a density of distribution along the track to be ~ 10–$12\,\mathrm{cm}^{-1}$ (Dajon *et al.*, 1967).

In 1968, Boris A. Dolgoshein and his team at MEPhI built a record-sized streamer chamber with a working volume of $8 \times 2 \times 1\,\mathrm{m}^3$ to search for W-bosons. The camera worked for many years at the U-70 proton synchrotron in Protvino. Using this camera, fast muons were first identified and their spectrum and polarization were measured. Unfortunately, the W-bosons were not discovered with this remarkable device for the reason that the energy of the U-70 (the highest energy accelerator at the time when it was launched in 1967) was not enough to produce particles whose mass turned out to be $\sim 80\,\mathrm{GeV/s}^2$. The W-boson was discovered in 1983 at CERN, for which Carlo Rubbia and Simon van der Meer were awarded the Nobel Prize in Physics in 1984 that happened in the record short time after the discovery.

The successful development of the streamer method for detecting tracks of relativistic particles in noble gases was awarded the Lenin Prize in 1970 with the wording "for creating a new type of track detector capable of detecting complex events in the interaction" of elementary particles. An attempt to develop a controlled track chamber with a working medium denser than gas at room temperature and a pressure of 1 bar, prompted the idea to try to increase the density either by lowering the temperature of the gas working medium, or by using liquefied noble gases instead of gas under normal conditions. In the first approach, experiments were conducted at ITEP in the laboratory of Valentin A. Lyubimov, where Igor V. Sidorov built a cryogenic streamer chamber filled with helium or a neon–helium mixture (70% Ne + 30% He), or hydrogen at temperatures of 80–100 K (Gorodkov *et al.*, 1974; Sidorov *et al.*, 1975). In these experiments, a surprising phenomenon was discovered: the visualization voltage (the threshold for the development of streamers) practically did not change with a decrease in temperature and an increase in the density of the gas medium, but the tracks became more compact, dense and bright. However, the cryogenic streamer camera turned out to be a rather cumbersome device, which proved to be difficult to fit into the configuration of modern accelerator experiments with detector geometry close to 4π, and therefore this technology has not been further developed.

About that time, Boris A. Dolgoshein *et al.* at MEPhI studied liquid argon as a possible working medium for a liquid streamer chamber. They did not succeed in providing a streamer discharge in liquid argon, but they have shown that the electrons resulting from the ionization of liquid argon by high-energy particles can be pulled out from the liquid by a fairly moderate ($\sim 1\,\mathrm{kV/cm}$) electric field into the equilibrium gas phase, where they can be detected using well-developed methods for obtaining track information in a gaseous medium (Rodionov, 1969).

1.3. Emission Spark Chamber

The first working emission detector was a liquid argon spark emission chamber (Dolgoshein *et al.*, 1970). It was a two-electrode plane-parallel chamber with 1.6 cm gap between the electrodes, with a disk-shaped cathode immersed in liquid argon, in the center of which an alpha source with a diameter of 3.5 cm was installed, and with a multi-wire anode (0.2 mm wire diameter made of Nichrome with a pitch of 0.6 mm) installed in the equilibrium gas phase over 1.4 cm thick layer of liquid argon. The gas phase consisted of a mixture of 50% Ar + 50% Ne. Electrons arising from the ionization of liquid argon were drawn into the gas phase by an electric field of 3 kV/cm. Using a pulse voltage generator, a voltage pulse with an amplitude of 40 kV and a duration of 100 ns was applied to the anode. If at the moment of the pulse supply near some wire there were electrons drawn by a constant electric field from the liquid, then a spark discharge developed near the wire. An optically transparent window was installed above the anode and spark discharges near the wires were recorded using a camera. The image of the 2D distribution of the activity of the alpha source was formed as a result of the superposition of images of many spark discharges at the anode wires.

This development did not find practical applications, but had shown that there is a chance to create a two-phase streamer emission chamber.

1.4. Emission Streamer Chamber

To implement the idea of a two-phase emission streamer camera, Boris U. Rodionov *et al.* at MEPhI built a special detector. Solid krypton with a diameter of 12.5 cm and a thickness of 5 mm at a temperature of 78 K was used as a working medium in this detector. A gas gap of 1 cm wide filled with neon at a pressure of 1 bar was used for visualization in streamer mode of electronic images of tracks of relativistic particles extracted from solid krypton by an electric field of 1.5 kV/cm strength. Particles passing through solid krypton were extracted from a beam of relativistic particles using a telescope of external scintillation counters. The counters generated a trigger signal that triggered the Arkadyev–Marx high voltage pulse generator to generate a high voltage pulse with amplitude of 100 kV and duration of 60 ns applied to the anode suspended in the center of the optical window above the solid krypton.

At the end of the 1970s, in a special experiment on the secondary beam of the ITEP proton synchrotron, the operability of this detector was demonstrated and it was shown that it is really capable of visualizing pion tracks with a pulse of 3 GeV/s passing through the solid krypton in the form of a chain of streamers (each with a diameter of ~0.5 mm and 2 mm long) and provides a high density of streamers along the track ($\sim 1\,\mathrm{mm}^{-1}$) and improved spatial resolution compared to gas-filled streamer cameras (Bolozdynya *et al.*, 1977a).

The study of the properties of the emission streamer chamber confirmed the observation, first made in the study of the properties of a cryogenic streamer gas chamber (Sidorov *et al.*, 1975), that despite the increased density of the cold working gas compared to gas at room temperature, the threshold field strength for the creation of streamers remained unchanged at the level of 2 kV/cm. It has been hypothesized that this effect is associated with increasing concentration of noble dimeric molecules with decreasing temperature. Since the ionization potential of dimers is reduced by 1–2 eV compared to atoms, the average ionization potential of the medium decreases in proportion to the increase in gas density, as a result of which the intensity of the visualizing field remains virtually unchanged.

An analysis of the results of exposure of this chamber with a
beam of high-energy pions revealed anomalous tracks with a very
low ionization density: one streamer per 1–2 cm of track length. More
detailed studies showed that this was a two-phase medium memory
effect associated with a capture of a part of some electrons at the
interphase: anomalous tracks have been detected exactly in the same
place where the track of a relativistic particle was recently detected
at normal ionization density. However, the idea of detecting particles
with an anomalously low ionizing ability (down to single electrons at
a few centimeters of the track length) using emission chambers was
first expressed at that time (Bolozdynya et al., 1980).

In the 1980s, the Nadezhda emission chamber was built at ITEP,
the working fluid of which was to be liquid krypton with a diameter
of 50 cm and a thickness of 20 cm, and the traces should be visualized using a cryogenic streamer chamber with a diameter of 1.5 m
(Fig. 6.6, Bolozdynya, 2010). It was supposed to use this device to
study the multiple productions of neutral pions during the annihilation of antiprotons in heavy nuclei. The ability to visualize high
ionization density tracks with a streamer camera was experimentally
tested using an LG-21 ultraviolet laser. The emission section was
used to accurately measure the radioactivity of krypton and to study
the possibility of obtaining efficient electron emission from large
volumes of liquid krypton (Anisimov et al., 1989a, 1989b). However,
the detector was not tested in the full assembly due to problems with
financing scientific work in the USSR in the early 1990s.

In parallel with the study of the possibility of visualizing tracks of
high-energy particles in dense working media, the electronic methods
of detecting particles in condensed working media began to actively
develop using condensed phase scintillation, electroluminescence, and
gas amplification in the gas phase. Significant efforts have been
expended in the search for other possible working environments for
emission detectors in addition to condensed noble gases.

1.5. Ionization Emission Detectors

In ionization emission detectors, the interaction of ionizing particles
with a condensed medium is recorded by analyzing the shape of

the ionization signal taken from the electrode system of detectors. Such detectors were used in the early stages of development of emission detectors when registering sufficiently intense ionization signals generated by alpha particles (Hutchinson, 1948), electron accelerators (Boriev *et al.*, 1978) or X-ray tubes (Guschin *et al.*, 1982b, 1982c; Bolozdynya and Stekhanov, 1984) in order to study the electron emission properties of various condensed dielectrics, including saturated liquid hydrocarbons such as hexane, isooctane, and tetramethylsilane (Bolozdynya, 1986a, 1986b), considered for the possibility of creating emission detectors operating at room temperature.

At that time ionization emission detectors have been used mostly for investigation emission properties of possible working media for two-phase emission detectors including condensed heavy noble gases (argon, krypton, xenon) and saturated hydrocarbons (methane, hexane, isooctane, TMP, TMSi). Despite the positive results of this research, saturated hydrocarbons did not find any real applications in experimental particle or nuclear physics.

1.6. Gas Gain Emission Detectors

The first emission detectors were built during the time of the development of multi-wire proportional chambers (MWPCs) invented by Charpak (*et al.*, 1968) for high-energy physics applications. Probably, for this reason, the first emission chambers most often had used multi wire anodes, which made it possible to explore some methods of gas gain of ionization signals. If the registration of emitted electrons was accompanied by a spark discharge, then such events could be photographed through the wire anode. If the anode wires could be capable of proportional gas amplification, then it would be possible to detect useful events electronically.

One of the first emission detectors of this type was a two-electrode liquid-argon ionization chamber similar to the device used to spark visualization of alpha particle tracks, as described above (Fig. 1.2). The electronic signal was recorded from the wired anode (the diameter of the wires varied in the range from 50 to 200 microns). Initially, attempts were made to obtain gas amplification at the wire

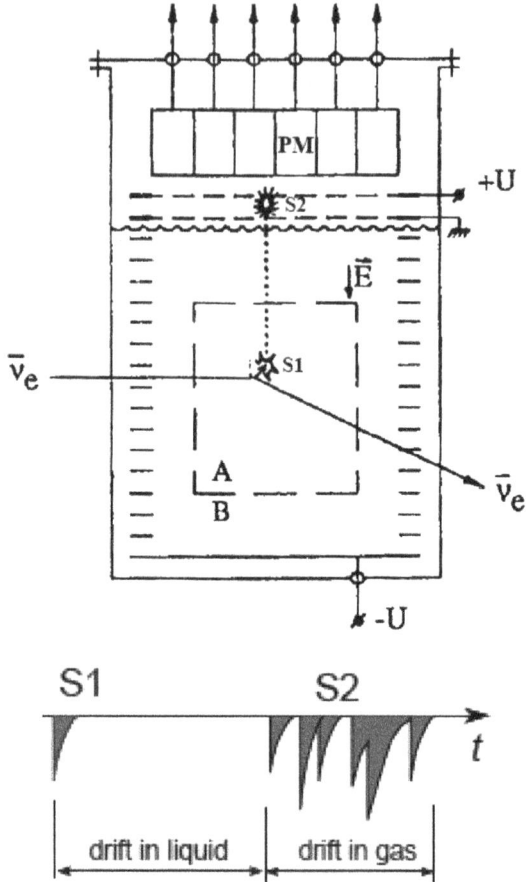

Figure 1.1: Schematic drawing of the "wall-less" two-phase emission detector for registration of rare point-like events (top) and readout signals in case of photon registration (bottom): PM — array of photo sensors; S1 — scintillation flash at the moment of interaction of the detected particle with condensed working medium; S2 — electroluminescence flash in the gas phase; +U — anode; –U — cathode; A and B — sensitive and screening nominal volumes of the working medium; E — electric field strength; ν_e — detected particle. Redrawn from Bolozdynya *et al.* (1995) and Chepel and Araujo (2013).

anode in the equilibrium gas phase, but in this mode it was not possible to obtain stable gas amplification with a coefficient of more than 500. To create conditions for more stable gas gain, the wire anode was immersed in liquid argon and the anode wires had been

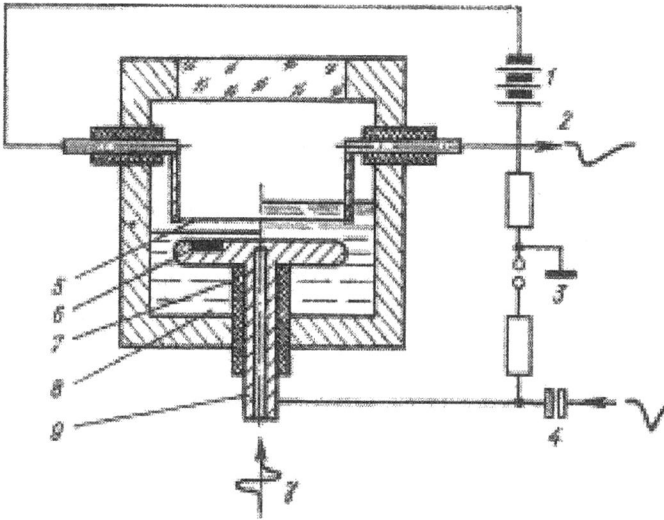

Figure 1.2:   Emission wire spark and proportional chamber: 1 — battery for heating anode wires; 2 — output anode signal; 3 — DC high voltage supply; 4 — pulsed HV supply; 5 — 100 micron diameter anode wires; 6 — alpha-particles source; 7 — channel for gamma-rays; 8 — liquid argon; 9 — cathode with collimator for gamma-rays. Redrawn from Dolgoshein *et al.* (1973).

heated by electric current of 0.1–1 A strength. The development of avalanches has been limited by the size of bubbles appearing on wires due to boiling of the liquid. In such an approach, it was possible to obtain gas amplification from electrons emitted from liquid argon into bubbles with a multiplication factor up to 10^4. However, it was observed that the dead time was significantly increasing (10 ms vs. 0.1 ms in a gas) due to localization of positive ions inside bubbles. Using pulsed high voltage to create a discharge around wires, the gas gain was raised up to 10^6 (Dolgoshein *et al.*, 1973). Nevertheless, this technique did not find practical application, since it required a huge expenditure of energy for heating the anode wires and boiling liquid argon around them.

Attempts to achieve high and stable gas amplification in emission detectors using organic working media were more successful. In particular, using liquid isooctane as emission detector medium, a gain of 3×10^4 has been achieved at the wire anode (Bolozdynya *et al.*, 1978).

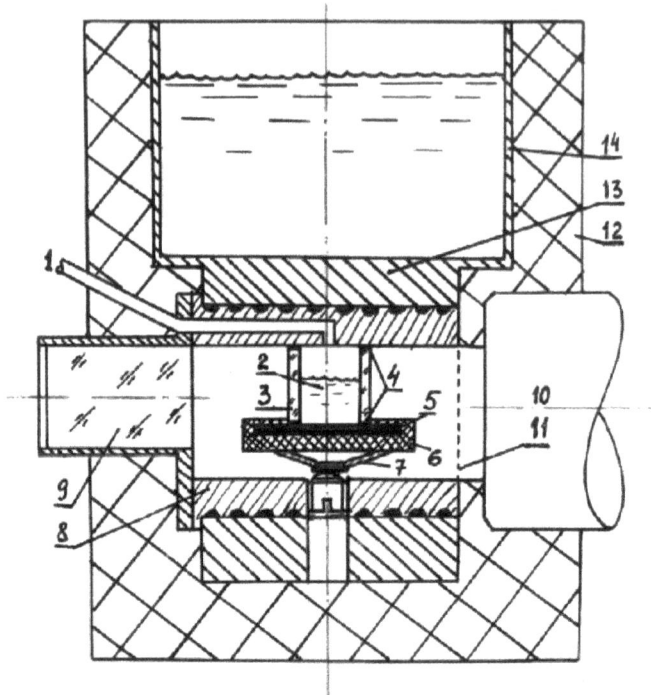

Figure 1.3:   Control chamber (CoC) luminescent ionization detector for investigation of electron emission properties of condensed noble gases and methane:
1 — gas input (output); 2 — $\sim$1 cm^3 condensed sample; 3 — glass cylinder of
15 mm diameter with inside surface coated by p-terphenyl wave-length shifter;
4 — Teflon foil made gaskets of 0.2 mm thickness; 5 — anode coated by alpha-
particles source material; 6 — Teflon made HV insulator; 7 — sealing system
based on spring washer; 8 — grounded cylindrical copper holder with a heater
coil; 9 — plexiglass window for visual control of the sample; 10 — photoelectron
multiplier FEU-110; 11 — screening greed; 12 — foam thermal-insulation; 13 —
heat conductor; 14 — liquid nitrogen bath. Redrawn from Bolozdynya *et al.*
(1977b).

In a similar experiment using liquid 2,2,4,4-tetramethylpentane, a gas
gain of $\sim$10^3 has been observed (Anderson *et al.*, 1987).

For a long time in emission detectors using pure noble gases
only spark discharge had been used for amplification of signals.
Any addition of organic impurities, which increases the gas gain
and provides a stable gas amplification in gas detectors, in emission
detectors based on condensed noble gases proved to be very difficult,

(a)

(b)

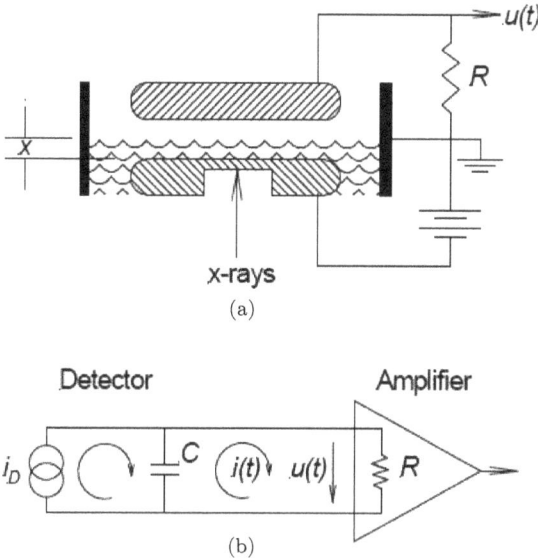

Figure 1.4:   Control Ionization Roentgen chamber of 3 cm diameter filled with noble liquid of $x = 1/3$ mm thickness (a) and readout schematic drawing (b). Redrawn from Bolozdynya (1986b).

since it inevitably led to the cooling of drifting electrons in the condensed phase and, accordingly, to the reduction of the probability of electron emission into the gas phase.

A new round in the development of technology for two-phase emission detectors was made due to the invention of gas electron multipliers (GEMs) by Fabio Sauli in 1997 (Sauli, 2016). The study of the possibility of using GEM for detecting electrons in two-phase emission detectors has shown that this technology combines well with argon as a working medium, and when using triple-GEM structures, the amplification of the electronic signal can be achieved up to 5000. Unfortunately, in two-phase xenon emission detectors, it was not possible to obtain gas gain more than 200 (Bondar *et al.*, 2006).

The next step in the development of gas signal amplification technology in two-phase emission liquid-argon detectors was the development of mechanically strong "thick" (ThGEM) (Chechik *et al.*, 2004; Bondar *et al.*, 2008) and "large" (LEM) (Cantini *et al.*, 2015) gas electron multipliers, which are now successfully used to amplify signals in multi-tone two-phase emission detectors based on liquid argon and designed to perform experiments such as the DUNE underground experiment on investigation long-base high-energy neutrino oscillations (Adams *et al.*, 2019).

1.7. Electroluminescence Emission Chambers

The development of the technology of electroluminescent emission detectors began with the creation of miniature detectors with a working medium volume of $\sim 1\,\text{cm}^3$ for studying the scintillation and emission properties of condensed argon, krypton, xenon, methane and their mixtures (Lansiart *et al.*, 1976; Bolozdynya *et al.*, 1977b). A qualitatively new stage in the development of emission detector technology was the development of a position-sensitive electroluminescent gamma camera with a hexagonal matrix of nineteen FEU-110 photomultiplier tubes (Egorov *et al.*, 1983), which was proposed to be used to visualize gamma radiation fields in nuclear medicine.

The development of this technology led to the idea of using the registration of two signals from one event: a scintillation signal (it is often called the S1 signal), which occurs at the moment of the interaction of the detected particle with the condensed working medium of the detector, and the subsequent electroluminescent signal (signal S2), which arises during drift through the gas phase of ionization electrons extracted by the electric field from the condensed phase. Using these two signals in recording quasi-point events allows us to determine the position of the interaction point in 3D space in order to create a "wall-free" detector for recording rare events such as interactions with neutrino baryonic matter and exotic particles that can make up the dark matter of the Universe (Bolozdynya *et al.*, 1995). This possibility of reconstructing a 3D picture of particle interactions in a dense noble gas was first demonstrated using compressed xenon (Bolozdynya *et al.*, 1997a). At about the same time, it was proposed to use a two-phase electroluminescent emission chamber of this type to search for rare decays (Bolozdynya *et al.*, 1997b). The idea of such a detector turned out to be very fruitful for experiments on the search for cold dark matter in the form of weakly interacting massive particles (WIMPs). To date, the best world results in the search for WIMPs have been obtained using this type of detectors. By the time of writing this book, two-phase emission WIMP detectors have reached several tons of working medium mass (liquid xenon as a rule) (Aprile *et al.*, 2018a; Akerib *et al.*, 2020a). The next generation of emission detectors searching for WIMPs is under consideration with liquid xenon mass up to 50 tons (Aalberts *et al.*, 2016) and with liquid argon mass up to 300 tons (Aalseth *et al.*, 2018).

Thus, the technology of two-phase emission detectors over 50 years of development has gone the way from miniature detectors for methodological studies with a mass of working substance of several grams to detectors with a mass of working substance of many tons for solving fundamental problems: the search for dark matter in the Universe, a study of the fundamental properties of neutrinos. At the next stage of this development the technology may find important practical applications in the field improving the safety of nuclear energy production and maintaining international efforts on non-proliferation of nuclear weapons as well as nuclear medicine.

Chapter 2

Particle Interactions and Energy Transfer Mechanisms in Noble Liquids

2.1. Structure of Noble Liquids

The energy transfer processes from an ionizing particle to the medium as well as those leading to formation of the observable signals in radiation detectors are deeply related with the medium thermodynamic state. A number of significant, from the point of view of particle detection, differences between dilute rare gases and rare gas liquids and solids exist. As the density increases, the interaction between the atoms becomes stronger that leads to formation of local regular structures in liquids, becoming macroscopic in solids. The valence and conduction bands are known to form already in the liquid state of heavy noble gases. The electron transport is also altered as the distance between the atoms becomes smaller than the electron wavelength so that the approximation of electron scattering on isolated atoms is no longer valid. In this section, we consider shortly the structural aspects of noble gas liquids.

Noble gas atoms have closed electronic shells and interact with each other through the weak van der Waals force. A pair of atoms can form a bound state with the binding energy of the order of 0.01 eV, comparable with the thermal energy, which makes the dimers unstable and easily breakable in collisions with other atoms even at cryogenic temperatures. Nevertheless, at any given time there is a certain fraction of dimers in the gas, which increases with gas density and with decrease of the temperature. For example, in xenon with the number density of $n = 2.6 \times 10^{19}\,\mathrm{cm^{-3}}$ (corresponds to pressure of 10 bar at room temperature) the fraction of atoms forming dimers is about 4%, according to Bernardes and Primakoff (1959).

The interaction between a pair of atoms is described by a pair potential, attractive at long distances and repulsive when the atoms are too close to each other. For neutral atoms, the attractive part is attributed to interaction between the electric dipole moments (and other multipoles, to a lesser extent) either existing or induced. As noble gas atoms are spherically symmetric, they do not have electric moments, in average. However, the fluctuations in the electron density distributions result in instantaneous asymmetry in the distribution of positive and negative charges in the atom which in turn affects the charge distribution in the other atom and *vice versa*. This attractive force, frequently called London dispersion force, rapidly decreases with the distance between the atoms as $O(1/r^6)$ and can be approximately described with the expression (Schmidt, 1997)

$$V_{\mathrm{disp}}(r) = -\frac{3}{4}\frac{\alpha_v}{r^6}I,\qquad(2.1)$$

where α_v represents atomic polarizability volume[1] and I is the ionization potential of the atom (see Table 2.1).

London dispersion force is responsible for formation of condensed state in rare gases. As the atomic polarizability increases with atomic number, heavier rare gas atoms are more prone to form bound states — dimers, trimers and clusters consisting of many atoms. This explains the general trend of increasing the phase transition temperature with the atomic number.

---

[1] Atomic polarizability is defined as the coefficient of proportionality between the dipole moment induced on the atom by an external electric field and the field strength $p = \alpha E$. In SI, it is measured in $\mathrm{C \cdot m^2/V}$, while in CGS in units of volume. In order to distinguish the two values, the latter is sometimes called "atomic polarizability volume". If we denote it as α_v, the relationship between the two is $\alpha = 4\pi\epsilon_0\alpha_v$.

Table 2.1:   Some parameters of noble gas atoms, diatomic molecules and similar quantities for condensed phase.

	Ne	Ar	Kr	Xe
Atomic parameters				
Ionization potential of an isolated atom (eV)	21.56	15.76	14.00	12.13
First excitation level of an isolated atom (eV)	16.62	11.55	9.92	8.32
Radiative lifetime of the lowest excited state of an isolated atom (ns)	25	10	4.4	3.6
Atomic polarizability volume, $\alpha_v(\text{Å}^3)$	0.397	1.64	2.48	4.01
Homonuclear diatomic molecule				
Equilibrium internuclear distance, $r_e(\text{Å})$[a,b]	3.1	3.76	4.02	4.36
Dissociation energy (meV)[b]	2[a]	12.3	17.2	24.2
Ionization energy (eV)	—	14.5	12.87	11.13
Excitation energy of homonuclear dimer relative to 0 vibrational level of the ground state (eV)				
$^1\Sigma_u^+$	—	11.56	—	—
$^3\Sigma_u^+$	—	11.46	9.87	—
Condensed phase parameters				
Minimum ionization energy (eV)				
liquid	21.6[c]	14.4[c]	11.56[d]	9.20[d]
solid[e]	—	14.16	11.61	9.33
Static dielectric constant, κ[f]				
liquid at triple point	—	1.52	1.67	1.89
liquid at 1 bar	—	1.51	1.66	1.88

Source: [a]Radzig and Smirnov (1985) if not stated otherwise; [b]Hübner (1998); [c]estimated by Schmidt and Yoshino (2005); [d]Steinberger (2005); [e]Schwentner *et al.* (1985); [f]calculated with Clausius–Mossotti relation from atomic polarizability, agrees with experimental values by Amey and Cole (1964) within 1% and by Marcoux (1970) (except for Xe, for which the difference is about 4%).

The repulsive part of the interatomic potential is largely related with the Pauli exclusion principle due to overlapping of the electron wave functions of the two atoms at short distances (also called quantum exchange interaction, which is repulsive for fermions but attractive for bosons). While the attraction energy varies slowly with the distance, the repulsion potential is a very steep function of distance that makes its mathematical description rather difficult.

The shape of the potential energy as a function of distance between the two rare gas atoms can be approximated with the 6–12 Lennard-Jones potential, which has a simple analytical form

$$V(r) = 4\varepsilon \left[\left(\frac{\sigma}{r} \right)^{12} - \left(\frac{\sigma}{r} \right)^{6} \right]. \tag{2.2}$$

Parameter ε in this expression represents the depth of the potential well and σ corresponds to the distance for which $V(\sigma) = 0$. The equilibrium distance between the atoms (corresponding to the minimum of energy equal to $-\varepsilon$) can be determined from σ as $r_m = \sqrt[6]{2}\sigma$. One readily recognizes in the Lennard-Jones potential the $O(1/r^6)$ attractive part and strong $O(1/r^{12})$ repulsion at short distances. In practice, the two parameters σ and ε are adjusted to result in the best agreement with experimental data, with the second virial coefficient, in the first place, as it is directly related to the pair interactions. The values of σ and ε for rare gas atoms are shown in Table 2.2. Using this simple potential in calculations results in predictions of thermodynamic and structural properties of noble liquids with quite reasonable accuracy. At short distances, the Morse potential in the form $V(r) = Ae^{-\beta r}$ is also used.

In search for better accuracy, a number of other analytical forms for the potential have been proposed. One can call these potentials semi-empirical as they do not result from a rigorous theory but rather from some theoretical models, each valid in a limited range of distances, which are combined in a unique (or piecewise) expression with the empirically obtained coefficients. Polynomial

Table 2.2:   Parameters of the 6–12 Lennard-Jones potential ε and σ following Rutkai *et al.* (2016) and the equilibrium distance between the atoms of the dimer molecule r_m.

	ε (meV)	σ(Å)	$r_m = \sqrt[6]{2}\sigma$(Å)
Ne	2.9231	2.801	3.144
Ar	10.064	3.3952	3.8110
Kr	14.010	3.6274	4.0716
Xe	19.487	3.949	4.4326

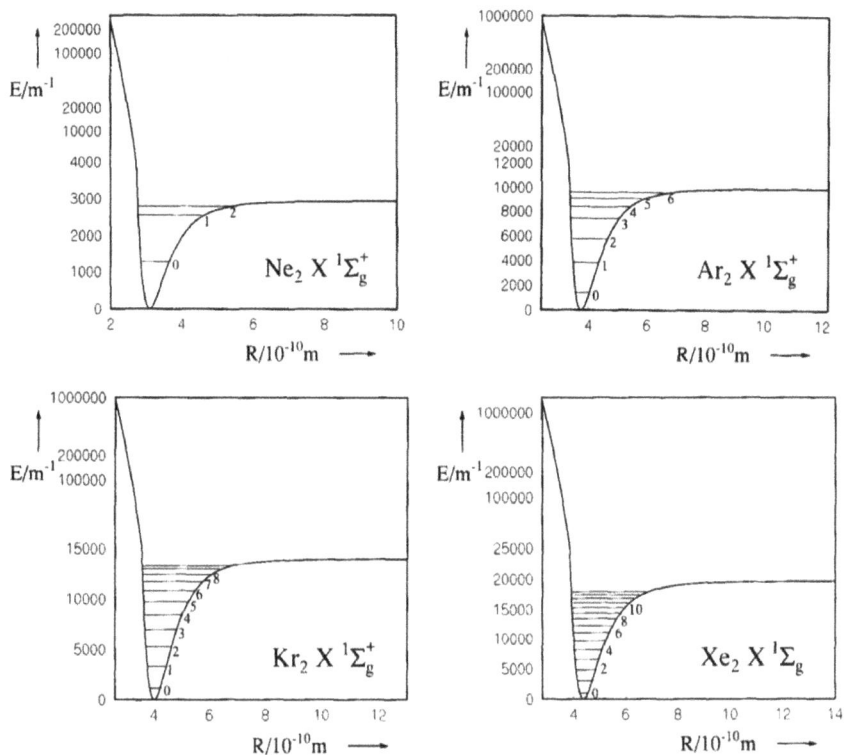

Figure 2.1: Pair potential energy as a function of internuclear distance for the ground state of rare gas dimers. The range of validity is $0.65 \leq R/r_m \leq 6.5$, where r_m is the distance corresponding to the minimum of the potential (here 3.091Å, 3.7565Å, 4.008Å, and 4.3627Å for Ne_2, Ar_2, Kr_2 and Xe_2, respectively). The vertical scale is the atomic energy units 10,000 m^{-1} = 0.0124 eV. Redrawn from Ogilvie and Wang (1992).

parametrization is quite a common approach (e.g., Ogilvie and Wang, 1992). The potential curves for the ground electronic state, obtained with the potential proposed in this work, are shown in Fig. 2.1. The vibrational excitation levels are also shown in the plots.

The presence of a third atom in the vicinity of a pair under consideration disturbs the charge density distribution in each of the two atoms and, in this way, affects the interaction between them. In the parametrization models, the many-body interactions are automatically taken into account in the potential as the model parameters are adjusted in order to reproduce the experimentally observable macroscopic thermodynamic properties of a particular fluid as precisely as possible. On the other hand, the applicability of such potential to the conditions different from those for which the parametrization has been performed cannot be guaranteed *a priori* and one should be careful in using it in those cases.

In principle, if the potential energy of an atom or a molecule resulting from its interaction with all other particles in the fluid is exactly known, all macroscopic properties can be calculated applying Newton's laws of dynamics to each particle. This is the approach of classical molecular dynamics (MD) which continuously gains more and more relevance in material science as the available computing power constantly increases. For a good overview (and also a tutorial) on molecular dynamics and its application to noble gas fluids we refer to Searles and Huber (2005); more specific details on the method can be found in the book of Allen and Tildesley (2017). Obtaining the exact shape of the potential energy curves theoretically is a difficult task even for a pair of atoms especially for those with many electrons. As an example of quantum mechanical *ab initio* computations for heavy rare gas dimers, one can refer papers by Laschuk *et al.* (2003) and Slaviček *et al.* (2003).

With the fluid density increasing, free atomic motion becomes more and more restricted by the presence of other atoms in the vicinity. Interactions between them promote formation of some sort of regular structure. Depending on density and temperature, this structure can be of lesser or higher degree of order which can be conveniently described by a pair correlation function $g(r)$. The pair correlation function, also called radial distribution function, represents a measure of the probability to find another particle at a certain distance from a given atom placed at the origin.[2] It also

---

[2] $g(r)$ can be seen as a histogram of the distances between all pairs of atoms normalized to the mean number density of the medium. At very short distances, $g = 0$ as the repulsive force prevents the atoms from approaching too close to each other. The first peak appears at the distance corresponding to the minimum of the potential energy of pair interaction (dimer bound state) and is present also

represents a distribution of the distances between any pair of atoms in a sufficiently large volume. Another closely related to $g(r)$ quantity is the static structure factor $S(k)$ which can be directly obtained from X-ray or neutron scattering experiments. Here, $|k| = \frac{4\pi}{\lambda} \sin\frac{\theta}{2}$ is the momentum transferred by a particle with the wavelength λ upon scattering to angle θ. The quantities $[S(k) - 1]$ and $[g(r) - 1]$ are related to each other through the Fourier transform so that

$$S(k) - 1 = \rho \int_V e^{-i\vec{k}\vec{r}}[g(\vec{r}) - 1]d\vec{r}. \tag{2.3}$$

Structure factors of LNe, LAr, LKr and LXe and the corresponding pair correlation functions are shown in Figs. 2.2–2.5. All liquids reveal a similar structure with the degree of order rapidly diminishing with the distance. While the nearest neighbor position is well defined at a distance close to the interatomic distance in the dimer molecule (r_m in Table 2.2), every next atomic layer is more diffuse and the order is practically disappearing at about 4–5 interatomic distances.

The correlation between the atoms in the liquid is weaker at higher temperatures but does not disappear even at the temperatures close to the critical temperature as one can see in Fig. 2.2 for LNe and Fig. 2.4 for LKr. In fact, some correlation already exists in dense gases. In Fig. 2.5, the radial distribution functions for liquid and solid Xe are compared. One can clearly see the difference — the solid, as expected, has much better defined structure than the liquid although the width of the peaks indicates that even in solid xenon the atoms are not strongly bound in the lattice as it would be in an ideal crystal and form a more amorphous structure.

Good agreement between the experimental data and calculations is noticeable. For example, Fig. 2.3(b) shows the pair correlation function for liquid argon obtained from neutron diffraction measurements (Henshow, 1957) and also that calculated from the pair

---

in gas. The following oscillations reflect the order existing (or not) in the medium: in a dilute gas, there would be no oscillations ($g = 1$, disorder); in an ideal crystal at zero temperature (i.e., no vibrations) the function would show series of sharp spikes at well-defined distances with no limit on r.

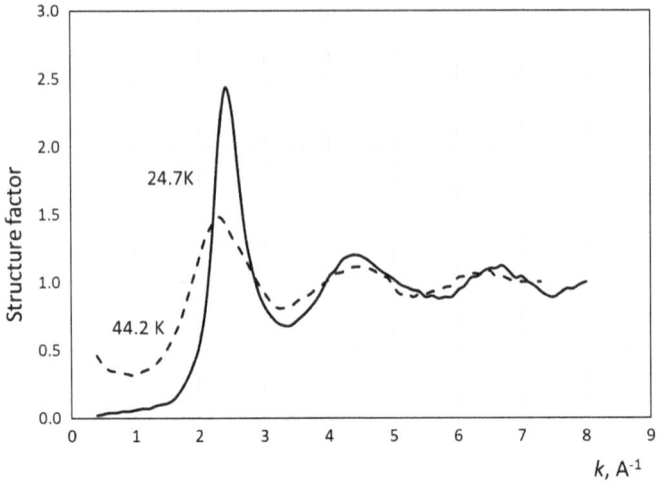

Figure 2.2:   Structure factor of liquid neon from X-ray scattering data at two temperatures (indicated in the plot) — near the triple point (24.56 K) and just below the critical temperature (44.49 K). Redrawn from Schmidt and Tompson (1968).

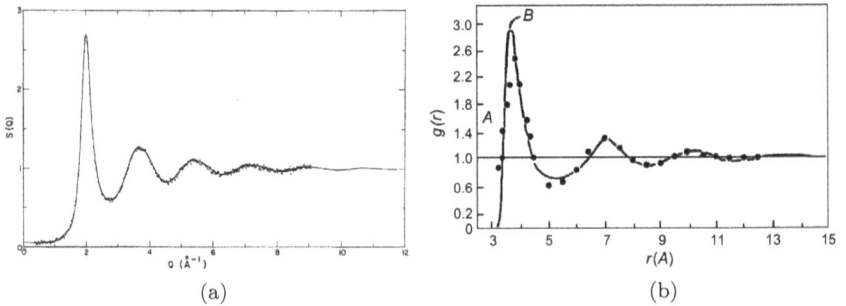

Figure 2.3:   Structure factor (a) and pair correlation function (b) of liquid argon at 84 K. The structure factor is from Yarnell *et al.* (1973) — experimental, from neutron scattering. For the pair correlation function, the data points are from neutron diffraction measurements (Henshow, 1957) while solid line represents calculations using the interatomic pair potential by Sakai (2005). Region A corresponds to the repulsive part of the potential; point B corresponds to its minimum. Redrawn from Yarnell *et al.* (1973) and Sakai (2005).

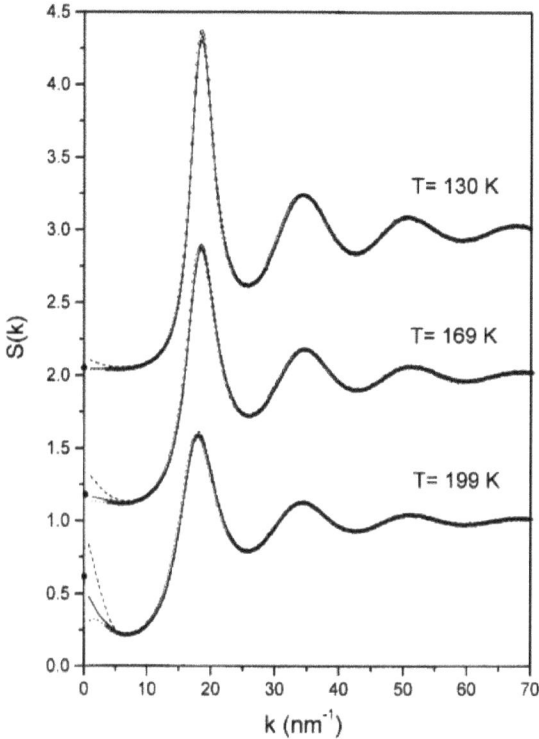

Figure 2.4: Structure factor of LKr at three different temperatures (indicated in the plot) according to Jakse *et al.* (2002). The experimental data points are from Barocchi *et al.* (1993). The curves for $T = 169$ K and $T = 130$ K are shifted upwards by an amount of 1 and 2, respectively. Redrawn from Jakse *et al.* (2002).

potential (Sakai, 2005). For the latter, the exclusion region (A) corresponds to the repulsive part of the potential and the first peak is at the position of the minimum of the potential energy between two Ar atoms, indicated as B in the plot. An excellent agreement is found also for LKr as one can see in Fig. 2.4 where the molecular dynamics computations by Jakse *et al.* (2002) are compared with the experimental data of Barocchi *et al.* (1993) obtained by neutron scattering.

At high fluid densities, the interatomic interactions result in some lowering of the atomic energy levels while overlapping of the electron wave functions of the nearby atoms leads to level splitting and,

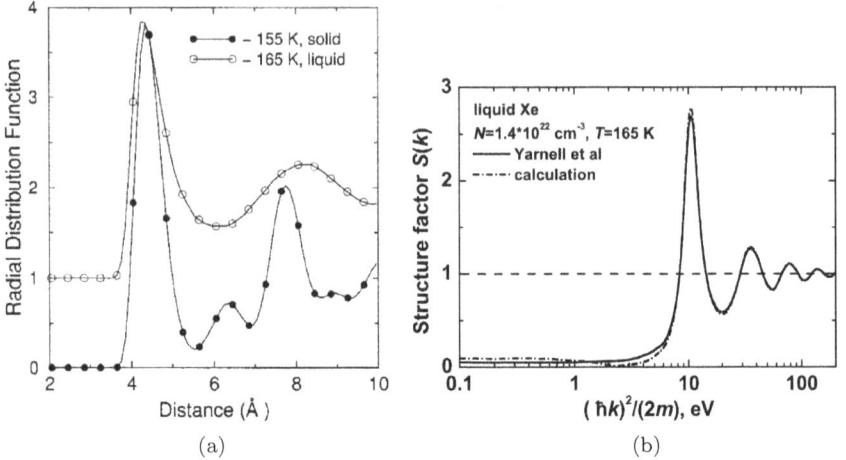

Figure 2.5:   Radial distribution function in liquid and solid Xe (a) and structure factor of LXe (b). Redrawn from Belonoshko *et al.* (2002) and Atrazhev *et al.* (2005), respectively.

at the limit of solid state, to band formation. It is experimentally proven that the valence and the conduction bands are also formed in heavy noble liquids in spite of the lack of global periodic structure. The band gap in liquid krypton and xenon have been measured to be 11.55 eV and 9.20 eV, respectively — a few electronvolts lower than the corresponding ionization potential of an isolated atom (Table 2.1), but no data are available for liquid argon (Steinberger, 2005). The band gap dependence on the liquid density has been observed to tend to increase as the density decreases, as expected — see Steinberger (2005) and references therein.

Band formation as well as, in general, formation of a regular structure in condensed rare gases has a profound impact on the electronic and optical properties of those media. Many properties of noble liquids are more correctly described in terms of solid state physics than considering liquid as a compressed gas. However, the latter approach is still tempting and frequently used given the existence of well-developed and relatively simple formalisms (such as a concept of electron scattering cross-section, for example). It provides a simpler and clearer picture for a variety of phenomena and we shall use "gas terminology" in this book, where appropriate, along with that from solid state physics.

2.2. Particle Energy Transfer in the Liquids

2.2.1. *Interaction of electrons and nuclear recoils with matter*

In the context of this book, we will be interested mainly in electrons, γ-rays and atomic ions, the latter being the result of elastic collisions of a hypothetical WIMP particle or a neutrino with the atomic nuclei of the detector medium. The interest in electrons and γ-rays is two-fold: first, these particles constitute the natural background in those detectors and, secondly, they can be conveniently used for calibration of the detector response in terms of the deposited energy.

In the energy range of interest, γ-rays interact with matter mainly through photoelectric absorption and inelastic (Compton) scattering, both resulting in ejection of an electron. Therefore, the energy transfer mechanism for γ-rays is essentially the same as that for electrons, although the energy of the photon is split between several secondary particles. In the case of the photoelectric absorption, fluorescence X-ray and Auger electrons are emitted apart from the photoelectron; Compton scattering results in one or more Compton electrons and ends up, again, with the photoelectric absorption. These processes are described, with different levels of details in many books, (e.g., Knoll, 2010; Leo, 1994). A concise review of the interaction processes with liquefied rare gases can be found in Lopes and Chepel (2005).

The energy spreading over several ionizing particles is especially relevant in the keV energy region where the photon energy is comparable with binding energies of the atomic electrons. As a consequence, the response of a detector to γ-photons and electrons of equal energy is not the same: both the scintillation and charge yields are known to be non-linear with energy, the effect being most significant for energies of $\sim 1\,\text{keV}$ to $\sim 10\,\text{keV}$ (see Sections 2.3.2 and 2.4.2).

A dark matter particle (WIMP) is expected to scatter elastically off atomic nuclei transferring a part of its kinetic energy to nuclear recoil. In the energy range considered ($\sim 1\,$keV to $100\,$keV), the initial velocity of the nuclear recoil is comparable to that of the atomic electrons, so that the recoiling atom conserves most of its electrons and moves through the liquid as a positive ion with low effective charge. The moving ion continuously exchanges electrons with other atoms along its way so its charge is not constant and even be zero (corresponds to a neutral atom). As the medium consists of atoms of the same kind, the primary recoil can transfer a significant fraction of its kinetic energy in each collision, thus losing rapidly its "projectile" identity and producing a cascade of secondary recoils of comparable energy which interact with the medium in the same way. Neutrinos, coherently scattered off the nuclei, also produce nuclear recoils behaving in the medium exactly as recoils from WIMPs. More information and references on that can be found in Chepel and Araújo (2013).

A charge particle moving through the medium interacts with the medium atoms via the electromagnetic force. In describing the energy–momentum transfer from the particle to an atom, the atom has to be considered, in general, as a single quantum mechanical system that is very difficult to implement in theory. A number of approximations have been developed, most successful for particles of high energies,[3] when the interaction time is too short for the atomic electrons to rearrange in the field of the impinging particle. In this case, the effect of the detailed atomic structure can be ignored as in the statistical Thomas–Fermi model of atom in which the set of the electron wave functions is replaced by a function describing the distribution of the electron density. Also, the projectile is considered as a point structureless charge. The well-known Bethe formula for the mean stopping power dE/dx is a result of such approximations. Discussion of those theories is outside the scope of

---

[3]In fact, particle velocity is a better parameter than energy being the Bohr velocity $v_B = \alpha c \approx 2.2 \times 10^6$ m/s a measure for applicability of an approximation. In terms of kinetic energy, it corresponds to ≈ 26.6 keV/amu.

this book; therefore the reader is referred to, for example, reviews by Ahlen (1980), Ziegler (1999), and books by Balashov (1997) and Sigmund (2004, 2006).

At lower ion velocities, $\beta \lesssim 0.2$, shell corrections become increasingly important. Also, the impinging particle cannot be considered to be a point-like charge anymore and the mutual effects on the electronic density distribution of the projectile-target system should be included in the calculations. The electron exchange between the two particles, at high energies leading to the electron stripping from the projectile, at low energies works the opposite way — the mean charge of the traveling particle can now be less than $+1e$. Using a semi-empirical formula proposed by Nikolaev and Dmitriev (1968), one can estimate the mean charge of a projectile of 100 keV energy as $\sim 1\,e$ and for 1 keV as $\lesssim 0.1e$, independently of the mass. Thus, the effective charge of a slowing down particle is constantly decreasing along its path and most of the time nuclear recoils with the initial energy <100 keV travel as neutral atoms. Very similar estimates for effective charge are obtained with simple formula $Z^* = Z^{1/3}(v/v_B)$ proposed by Bohr (here Z is the atomic number of the projectile and v its velocity).[4] The calculated width (r.m.s.) of the equilibrium charge distribution at any given instant is quite significant and can be as large as $\sim 100\%$.

The stopping power (or linear energy transfer — LET, looking to the problem from the point of view of energy depositions in the medium) can conveniently be split into two parts

$$\frac{dE}{dx} = \left(\frac{dE}{dx}\right)_e + \left(\frac{dE}{dx}\right)_n, \qquad (2.4)$$

where the first term accounts for all electronic excitations (including ionization) caused by the travelling particle, while the second term

---

[4]One must note that the semi-empirical expressions obtained by Nikolaev and Dmitriev (1968) were adjusted to the experimental data for energies $\gtrsim 5$ MeV and for solid state targets. At lower energies, there is no sufficient data to allow reliable parametrization. Other approaches have been developed (e.g., Brandt and Kitigawa, 1982) although for even higher energies. A concise review on this subject can be found in the book by Ziegler *et al.* (2008).

takes into account the energy lost in elastic collisions with atomic
nuclei. Such division means to separate the energy "visible" by a
conventional detector, which measures the ionization or scintillation
(or both), from the energy lost to heat and undetectable by those
detectors. For electrons, the second term is negligible given the
large difference in masses on the projectile and the target (this
does not mean, however, that no loss to heat takes place — see
Section 2.2.2). The situation is completely different when ions are
concerned — these particles indeed lose a significant part of their
energy in collisions which do not result in electronic excitations (the
second term is frequently referred to as "nuclear stopping power"
while the first one as "electronic stopping power"). The energy
partition between electronic excitations and scattering on nuclei for
ions is most frequently described using the Lindhard–Scharff–Schiøtt
(LSS) theory, developed by Lindhard *et al.* (1963a, 1963b, 1968) and
allowing a simple parametrization, in spite of the limitations at low
energies which have been discussed by several authors[5] (e.g., Tilinin,
1995, and in connection with dark matter search with two-phase
detectors — by Mangiarotti *et al.*, 2007; Mei *et al.*, 2008; Bezrukov
et al., 2011). Alternatives exist, among which the approach of Ziegler
et al. (1985), strongly assisted by the existing experimental data,
has been successful in a wide range of energies and projectile/target
species (Ziegler *et al.*, 2008, 2010).

Lindhard *et al.* (1963a) expressed the fraction of the ion energy
spent in electron collisions in the form

$$L(\epsilon) = \frac{kg(\epsilon)}{1 + kg(\epsilon)}, \qquad (2.5)$$

where k is the coefficient of proportionality between the stopping power dE/dx and the ion velocity, and $g(\epsilon)$ is a smoothly
increasing function of the dimensionless projectile energy defined as

---

[5] As noted by Lindhard *et al.* (1968), the Thomas–Fermi treatment, used in the
theory, may be not valid below a certain energy limit which was estimated to be
~ 1 keV for argon and ~ 10 keV for xenon.

$\epsilon = 11.5\,E/Z^{7/3}$ with E expressed in keV (the numerical coefficient is calculated for a symmetric projectile-target system which is of interest for DM search and coherent neutrino scattering experiments). No analytical expression has been given by the authors for $g(\epsilon)$ but a simple parametrization $g(\epsilon) = 3\epsilon^{0.15} + 0.7\epsilon^{0.6} + \epsilon$ was proposed by Lewin and Smith (1996). The coefficient k is related with the atomic and mass numbers of the projectile as $k = 0.133\,Z^{2/3}/A^{1/2}$; for xenon it is equal to 0.166 and for argon 0.144. A somewhat different value of $k = 0.110$ was obtained by Hitachi (2005) for xenon. The function $L(\epsilon)$ in Eq. (2.5) is a saturation function of ϵ tending to 1 for high energies and to 0 when the recoil energy $\to 0$.

It should be noted that while the term *linear energy transfer* is good for describing energy losses by a charged particle moving in a medium (even if its trajectory is not a straight line), it does not necessarily describe correctly the distribution of energy depositions in the medium. The light particles (electrons and positrons) are subjects of multiple scattering, especially significant at low energies and in high Z materials, so that the track not always can be seen as a linear sequence of excitations but more like a blob. Production of secondary particles capable of ionizing, such as δ-electrons or fluorescence photons followed by photoelectrons or Auger electrons, also makes the distribution of the energy depositions in the medium more complex.

In the case of electrons with the energy of $\sim$1 MeV, most of the track can be well pictured as a linear sequence of ionizations and excitations as illustrated for LXe by Fig. 2.6, obtained by a Monte Carlo simulation with the PENELOPE code (Salvat *et al.*, 2011). The figure shows only the initial part of the track which is to be compared

Figure 2.6: The initial part of 1 MeV electron track in LXe with energy depositions shown as bubbles. The area of each bubble is proportional to local energy deposition; the scale (100 eV) is shown as a separated bubble on the left. The electron propagates from left to right. Simulated with the PENELOPE code.

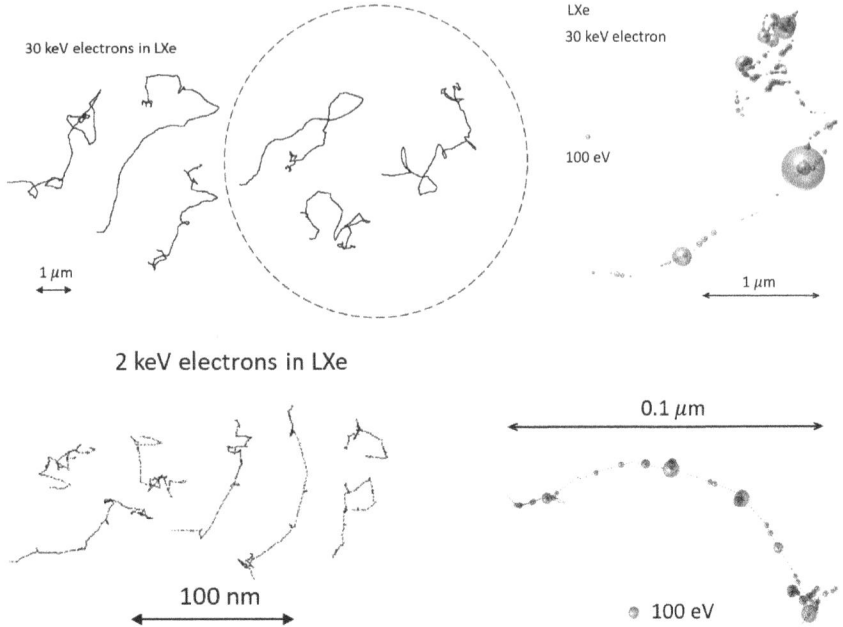

Figure 2.7: Examples of low energy electron tracks in LXe. A circle of 4.5 μm radius corresponds to the mean thermalization distance according to Mozumder (1995b). The tracks are simulated with PENELOPE. At the right, an example of energy deposition pattern is shown. The area of each bubble is proportional to local energy deposition; the scale (100 eV) is shown as a separated bubble.

with tracks of electrons of lower energies (Fig. 2.7).[6] In fact, the $\sim$1 MeV electrons also suffer quite a significant scattering in LXe and somewhat less in LAr, which is noticeable in the scale of their range (a fraction of mm), however in the scale comparable with the range of the $\sim$keV electrons the track pattern of 1 MeV electron clearly differs from that of low energy electrons. For electrons, $\frac{dE}{dx} \approx \left(\frac{dE}{dx}\right)_e$ and the number of electron–hole pairs per unit of length can be expressed as $\frac{dN_i}{dx} = \frac{1}{W}\frac{dE}{dx}$, from which an estimate for the average distance

---

[6]One must be aware that calculations of the energy losses for particle energies $\lesssim$10 keV can be not as precise as for higher energies due to difficulties of taking into account the effects related with the atomic structure and possible collective effects.

between the positive charges in the track can be obtained: $\sim$40 nm for LXe and $\sim$90 nm for LAr. These distances are close to the Onsager radius for the respective liquid: 54 nm for LXe, 84 nm for LKr, and 126 nm for LAr (see Section 2.4.1). The local energy depositions, represented in the images by bubbles with the area proportional to deposited energy, are subject to significant fluctuations, however.

Nuclear recoils undergo multiple collisions with the nuclei producing secondary recoils, indistinguishable from the primary projectile. This results in an extended tree-like energy deposition pattern as one can see in Fig. 2.8, that shows several examples of Monte Carlo simulations of Xe recoils in LXe with the SRIM code (The Stopping and Range of Ions in Matter (Ziegler *et al.*, 2010).

Each of the images in Figs. 2.6–2.8 represents a spatial pattern of energy deposition in the liquid resulting from the interaction of the respective particle with it and, simultaneously, a distribution of the positive ions/holes and excitons along the track. The distribution of the ionization electrons coincides with this pattern, too, but only at the initial instant when their kinetic energy becomes insufficient to ionize or excite the atoms of the medium, i.e., just below $\sim$10 eV. At such a high energy, the probability for an electron to recombine with a positive ion is low. Therefore, the electrons will continue to

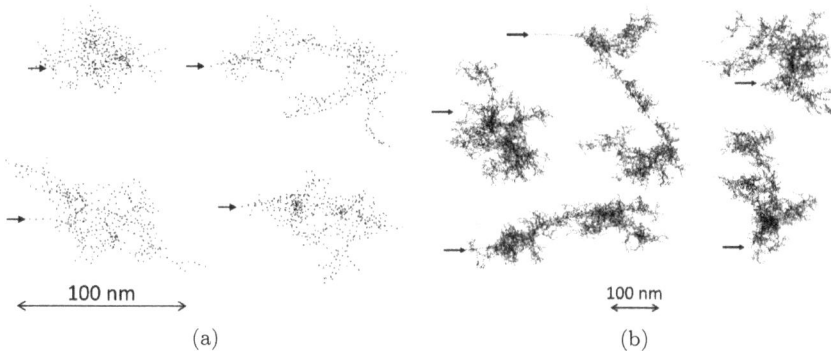

100 nm

(a)   (b)

Figure 2.8:   Examples of xenon recoil tracks in LXe for 1 keV (a) and 10 keV (b) initial energy. Each dot represents an energy exchange location due to either the primary or secondary recoils. An arrow indicates the initial position and direction on the primary recoil. Simulated with the SRIM code.

move in the liquid, colliding with atoms elastically, until they reach thermal equilibrium with the liquid. The distance from the origin of an electron (i.e., position of its sibling ion) to the "point" where its energy decreases down to the thermal energy is called *thermalization length*. The thermalization length in liquefied noble gases is rather large, $\sim$1700 nm in LAr (Mozumder, 1995a) and $\sim$4500 nm in LXe (Mozumder, 1995b), as shown by a dashed line circle in Fig. 2.7 for liquid xenon. The time required for that is in the nanosecond scale (Sowada *et al.*, 1982). This is when one can roughly consider the recombination process to begin although indications exist that for high ionization density tracks (e.g., α-particles or heavy ions) the recombination starts earlier. We shall consider these processes in more detail in Section 2.4.1.

2.2.2. *Energy dissipation in noble gas liquids*

In general terms, the energy transferred by a particle to the medium is dissipated through the following three channels — ionization, excitation and heat (production of phonons in condensed matter). Ionization and excitation of the atoms and molecules, constituting the medium, are often jointly called *excitation*. It is important to distinguish (at least in theory) the direct effects caused by the primary particle, those due to the secondary particles, and the energy interchange between the channels (e.g., recombination of electrons with positive ions, resulting in formation of excited dimer molecules with subsequent light emission). Each of these processes has its own time scale. The time required for the energy of all particles, set to motion, to decrease to the sub-excitation level is of the order of 10^{-12} s. From this moment on, no new ionized and excited species are generated directly (although transformation is possible through the above mentioned mechanism) and their total number can only decrease.

It is also important to distinguish the effects due to light charged particles and photons (these interact with matter through photoelectric absorption, Compton scattering and electron–positron pair production, in all cases resulting in ejection of an electron or a pair of electron and positron) from those due to heavy particles like ions. If the light particles transfer their energy almost entirely through the electronic excitation of the medium (including ionization), heavy particles also lose a significant part of their energy in nuclear collisions resulting in a cascade of nuclear recoils. This is the case of heavy ions and also of neutral particles like neutrons and WIMPs which interact with the medium via elastic scattering on the medium nuclei.[7]

The energy sharing between the three energy dissipation channels is formalized in the well-known Platzman energy balance equation written initially for gases and for light particles, which do not produce nuclear recoils (Platzman, 1961)

$$E_0 = N_i \bar{E}_i + N_{ex} \bar{E}_{ex} + N_i \bar{\varepsilon}. \tag{2.6}$$

According to this equation, the energy transferred to the medium E_0 is spent to produce N_i singly ionized atoms at an average energy expenditure of $\bar{E}_i$, N_{ex} excited atoms at an average energy expenditure of $\bar{E}_{ex}$, and N_i sub-excitation electrons having average kinetic energy $\bar{\varepsilon}$ immediately after the last collision to result in either ionization or excitation. The sub-excitation electrons can only participate in elastic collisions and, therefore, their kinetic energy will be ultimately transferred to the thermal motion of the medium atoms. The Platzman equation reflects the energy partition at the (idealized) track formation instant, i.e., when all primary excitations and ionizations have already occurred but no secondary processes started due to interactions between the created species (free electrons, ions, excited atoms) and between them and the atoms.

It is customary to characterize the radiation detector media with average energy expended per one ionization in the particle track, defined as $W = E_0/N_i$ — the so called W-value. Using this definition, the Platzman equation can be re-written in the following form:

$$W = \bar{E}_i + \bar{E}_{ex} \frac{N_{ex}}{N_i} + \bar{\varepsilon}. \tag{2.7}$$

---

[7]Neutrons also can interact inelastically through nuclear reactions or capture. In the first case, one has to consider the energy transfer from the reaction products; in the case of neutron capture by a nucleus, this can disintegrate or relax to the ground state with emission of a cascade of γ-rays.

All values in this equation are characteristics of the medium, in the first place, although may depend on the particle type (as it is indeed verified for N_{ex}/N_i) and, to a lesser extend, on particle energy. The ratio N_{ex}/N_i was assumed to be independent of energy by Platzman. More recent work of Dahl (2009) supported this assumption (at least, for liquid xenon).

The quantities in Eqs. (2.6) and (2.7) cannot be easily accessed experimentally but some obvious boundaries can be set: in a diluted gas it must be $\bar{E}_i \geq I$ — ionization potential of the atoms. Indeed, $\bar{E}_i$ was found to be higher then I by $\sim$10% thus reflecting the fact that multiple ionizations and formation of ions in an excited state can occur. In liquid or solid, $\bar{E}_i$ should be larger than the band gap E_g. Similar considerations are valid for E_{ex}. The values from the literature of the parameters entering Eqs. (2.6) and (2.7) are shown in Tables 2.3 and 2.4.

The ratio N_{ex}/N_i can be theoretically estimated from the oscillation strengths as well as by studying the energy transfer between the excited states and ionizations. Thus, observation of the Penning effect in liquid argon doped with xenon, in which the excitation energy of argon excitons R^* is transferred to Xe atoms, results in the N_{ex}/N_i value of 0.19 ± 0.02 for LAr in good agreement with previous calculations for solid argon yielding the value of 0.2 (Kubota *et al.*, 1976). For liquid xenon, the theoretical estimates give 0.06 while measurements of the recombination dependence on the electric field suggest a much higher value of about 0.13 (Doke *et al.*, 2002; Doke, 2005; Suzuki and Hitachi, 2011). There is no complete clarity on that, yet. Thus, an even higher value of 0.20 has been reported for liquid xenon although with an uncertainty of about 60% (Doke *et al.*, 2002; Aprile *et al.*, 2007). For heavy particles like nuclear recoils, much larger values than for electrons have been reported. For xenon recoils of 10–100 keV in LXe, N_{ex}/N_i in the range from 0.9 to 1.1 was obtained for LXe (Dahl, 2009; Sorensen and Dahl, 2011; Angle *et al.*, 2011 — see Section 2.4.3).

By measuring the amount of charge produced by a particle of a given energy in a detector, one can determine the W-value and therefore to set further constraints on the variables in the Platzman

Table 2.3: Ionization energy parameters of liquefied rare gases (all in eV except Fano factor).

Liquid	I (gas)	E_g (liquid)	E_g (solid)	$\bar{E}_i$	W	F
Ne	21.56	—	—	—	—	—
Ar	15.76	(14.4)	14.16	15.4	23.6 ± 0.3	0.107
Kr	14.00	11.55	11.61	13	20.5 ± 1.5	0.057
Xe	12.13	9.20	9.27	10.5	15.6 ± 0.3	0.041

Source: Atomic ionization potential I from Radzig and Smirnov (1985); band gap E_g from Steinberger (2005) (minimum ionization energy for LAr as estimated by Schmidt and Yoshino, 2005); mean ionization energy in the Platzman equation $\bar{E}_i$ from Takahashi *et al.* (1975); mean energy expended per ion pair W and the Fano factor F are from the monograph by Aprile *et al.* (2006a). For liquid argon, the value of E_g is in parentheses because the band structure in this liquid is not confirmed; the minimum ionization energy is given instead.

Table 2.4: Excitation energy parameters of liquefied rare gases (all in eV, except N_{ex}/N_i).

Liquid	E_{ph}	$\bar{E}_{ex}$	$W_s^{\min}$	N_{ex}/N_i	$\bar{\varepsilon}$
Ne	15.8 (1.2)	—	—	—	—
Ar	9.80 (0.6)	12.7	19.5 ± 1.0	0.21	5.15
Kr	8.37 (0.6)	10.5	15	0.08	5.50
Xe	7.10 (0.5)	8.4	13.8 ± 0.9	0.06	4.45

Source: Energy of the emitted photons E_{ph} (width of the spectrum in brackets) are from Morikawa *et al.* (1989); mean energy to create an excited specie $\bar{E}_{ex}$, N_{ex}/N_i and the mean energy of sub-excitation electrons $\bar{\varepsilon}$ in the Platzman equation — from Takahashi *et al.* (1975); mean energy expended per scintillation photon W_s — from the monograph by Aprile *et al.* (2006a) (depends on the ionization density; the minimum values are given).

equation. In practice, however, one does not have access to the value of N_i but only to the number of electrons extracted from the particle track under the applied electric field. The number of extracted electrons N_e is always smaller than N_i because some electron–ion (electron–hole) pairs recombine at any practically achievable field. The W-value is therefore determined by interpolation of the ratio E_0/N_e, measured as a function of field, to an infinite field that requires a recombination model to be chosen. Recombination is a complex process depending strongly on the spatial density of the ionization that will be addressed in Sections 2.3 and 2.4 of this chapter. As an example, one can refer here two extreme cases — fast electrons with the energy of $\sim$1 MeV, for which >90% of the electrons created along the track can be collected in fields of a few kV/cm, and fission fragments for which only $\sim$1% of the total charge was measured at 10 kV/cm (e.g., Suzuki and Hitachi, 2011; Aprile *et al.*, 2006b). This shows that the determination of the W-value by measuring the collected charge is nearly impossible for highly ionizing particles.

The generally accepted values of W for liquefied rare gases are shown in Table 2.3 together with the best estimates for other parameters in the Platzman equation (also in Table 2.4). The W-value is assumed here to be independent on the particle energy and type although some variations have been reported below $\sim$1 keV.[8] Using these values, one can get a general picture of the energy dissipation through different channels: about 65% of the particle energy is spent for ionization, 20–30% is taken by the sub-excitation electrons, i.e., converted to heat, and only 3–11% of the particle energy is spent in direct excitations. In practice, however, one does not measure energy (except for the phonon channel) but the number of particles — electrons and photons — produced along the particle track. Therefore, the ratio N_{ex}/N_i, estimated to be between 0.06 and 0.20 for electrons and γ rays, is a more adequate parameter to determine the initial share of each of the two energy dissipation channels in the total signal: $1/(1 + N_{\text{ex}}/N_i)$ for the charge channel and $(N_{\text{ex}}/N_i)/(1 + N_{\text{ex}}/N_i)$ for the light channel, although one has to keep in mind that not all ionizations result in free electron–hole pairs (which can be measured) but can be converted into excitations via recombination as will be discussed next.

---

[8]In liquid water, for example, the W-value is practically constant for electrons with the energy above 1 keV but increases below that energy (Plante and Cucinotta, 2009). Small variations have also been reported below $\sim$1 keV for gases (e.g., Pansky *et al.*, 1997). In Xe gas, the W-value varies significantly for electrons with the energy <100 eV (Samson and Haddad, 1976).

2.3. Primary Scintillation in the Liquids

2.3.1. *Scintillation mechanism*

There is much similarity between gas and liquid and also between different rare gases in what the evolution of the excited states is concerned. In gases, the luminescence mechanism is well understood and, in a simplified form, can be represented as a sequence of the following reactions:

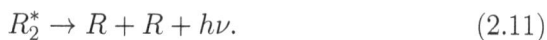

$$e^- + R \to R^* + e^-, \tag{2.8}$$

$$R^* + R \to R_2^{*,\nu}, \tag{2.9}$$

$$R_2^{*,\nu} + R \to R_2^* + R, \tag{2.10}$$

$$R_2^* \to R + R + h\nu. \tag{2.11}$$

Here, $R =$ Ar, Kr, or Xe; the superscript ν is used to distinguish electronically excited states with vibrational excitation ($R_2^{*,\nu}$) from the electronically excited state with the lowest vibrational energy $\nu = 0$ (R_2^*). Reaction (2.8) represents formation of an excited by electron impact atom R^*, most probably in one of the highly excited states which rapidly relax to the lowest excited level through collisions with other atoms. The atomic radiation emitted in a transition from the first excited level to the ground state is resonantly absorbed by the neighbors at short distances and re-emitted again so that in practice the atomic emission can only be observed at gas pressures below a few hundreds Torrs (Policarpo, 1981). At higher pressures, this process is interrupted by formation of an excited diatomic molecule R_2^* (excimer) through reaction (2.9). Vibrational relaxation promptly occurs, mostly through non-radiative channel with the energy transferred to the kinetic energy of surrounding atoms, as reaction (2.10) shows, although emission of near infrared radiation is also possible with a small probability (Belogurov *et al.*, 2000; Bressi *et al.*, 2001a; Buzulutskov *et al.*, 2011). As a result, an excimer is formed in the singlet $^1\Sigma_u^+$ or triplet $^3\Sigma_u^+$ state, both with $\nu = 0$. These states have almost equal and rather high energy of ~ 10 eV that prevents non-radiative de-excitation. A VUV photon is then emitted in a transition from one of those states to the repulsive ground state $^1\Sigma_g^+$ followed by dissociation of the dimer as indicated in reaction (2.11). The singlet–singlet transition $^1\Sigma_u^+ \to {}^1\Sigma_g^+$ is rather fast (lifetime of an isolated excimer molecule is in the nanoseconds scale) but the triplet–singlet transition is, strictly speaking, forbidden within the *LS*-coupling formalism. However, for high Z atoms ($Z \gtrsim 30$) the jj coupling description becomes more adequate and the excimer states cannot be considered as pure states with the electronic terms $^1\Sigma_u^+$ or $^3\Sigma_u^+$. The transition from the triplet state $^3\Sigma_u^+$ is then becomes possible due to mixing of this state with the singlet $^1\Pi_u$ state. The mixing is stronger for higher atomic numbers thus resulting in a noticeable reduction of the lifetime of the triplet state, more and more significant from Ne to Xe (see Table 2.5). The energies of the two transitions differ by less than ~ 0.1 eV (e.g.,

Table 2.5: Radiative lifetimes of the two lowest excited states of homonuclear rare gas dimers in different aggregation states of the medium.

	Ne$_2^*$	Ar$_2^*$	Kr$_2^*$	Xe$_2^*$
Radiative lifetime of the singlet state $^1\Sigma_u^+$ (ns)				
Free molecule[a]	2.8	4.2	3.4–5.2	4.6–6.2
Gas $\sim$1 bar[a,b]	—	3.3–4.8	3.4–5.2	4.5–6.9
		4.2[d]		
Liquid[b]	—	(1.65) 4.2–7.1	1.45–2.1	1.4–4.3
		7[d]		
Solid[a,b]	1	3.2–4.1	1.3	1.1–1.3
Radiative lifetime of the triplet state $^3\Sigma_u^+$ (ns)				
Free molecule[a]	5000–12000	3200	150–350	100
Gas $\sim$1 bar[a,b]	—	2800–4000	(150) 250–395	96–130
		3100[d]		
Liquid[b]	15000[c]	860–1660	80–110	19–32
		1600[d]		
Solid[a,b]	2000–4000	1100–1410	82–90	19–29

Note: Liquids and solids at temperatures near the triple point. Values in brackets are outside the generality of the results.

Source: [a]compilation by Schwentner *et al.* (1985); [b]compilation by Morikawa *et al.* (1989); [c]measurements by Nikkel *et al.* (2008); [d]compilation by Buzulutskov (2017).

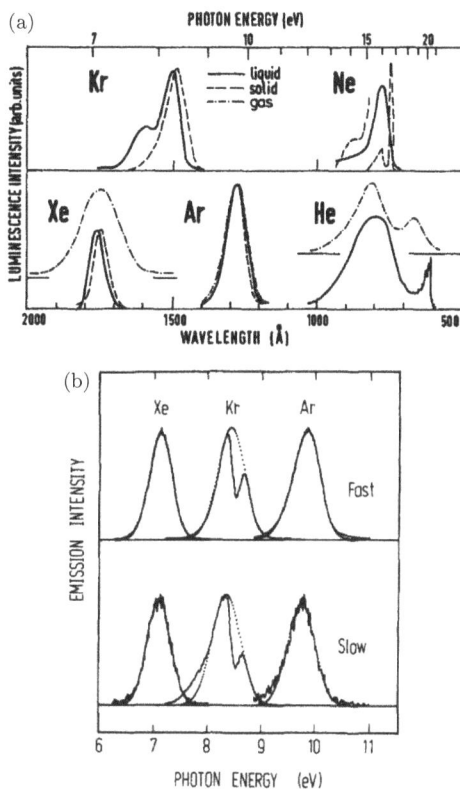

Figure 2.9: (a) Comparison of the emission bands in liquefied rare gases together with solid and gas phase spectra (from Schwentner *et al.*, 1985). (b) Emission spectra corresponding to the fast and the slow components in the liquefied rare gases as measured by Morikawa *et al.* (1989). A dip in the middle of the krypton spectra is explained with the absorption by <5 ppm xenon impurities. The separation of the two components was done by setting a proper time window after the excitation with synchrotron radiation. Redrawn from Schwentner *et al.*, (1985) (a) and Morikawa *et al.* (1989) (b).

Mulliken, 1970; Lorents, 1976) corresponding to the difference in the emission peak maxima $\Delta\lambda \sim 2\,\mathrm{nm}$, while the width of the emission spectrum, which has a form close to a Gaussian[9] in the liquid and solid states as well as in the gas at pressures $\gtrsim 1$ bar is of the order of 10 nm as illustrated by Fig. 2.9 (also Table 2.4).

---

[9]In krypton, absorption by xenon impurity atoms is observed.

In rare gas solids, much similarity is found with gases (Jortner *et al.*, 1965; Schwentner *et al.*, 1985). The general reaction scheme is still valid although the meaning of the symbols used in Eqs. (2.8)–(2.11) is different. It is not correct to consider rare gas atoms as mostly non-interacting. The environment has a strong effect on the atomic energy levels resulting in their lowering and overlapping. The well-confirmed band structure in rare gas solids makes evident that all the above processes have to be treated as collective and no complete analogy with gas should be expected *a priori*. In the solid state, the initial excitation by the electron impact — R^* in Eq. (2.8) — should be regarded as exciton, a state lying within the band gap. At least three narrow exciton bands were confirmed to exist just below the bottom of the conduction band, each of which can be excited by impinging electrons or through photon absorption. The relaxation to the lowest state with $n = 1$ then occurs, mostly non-radiatively with the energy transferred to phonons, within the time ranging from a few picoseconds for Xe to 1.5 ns for Ar (Schwentner *et al.*, 1985, Table 6.7; Suzuki and Hitachi, 2011, for more references). At the same time, the free exciton (excited atom in gas) can form a bound state with one of its neighbors — the so-called self-trapped[10] exciton R_2^*. The trapping time constant is estimated to be $\sim 10^{-12}$ s. The behavior of self-trapped excitons in solid rare gases and excimers in gas is very similar, so that the molecular spectroscopy terminology is frequently used even for solids by designating the lowest excited states with symbols $^1\Sigma_u^+$, $^3\Sigma_u^+$, and $^1\Sigma_g^+$ for the ground state. However, one has to keep in mind that formation of an extra correlated pair of atoms in a crystal implies some structural change (does not exist in gas) meaning that the rare gas crystals cannot be considered as having a rigid lattice (Jortner *et al.*, 1965). The self-trapped exciton de-excitation occurs with emission of a VUV photon — reaction in Eq. (2.11) — with the energy very similar to that of the corresponding excimer molecule.

---

[10]The term "self-trapped" refers to formation of a bound state between two atoms of the same kind, which share the excitation energy, in contrast with the excitation trapping to an impurity level in a crystal.

The emission spectrum is narrower and decay times are in general shorter than in gas, especially for the triplet state and for heavier rare gas solids (Table 2.5).

The fact that band formation is verified in heavy rare gas liquids makes them resembling more solids than compressed gas and therefore solid state physics terminology applies better. The exciton physics in liquefied rare gases is much less studied than in solids. Many authors, therefore, assume that the excitons in liquids are essentially the same as in solids that gives rather good predictions for the properties of the liquefied rare gases relevant for radiation detection (e.g., Doke, 2005). Spectroscopic data and time resolved studies do exist providing valuable information on the exciton properties and dynamics in rare gas liquids. Formation of self-trapped excitons is confirmed (Jortner *et al.*, 1965); the emission spectra are similar to those of excimers in gases and excitons in solids (Fig. 2.9). The decay times are closer to those observed in the solid state than in gas except probably for neon where very few measurements exist (Table 2.5). As well as in gas, we notice a very significant variation of the triplet lifetime with a clear trend to increase from xenon to neon. In liquid neon, lifetime as long as 15 μs has been measured by Nikkel *et al.* (2008), while in liquid helium it was found to be $\sim$13 s (McKinsey *et al.*, 1999).

Relaxation from the higher excited states takes place mostly through non-radiative channels with transfer of energy to phonons, but emission of infrared photons can also occur with a small probability. Similarly to gas, emission of near infrared light in liquid argon and xenon has been reported (Bressi *et al.*, 2000, 2001a, 2001b; Belogurov *et al.*, 2000; Heindl *et al.*, 2010; Buzulutskov *et al.*, 2011). For LAr, emission in the range 400–1000 nm represents $\sim$1% of the yield in VUV (Heindl *et al.*, 2010); a value of 0.51 infrared photons with wavelengths between 690 nm and 1000 nm per keV was reported by Buzulutskov *et al.* (2011). Controversially, no infrared light has been detected in pure LAr in the range of 0.5–3.5 μm in the spectroscopic measurements of Neumeier *et al.* (2015). On the other hand, a rather intense emission at $\lambda = 1.18\,\mu$m was observed in LAr doped with a few ppm of Xe yielding about 13 IR photons/keV under the excitation with a 12 keV electron beam injected through a thin window. For pure LXe, an estimate of $\gtrsim 0.2$ photons/keV can be obtained combining the results published by Belogurov *et al.* (2000) and Bressi *et al.* (2001a).

Free excitons can also be formed through the recombination of electrons with positive ions (holes) as represented by the following reactions (Doke, 1981; Doke *et al.*, 2002)

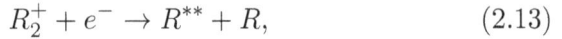

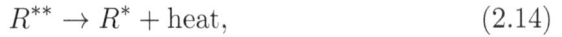

$$R^+ + R \rightarrow R_2^+, \tag{2.12}$$

$$R_2^+ + e^- \rightarrow R^{**} + R, \tag{2.13}$$

$$R^{**} \rightarrow R^* + \text{heat}, \tag{2.14}$$

after which the free exciton R^* follows the same path as the excitons formed by electron impact, shown by reactions in Eqs. (2.9)–(2.11). The formation of molecular ions R_2^+ (self-trapped hole; Eq. (2.12)) is estimated to be $\sim 10^{-12}$ s (Hitachi *et al.*, 1983).

Although recombination leads to formation of self-trapped excitons in the same final state as direct excitation by electron impact, it does affect the observed decay time. The three scintillation components, corresponding to the singlet and triplet transitions of the directly formed excitons and that due to recombination, can be clearly distinguished in time resolved measurements as those shown in Fig. 2.10 for liquid xenon excited by ~ 1 MeV electrons, α-particles and fission fragments with the energy of ~ 1 MeV/amu. The recombination process is extremely sensitive to ionization density along the particle track which is the lowest for the electrons, among these three particles (this will be discussed in more detail in Sections 2.3 and 2.4). The recombination time is longer than the exciton lifetimes and therefore it determines the observed decay time in this case. The decay curve does not follow an exponential as it should be for exciton decays but is well described by an inverse parabola $I \propto t^{-2}$ thus confirming the recombination dominated nature of the emission process for electrons (Hitachi *et al.*, 1983). We also notice that emission from the directly formed excitons is hardly noticeable in the logarithmic scale plot. In fact, its contribution to the total amount of the emitted light for ~ 1 MeV electrons is about 30% as

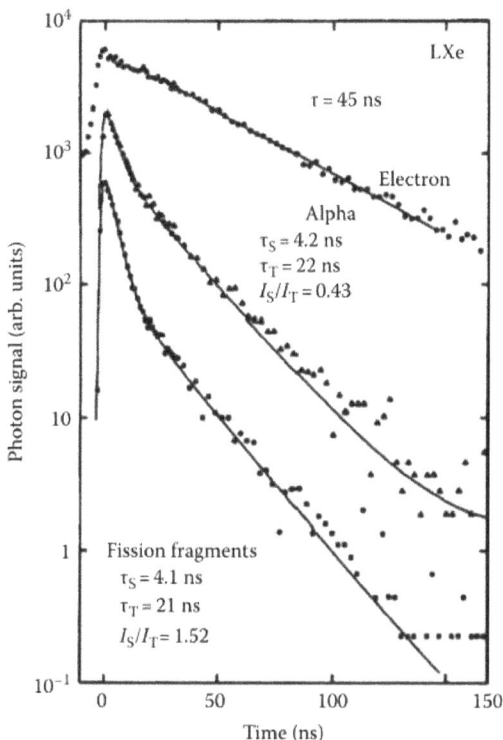

Figure 2.10:   Scintillation decay curves for liquid xenon excited by electrons ($\sim$1 MeV), α-particles ($\sim$5 MeV) and fission fragments ($\sim$1 MeV/amu). Redrawn from Hitachi *et al.* (1983).

one can see from the dependence of the scintillation intensity on the electric field which allows to suppress recombination (Fig. 2.11(a)). In liquid xenon, the 45 ns recombination component disappears when a sufficiently strong electric field is applied.

A diagram of the energy distribution between the recombination and the direct excitation channels in liquid argon and xenon for fast electrons is shown in Fig. 2.12 (Chepel and Araújo, 2013). Although there is some discrepancy between the exact values obtained by different authors and in different experiments, the general picture is quite consistent. It is worth noting that the presence of even a small amount of impurities can affect the exciton dynamics, especially the observable decay time. Measurement of the singlet and triplet decay

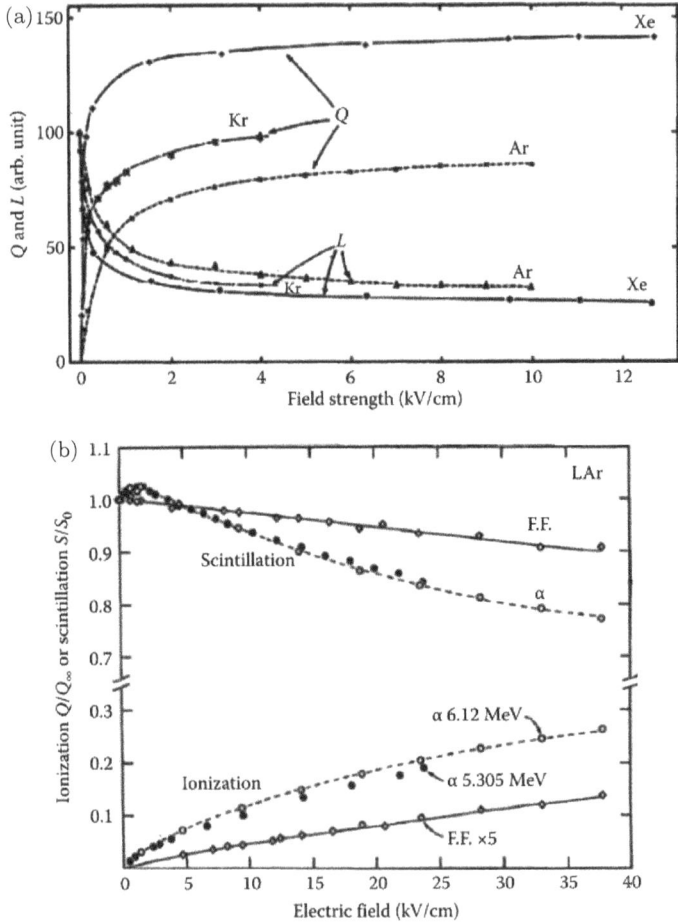

Figure 2.11: Dependence of the relative charge (Q) and scintillation light (S) yields as a function of electric field for electrons (a) and α-particles and fission fragments (b). Redrawn from Suzuki and Hitachi (2011); original data from Kubota *et al.* (1979) and Hitachi *et al.* (1987).

times at zero electric field is complicated by the presence of the recombination component. One also notices some difference between liquid argon and xenon in what the recombination component is concerned: faster recombination and longer self-trapped exciton lifetimes in liquid argon result in practically the same emission times with and without field.

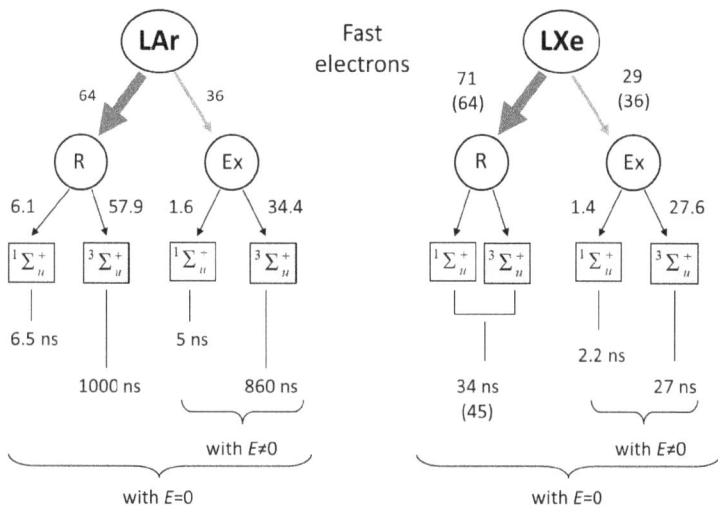

Figure 2.12: Diagram of the observable distribution of luminous energy (in percent) for fast electrons (from Chepel and Araújo, 2013). Symbols R denotes the recombination channel while Ex stands for direct excitation by electron impact. Original data are from Kubota *et al.* (1979) (energy share between the two channels) and Kubota *et al.* (1978) and Kubota *et al.* (1982a) (decay times and relative intensities).

Note: Somewhat different values has also been reported. For singlet/triplet intensity ratios: in LAr — 0.31/0.69 and 0.23/0.77 (Kubota *et al.*, 1979; Hitachi *et al.*, 1983); in LXe — 0.36/0.64 (Kubota *et al.*, 1979). For time constants: in LAr — 7.0 ns for the fast component (Hitachi *et al.*, 1983), 1600 ns, 1463 ns, and 1260 ns (Hitachi *et al.*, 1983; Lippincott *et al.*, 2008; Acciarri *et al.*, 2010, respectively); in LXe — 4.3 ns and 22 ns for singlet and triplet decay times and the apparent decay time of 45 ns (Hitachi *et al.*, 1983). All values at zero electric field. Redrawn from from Chepel and Araújo (2013).

The ionization density in the tracks of α-particles and fission fragments is much higher than for 1 MeV electrons. The decay curves observed for these particles reveal more exponential-like variation of the light intensity with time (Fig. 2.10) with two decay constants — fast (4.2 ns) and slow (22 ns) — attributed to the singlet–singlet and triplet–singlet transitions (Hitachi *et al.*, 1983). This data indicates that the recombination time along the tracks of α-particles is shorter than the self-trapped exciton lifetimes. Experimentally, separation of the recombination component from that due to direct

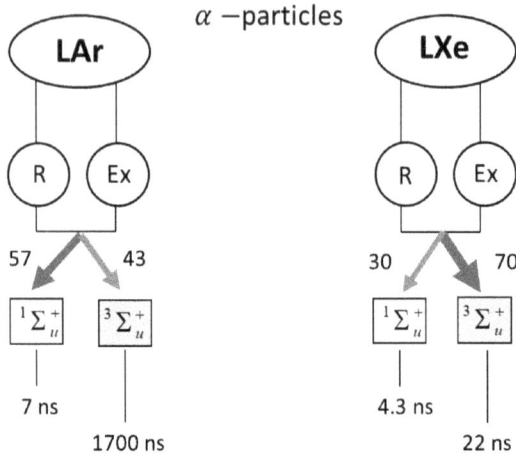

Figure 2.13: Diagram of the observable distribution of luminous energy (in percent) for α-particles based on the data from Hitachi *et al.* (1983). For xenon, different values of the fast/slow intensity ratios have been also reported: 0.53/0.47 and 0.77/0.23 (Kubota *et al.*, 1982a; Kubota *et al.*, 1980, respectively). Redrawn from Chepel and Araújo (2013).

excitation is difficult as very high electric fields are required to suppress recombination along the α-particle tracks appreciably (in practical fields of <10 kV/cm only a few percent of the total charge can be extracted from the track). These facts are reflected in the diagram of the luminous energy sharing and timing for α-particles (Fig. 2.13) by joining the two channels. We also notice that the energy sharing between the singlet and triplet exciton states for α-particles is quite different from that observed with fast electrons — the population of the singlet states is very much enhanced for α-particles. For fission fragments, which produce even higher density of excitations/ionizations, this trend is also confirmed — more light is emitted in the fast component by a factor of about 3.5 compared to the α-particles (Hitachi *et al.*, 1983; Fig. 2.10). The physical nature of this mechanism remains unclear although some hypotheses have been discussed (Hitachi *et al.*, 1983). Thus, Lorents (1976) pointed out that if the free electron density is high, collisions between excimers and electrons would result in frequent transitions between the singlet and triplet states $^3\Sigma_u^+ \rightleftarrows {}^1\Sigma_u^+$, the energy of which differs by only

~0.1–0.2 eV (the singlet state has a slightly higher energy), so that the two states are, in effect, mixed. As the lifetime of the singlet state is much shorter, the net result will be the energy pumping from the slow decay channel to the fast one.

2.3.2. *Light yield for different particles*

The excitation/ionization density along the particle track also has a significant impact on the observed scintillation light yield as shown in Fig. 2.14 (Chepel and Araújo, 2013). Here the light yield is presented as a function of LET dE/dx in liquid argon and xenon for different particles at zero electric field. The LET dependence of the scintillation yield of liquid argon and xenon has been thoroughly discussed by Doke *et al.* (1988, 1990, 2002); Doke and Masuda (1999) and Tanaka *et al.* (2001). The plot in Fig. 2.14 reveals a rather complex behavior which cannot be described solely in terms of LET but depends also on the particle type and energy. The maximum light yields are observed for the light and average mass ions with LET in a wide range of dE/dx ~10^1–10^3 MeV·cm^2/g for liquid xenon and ~10^2–10^4 MeV·cm^2/g for argon. If we assume that in this "flat-top response" region all excitations, both direct and those due to recombination, result in emission of a VUV photon (reactions in Eqs. (2.8)–(2.11)) and that all electron–ion pairs recombine, each resulting in formation of an exciton through the processes in Eqs. (2.15), (2.13) and (2.14), then the number of the emitted photons is at its maximum possible value of $N_{\mathrm{ph}} = N_{\mathrm{ex}} + N_i$.

In similarity with the W-value for charge, the average energy expended per one photon can be found as $W_s = E_0/N_{\mathrm{ph}}$ (here E_0 is the deposited energy as before). Under the above "flat-top response" assumptions, W_s is at its minimum possible value, which can be related to the W-value for charge as

$$W_s^{\mathrm{min}} = \frac{W}{1 + N_{ex}/N_i}. \tag{2.15}$$

The experimental values of W_s^{min} are shown in Table 2.4. One should stress that W_s^{min} is defined for zero field $E = 0$, contrary to the W-value, which is measured by extrapolation to an infinite electric field.

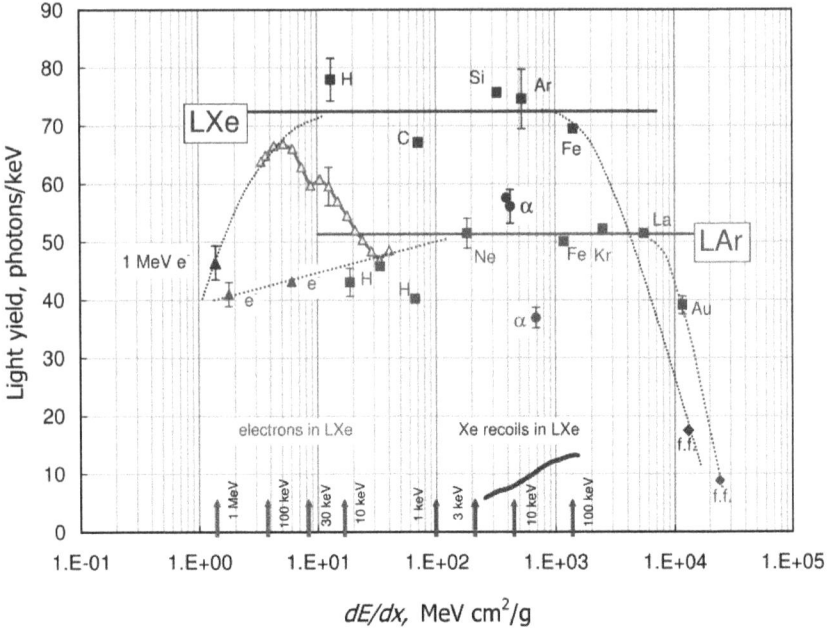

Figure 2.14:   Scintillation yield of LXe and LAr as a function of LET for various particles. Data points (in blue for LXe, and in green for LAr) are after Doke *et al.*, 2002 (figures 2, 4 and Table III); not all data points are shown for clarity. For the same reason, only a typical error bar is shown for each dataset. Dashed lines are for guiding the eye only, no underlying model is assumed; the top plateau corresponds to a minimum energy to produce a scintillation photon $W_s^{min} = 13.8$ eV for LXe and 19.5 eV for LAr according to Doke (2005) and Doke *et al.* (2002). Red arrows indicate average dE/dx values for electrons in LXe calculated as the initial particle energy (indicated next to arrow) divided by the range from ESTAR (ESTAR); blue arrows indicate $(dE/dx)_e$ for Xe recoils in LXe calculated in a similar way using stopping power and range tables from SRIM (Ziegler *et al.*, 2008) and the Lindhard partition function from Lindhard *et al.* (1963a). The red yellow-filled triangles represent the relative measurements with Compton electrons reported by Aprile *et al.* (2012a) (the energy is re-scaled to dE/dx using ESTAR as above, and the response re-normalized to that of γ-rays at $\approx$120 keV using the evaluated curve from Szydagis *et al.*, 2011). Redrawn from Chepel and Araújo (2013).

The validity of the assumption $N_{ph} = N_{ex} + N_i$ for the medium dE/dx values can be tested using the relationship between W_s^{min} and W (Eq. (2.15)) which can be measured more easily. The value of W_s^{min} for liquid argon was calculated to be 19.5 ± 1.0 eV, in good agreement with the experiment (Doke *et al.*, 2002). In those calculations, $N_{ex}/N_i = 0.2$ was used. For liquid xenon, an estimate based on the theoretical value of $N_{ex}/N_i = 0.06$ and $W = 15.6$ eV (Takahashi *et al.*, 1975) results in $W_s^{min} = 14.7 \pm 1.5$ eV (Doke *et al.*, 2002) while measurements of the light yield point to a lower value between 13.0 eV and 13.8 eV, which translates into N_{ex}/N_i between 0.2 and 0.14, respectively (for the same W-value). Doke *et al.* (2002) consider $W_s^{min} = 13.8 \pm 0.9$ eV to be the most probable value for liquid xenon.

The reduction of the scintillation yield at low LET values is attributed to a higher probability for electrons to escape recombination even at zero field (see Section 2.4). As estimated by Doke *et al.* (1985), the probability for an electron to escape from the track of 1 MeV electron in liquid argon ($dE/dx \sim 1$ MeV cm^2/g) is as high as 0.35 while for high energy Ne and Fe ions it is merely ~ 0.003 ($dE/dx \sim 200$ and ~ 1000 MeV cm^2/g, respectively).

The LET dependence of the scintillation yield in liquid argon has been analyzed in the paper by Doke *et al.* (1988). The authors noted that the recombination component of scintillation might be, in fact, due to two distinct mechanisms — volume recombination and geminate (Onsager) recombination (see Section 2.4), which have different dependences on dE/dx. By imposing a finite recombination time for the volume component in order to account for the escape electrons, the authors came to the following expression for the scintillation yield

$$\frac{dL}{dE} = A\frac{dE/dx}{1 + B(dE/dx)} + \eta_0, \qquad (2.16)$$

where the first term describes the contribution of volume recombination and η_0 is a constant, which takes into account the light resulting from the direct excitation and geminate recombination.

At the other end of high LET radiation, such as slow heavy ions and fission fragments, the scintillation yield also decreases but for different reasons. At high ionization and excitation densities, exciton quenching phenomena are observed in solid and liquid scintillators (e.g., Birks, 1964). For noble liquids, a bi-excitonic quenching

mechanism has been proposed (Hitachi *et al.*, 1992b). At high density
of the excited species in the track, the probability of collision and
energy exchange between two excitons becomes not negligible. The
sum of the energies of the two excitons is sufficient to ionize one of
the atoms and therefore the following process may occur:

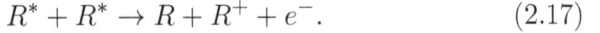

$$R^* + R^* \rightarrow R + R^+ + e^-. \tag{2.17}$$

As a result, at most one photon can be expected after the recombination of the positive ion with an electron instead of two that would
be observed in the absence of exciton–exciton collisions.

It is clear from Fig. 2.14 that low energy electrons and nuclear
recoils ($\lesssim$100 keV) do not follow the same trend as other particle in
function of LET. It has to be noticed, that the very concept of LET
presumes that the track can be considered as linear which is not so
at low energies due to strong scattering of electrons and significant
energy losses in nuclear collisions in the case of nuclear recoils. The
motion of nuclear recoils through the liquid results in a cascade
of a large number of secondary recoils so that the ionization and
excitation pattern is far from being linear (see Figs. 2.7 and 2.8). The
track structure (i.e., spatial distribution of positive ions/holes and
thermalized electrons) and its effect on the observable scintillation
and ionization signals will also be discussed in Section 2.4. Here,
we concentrate on the experimental data on the scintillation yield
for electrons and nuclear recoils in the energy region of interest for
dark matter search and coherent neutrino scattering experiments,
i.e., <100 keV.

Great experimental efforts have been undertaken to measure the
scintillation yield of liquid xenon and argon for electrons with the
energies from ∼100 keV down to ∼1 keV. The results for LXe are
summarized in Fig. 2.15. Some of the measurements have been done
with dedicated setups using Compton scattering of γ-rays from an
external radioactive source (Aprile *et al.*, 2012a; Baudis *et al.*, 2013;
Goetzke *et al.*, 2017), other data were obtained from calibrations of
operational DM detectors with γ-rays from internal sources diluted in
the liquid (e.g., Akerib, *et al.*, 2016a, 2017a,b). Among the convenient

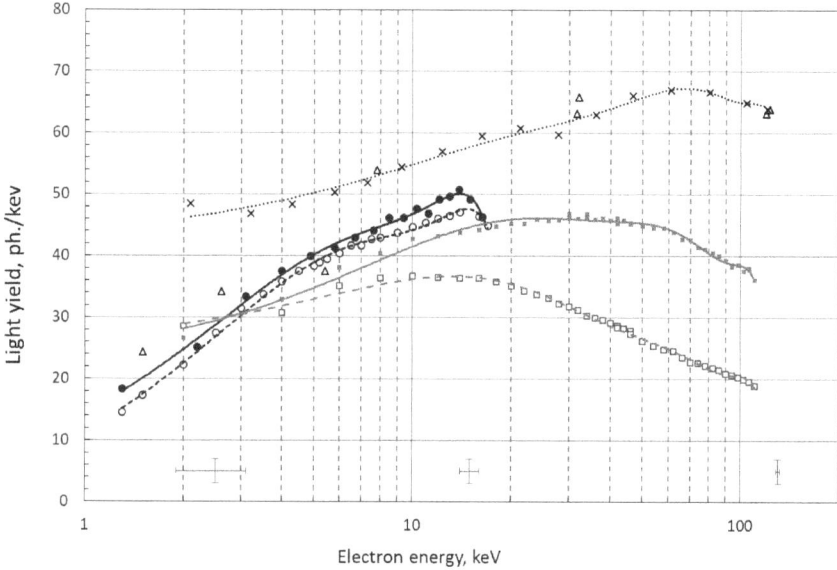

Figure 2.15: Scintillation yield in liquid xenon as a function of electron energy and at different electric fields. Aprile *et al.* (2012a) ($\times$) and Baudis *et al.* (2013) (Δ) — relative measurements at zero electric field; both are converted absolute values using $W_s = 15.66$ eV at 122 keV from Szydagis *et al.* (2011). Measurements with non-zero field: Akerib *et al.* (2016a) ($\bullet$ — 105 V/cm, $\circ$ — 180 V/cm), Goetzke *et al.* (2017) ($\blacksquare$ — 190 V/cm, $\square$ — 2320 V/cm). The three isolated points with error bars indicate approximate uncertainties. Lines are polynomial fits for eye guidance only.

calibration sources are ^{83m}Kr, ^{127}Xe (fluorescence X-rays following the electron capture), or tritium β^--decay. In the latter case, the β-particle energy is measured via the luminous signal and the light yield is adjusted, in each energy bin, to fit the expected energy spectrum.

Absolute calibration of the light detection system in the number of photons, emitted at the source location, is rather challenging and represents an important source of uncertainties that may be partly responsible for some discrepancies. Nevertheless, the general trend is quite clear and confirmed by many measurements. The data in Fig. 2.15 reveal a significant non-linearity of the scintillation response

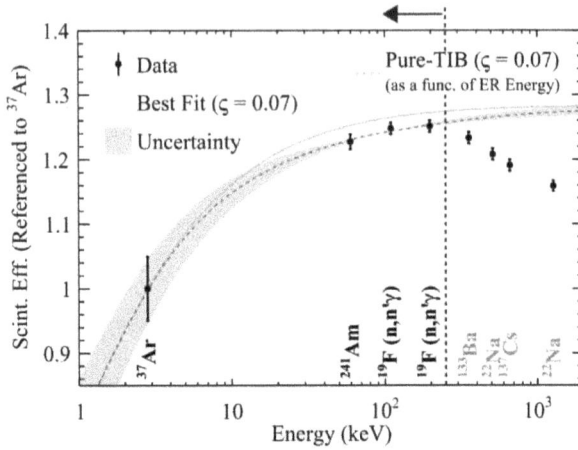

Figure 2.16: Relative scintillation yield in liquid argon as a function of γ-ray energy at zero electric field. TIB stands for Thomas–Imel box model which was used to fit the data. Redrawn from Kimura *et al.* (2020).

in the low energy region.[11] Another important observation is that the electric field affects the scintillation yield very differently in different energy regions being the effect almost nonexistent at the energies of a few keV.

A similar picture is observed in LAr — the light yield drops significantly at low electron energies as can be seen in Fig. 2.16, which represents relative response of a liquid argon scintillation detector to γ-rays of different energies normalized to that for 2.82 keV fluorescence due to electron capture in ^{37}Ar — a natural radioactive isotope of argon (Kimura *et al.*, 2020). The measurements were done at zero electric field. As well as in xenon, one observes a decrease of the light yield when the γ-ray/electron energy approaches $\sim$1 MeV (corresponds to the lowest LET value in Fig. 2.14). The curves in the plot result from adjustment of a combined recombination model to the data (the model will be considered in Section 2.4).

---

[11]Given that, measurements with γ-rays cannot be directly compared with those with electrons of the same energy.

For nuclear recoils, a significant reduction of scintillation yield with respect to all other particles has been observed both in liquid xenon and argon, as one can see in Fig. 2.14 for xenon. Most experimental measurements of the scintillation efficiency for nuclear recoils have been done using elastic scattering of monoenergetic ~MeV neutrons off a LXe or LAr target in a dedicated chamber equipped with photomultipliers. The recoil energy is determined kinematically by setting a detector of scattered neutrons at a fixed angle with respect to the target. The setup can be conveniently calibrated with 122 keV γ-rays from a ^{57}Co source. The relative scintillation efficiency for nuclear recoils at zero electric field is then defined as $L_{eff}(E_{nr}) = Y(E_{nr})/Y(122\,keV)$, where E_{nr} is recoil energy and Y denotes scintillation yield. In more recent experiments, lower energy references have been used, ^{83m}Kr dissolved in liquid xenon or argon, in particular. This isotope emits two photons in cascade with the energies of 32.1 keV and 9.4 keV summing 41.5 keV. One has to stress that inter-comparison of the data obtained with different reference sources has to be done with care given the nonlinearity of the scintillation yield for γ-rays/electrons, which has to be taken into account (Figs. 2.15 and 2.16). It is also important to keep in mind the difference in the energy transfer process for electrons and γ-rays. In the latter case, a cascade of secondary particles of lower energies is produced due to photoelectric absorption on different atomic shells, emission of fluorescence X-rays and Auger electrons. Therefore, the scintillation yield cannot be expected to be the same for γ-rays and electrons of the same energy.

In large detectors, the information on $L_{eff}(E_{nr})$ can be extracted from the calibration data obtained with a broadband neutron source, for example Am/Be. The measured scintillation signal distribution results from the convolution of the spectrum of nuclear recoils and their spatial distribution in the detector, the (unknown) scintillation efficiency $L_{eff}(E_{nr})$, and the detector spatial response function. In order to extract $L_{eff}(E_{nr})$, a detailed knowledge of contributions from the other two factors is necessary. These are obtained by performing a detailed simulation of the whole setup with one of the modern and highly reliable Monte Carlo codes, usually with

GEANT4 (Agostinelli *et al.*, 2003). Then, the function $L_{\text{eff}}(E_{\text{nr}})$ is fitted so that the simulated and measured scintillation spectra coincide. Although more complex, the advantage of this indirect method is that it can be applied to calibration data from real dark matter experiments thus providing a useful validation of nuclear recoil detection efficiencies.

More recently, monoenergetic neutron generators became to be used for calibration of the response of a dark matter detector to nuclear recoils as, for example, in Akerib *et al.* (2016b).

A significant bulk of data on scintillation efficiency for nuclear recoils in liquid xenon has been accumulated up to the date both using direct (Arneodo *et al.*, 2000; Akimov *et al.*, 2002; Aprile *et al.*, 2005a, 2009; Chepel *et al.*, 2006; Manzur *et al.*, 2010; Plante *et al.*, 2011) and indirect methods (Sorensen *et al.*, 2009; Horn *et al.*, 2011). A compilation of data on relative scintillation efficiency for nuclear recoils in liquid xenon is shown in Fig. 2.17. In liquid argon, the data summarized in Fig. 2.18 have been obtained with dedicated setups using scattering of neutrons of fixed energy (Regenfus *et al.*, 2012; Gastler *et al.*, 2012; Alexander *et al.*, 2013b; Creus *et al.*, 2015; Cao *et al.*, 2015; Agnes *et al.*, 2018a). Some data for liquid neon

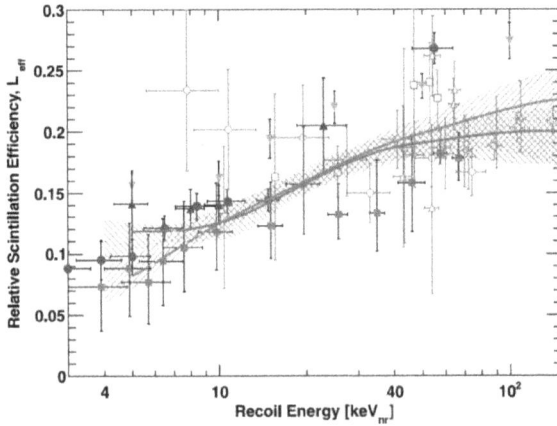

Figure 2.17:   Experimental data on the scintillation efficiency of LXe for nuclear recoils. Redrawn from Horn *et al.* (2011). See the original paper for references.

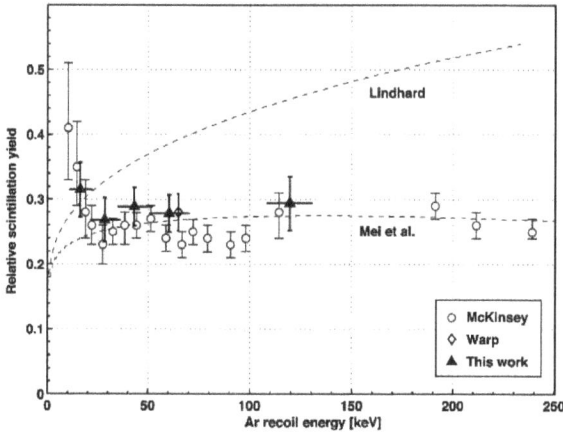

(a)

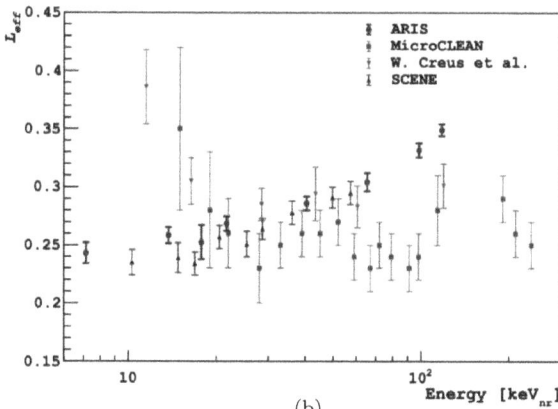

(b)

Figure 2.18: Scintillation efficiency of LAr for nuclear recoils. (a) — from Regenfus *et al.* (2012); (b) — as summarized by Agnes (2020). Data points labeled as "MicroCLEAN" and "McKinsey" are from Gastler *et al.* (2012); "WARP" (single point) — from Brunetti *et al.* (2005); "CENE" — from Creus *et al.* (2015) and Cao *et al.* (2015); "ARIS" — from Agnes *et al.* (2018). The curve marked "Lindhard" shows the fraction of the recoil energy transferred to the liquid in electronic collisions (see Eqs. (2.4) and (2.5)); the lower curve represents calculations with a model by Mei *et al.* (2008) combining the Lindhard's partition function with an empirical saturation formula. Redrawn from Regenfus *et al.* (2012) (a) and Agnes (2020) (b).

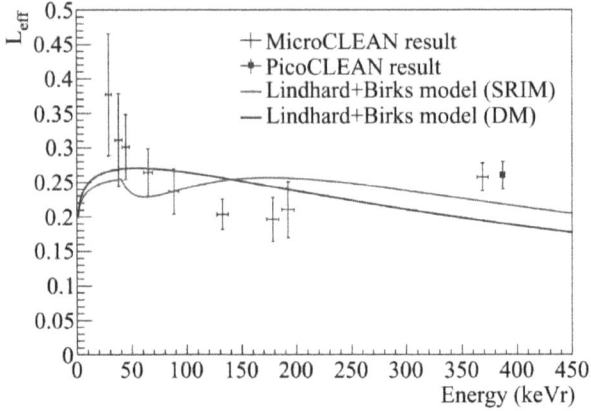

Figure 2.19:   Scintillation efficiency of liquid neon for nuclear recoils: DM stands for the combined model by Mei *et al.* (2012); SRIM — calculations using stopping power from SRIM (Ziegler *et al.*, 2008). Redrawn from Lippincott *et al.* (2012).

have also been published (Lippincott *et al.*, 2012). They are shown in Fig. 2.19.

In spite of some inconsistencies still existing in the data (in particular the upturn at low energies observed in the earlier experiments in argon as well as in neon and xenon but not confirmed by the more recent measurements), it is clear that the scintillation yield for nuclear recoils is much reduced compared to that for γ-rays and electrons. At the energies $\sim$100 keV, the scintillation efficiency for nuclear recoils is about $1/5$ of that for γ-rays of similar energy in LXe and LNe, and $\sim$1/4–1/3 in LAr. A lower light yield for nuclear recoils is explained, in the first place, by the fact that heavy charged particles moving in matter lose a significant part of their energy in nuclear collisions resulting in low energy secondary nuclear recoils and, at the end, production of phonons (see Section 2.2). The LSS and other theories predict a smooth decrease of the fraction of the recoil energy lost in electron collisions (e.g., Fig. 2.18(a)) and thus a reduction of the scintillation yield at low energy, qualitatively consistent with the experiments. The predicted reduction of the electronic losses, however, fails by a factor of $\sim$2 in describing the data which indicate that additional quenching mechanisms exist.

Similarly to slow heavy ions (high LET), quenching of excitons in collisions between them may become significant at high concentrations of the excited species along the track. Hitachi assessed the effect of scintillation quenching for nuclear recoils in LXe and LAr theoretically, taking into account both the Lindhard theory and the bi-excitonic quenching mechanism (Hitachi, 2005a, 2005b; Suzuki and Hitachi, 2011). The model predicts the experimentally observed quenching factors quite successfully. For liquid xenon, for example, at 60 keV recoil energy the model predicts a reduction of the scintillation efficiency by a factor of about 1.5 due to the exciton quenching added to another factor of about 3.2 due to energy losses in nuclear collisions, resulting in total reduction of the scintillation efficiency[12] to the level of ≈ 0.2. For liquid argon, the estimates give ≈ 0.26 for 100 keV recoils with gradual decrease to ≈ 0.15 for 5 keV.

One must comment in this respect that the two distinct processes — the loss of energy by a particle in nuclear collisions (and thus not contributing to excitation of the medium) and quenching of excitons — are sometimes misleadingly referred to as "nuclear quenching" and "electronic quenching".

Manzur *et al.* (2010) noticed that a better agreement with their measurements in LXe can be achieved if, in addition to nuclear energy losses described by the Lindhard theory and bi-excitonic quenching, a non-negligible electron escape probability is assumed for nuclear recoil tracks. This assumption was supported by a very weak dependence of the extracted charge on the applied electric field. Considering a constant ratio $N_{\text{ex}}/N_i = 0.06$, both for electrons and nuclear recoils, they arrived at rather high values for the fraction of escaping electrons, which increases with decreasing recoil energy from ~ 0.15 for 70 keV up to 0.7 ± 0.2 for 4 keV.

---

[12]It is important to underline that in the theory the quenching factor is defined with respect to the maximum possible light yield, i.e., when $N_{ph} = N_{ex} + N_i$ (this situation is realized at moderate dE/dx values — see Fig. 2.14), while experimentally L_{eff} is measured using a γ-ray source as a reference. For instance, for 122 keV γ-rays there is already a reduction of the scintillation yield by a factor of ≈ 0.8.

As the exact nature of the exciton quenching is not completely clear, some authors proposed using empirical parametrizations such as the Birks' formula, previously used with great success to parametrize the light yield of organic scintillators as a function of LET: $\frac{dL}{dE} = \frac{A}{1+kB(dE/dx)}$ (Birks, 1951).[13] Thus, Mei *et al.* (2008) proposed to generically describe the electronic quenching (i.e., quenching of excitons created along the track) with this formula and achieved reasonably good agreement with the experimental data on the scintillation efficiency for nuclear recoils in LXe (for LAr, see Fig. 2.18(a)). Agnes *et al.* (2018a) have found, however, that more terms in the Birks formula are necessary to describe their results for LAr. Thus, a quadratic term $(\frac{dE}{dx})^2$, with another adjustable coefficient, was added to the denominator (known also as the extended Birks formula).

2.4. Charge Yield in the Liquids

2.4.1. *Electron thermalization*

Track structure details are of crucial importance for understanding recombination along the particle track and, consequently, the scintillation and charge signal formation in particle detectors. Under the *track structure* term it is usually understood the spatial distribution of positive ions (holes), ionization electrons and excitations at the instant "right before" the recombination starts. Each of these species follows its own dynamics which is however not totally independent of the dynamics and distribution in space of other participants. The spatial distribution can be characterized by a volume or linear density (or both) depending on the spatial pattern of the track, that varies very much with particle kind and energy (Section 2.2.1). Some common features can nevertheless be identified. First of all,

---

[13]In the original form, the formula reads $\frac{dL}{dx} = \frac{A(dE/dx)}{1+kB(dE/dx)}$, where dL/dx is the scintillation yield per unit length of the particle track, and A and kB are some adjustable constants (Birks, 1951). Parameter k is interpreted as the probability of exciton capture to some quencher. The product kB is frequently called Birks' constant.

we note that the electrons, after leaving the sibling ions, tend to spread out over a larger volume than the holes end excitons given the much higher electron mobility. The energy of sub-excitation electrons is rather high, ~ 5 eV on average (see Table 2.4). Therefore, the Coulomb attraction to the positive ions has little effect on them until their energy is substantially reduced. The only mechanism of losing energy for those electrons is through the elastic collisions with the atoms — a very inefficient process because of a large mass difference between the colliding particles (the relative energy loss in each collision is $\frac{\Delta E}{E} = \frac{2m_e}{M_{at}} \sim 10^{-5}$), that makes the process slow. The time needed for electrons to reach thermal equilibrium with the liquid has been measured to be 0.9 ns for LAr, 4.5 ns for LKr and 6.5 ns for LXe by Sowada *et al.* (1982). This time can be compared, for example, with the characteristic time of formation of self-trapped excitons which is of the order of 10^{-12} s. This also indicates that the recombination would not start effectively until a few nanoseconds after the particle passage as confirmed by the observed scintillation rise time for relativistic electrons (Fig. 2.10). One can also see in the figure that the rise time for α-particles and fission fragments is faster than for electrons thus indicating that the recombination starts before the electrons are fully thermalized.

The mean electron thermalization distance from the sibling ion is also an important parameter found to be rather long in the rare gas liquids: $l_{th} \sim 1700$ nm in LAr (Mozumder, 1995a) and ~ 4500 nm in LXe (Mozumder, 1995b). This length can be compared with the Onsager radius defined as the distance between a positive ion and an electron at which the energy of thermal motion of the electron is equal to the energy of Coulomb attraction to the ion $r_c = e^2/(4\pi\epsilon k_B T)$, where $\epsilon = \kappa\epsilon_0$ is the dielectric constant of the medium. The Onsager radius is equal to 126 nm for LAr, 84 nm for LKr and 54 nm for LXe at the normal boiling point.

The exact distribution of thermalized electrons on distance r from the sibling ion is not known. Some authors found that their results on charge yield as a function of electric field in organic liquids are well fitted if the thermalization length distribution in the following form is assumed (Dodelet *et al.*, 1973)[14]

$$p(x) \propto \frac{4}{\sqrt{\pi} l_{th}} x^2 e^{-x^2}, \qquad (2.18)$$

where $x = r/l_{\text{th}}$ and l_{th} corresponds to the most probable thermal-ization distance.

When considering the electron thermalization process, one also has to bear in mind that the final electron temperature can differ from that of the environment if an electric field is applied. In the presence of electric field, the steady state kinetic energy of quasi-free electrons starts to deviate from the thermal energy at rather low field values — about 50 V/cm in liquid xenon and about 250 V/cm in argon (see Table 3.1 in Chapter 3).

Even at zero electric field, not all electrons recombine with a hole but some can escape recombination and diffuse from the track to the chamber wall or wander around and recombine after $\gtrsim 1$ ms, so that the emitted photon does not contribute to the measured luminous signal. These electrons, termed *escape electrons*, are responsible for the apparent contradiction between small N_{ex}/N_i ratio (Table 2.4) and quite significant contribution of the direct excitation component to the observed scintillation yield for relativistic electrons with the energy of ~1 MeV (Fig. 2.11). While the number of directly created excitons in the particle track is by a factor of 5 to ~17 smaller than the number of primary electron–hole pairs (this corresponds to N_{ex}/N_i between 0.21 and 0.06), their contribution to the total scintillation intensity is as large as 20–30% (Kubota *et al.*, 1979). Taking into account both components, one can write for the total number of emitted photons $N_{\text{ph}} = N_{\text{ex}} + rN_i$, where $r(E)$ is recombination probability as a function of electric field strength: $r = 0$ for an infinite field and $r = r_0$ for $E = 0$. Then, $\chi = 1 - r_0$ represents the escape probability, which can be estimated from the data of Kubota *et al.* (1979). Denoting scintillation yield at zero field

---

[14]These authors also found that a long range tail is somewhat enhanced corresponding to less than 10% of electrons, the fact they have taken into account by introducing an additional polynomial term at the distances $>2.4 l_{th}$.

as L_0 and that at the limit of $E \to \infty$ as L_∞, one obtains

$$\frac{L_0}{L_\infty} = \frac{N_{\text{ex}} + r_0 N_i}{N_{\text{ex}}}, \tag{2.19}$$

from which for the escape probability

$$\chi = 1 - \frac{N_{\text{ex}}}{N_i} \left(\frac{L_0}{L_\infty} - 1 \right). \tag{2.20}$$

An extrapolation is required to estimate L_0/L_∞ from the data on the scintillation yield as a function of field. Therefore, only a rough estimate can be obtained for the escape probability. Using N_{ex}/N_i from Table 2.4, one obtains $\chi \sim 0.45$ for LAr and ~ 0.8 for LKr and LXe. This can be compared with a similar estimate by Doke *et al.* (1985) for liquid argon yielding a value of ≈ 0.35. In general, the trend is consistent with the longer electron thermalization length in heavy rare gas liquids. One should note that the above estimate is rather sensitive to the value of the ratio N_{ex}/N_i. For liquid xenon, a much larger value of 0.2 has been reported by some authors for that ratio (although with the uncertainty of >50%) which would result in $\chi \sim 0.2$–0.4 in tension with what should be expected from the comparison of the thermalization lengths in LXe and LAr.

Much smaller values for the electron escape probability (~ 0.003) have been reported by Doke *et al.* (1985) for Ne and Fe ions with the energy of 0.6 GeV/amu and 0.7 GeV/amu, respectively. This is a result of much higher ionization density in the tracks of those ions and, consequently, much higher probability for electrons to recombine.

2.4.2. *Charge yield and recombination models (electrons)*

At present, there is no comprehensive theory of recombination applicable to the liquefied rare gases. The linear energy transfer to liquid argon, krypton and xenon is rather high even for minimum ionizing particles (about 1–1.5 MeV/g·cm^2). The track structure has a significant impact on the probability of recombination, which turns out to be different in different parts of the track. Several existing theories and models have been adjusted for noble liquids and tested against experimental data allowing, in general, a fair parametrization of the existing data on charge yield as a function of electric field. A complete and consistent physical picture of the recombination in heavy rare gas liquids is however still missing.

The theory of initial (geminate) recombination has been developed by Onsager (1935) and successfully applied for organic liquids in those cases when the ionizations are sparse (e.g., photoionization). The theory considers an isolated electron–ion pair and performs Brownian motion of the electron, thermalized at some distance r_{th} from the sibling positive ion, using the Smoluchowski equation. This motion is a result of scattering of the electron on neutral atoms under the effect of Coulomb attraction to the positive ion and also of the external electric field. For a pair of charges initially separated by a distance r, the escape probability in the absence of electric field is well described by a function $P_{esc} = \exp(-r_c/r)$, where r_c is the Onsager radius $r_c = e^2/(4\pi\epsilon k_B T)$ with $\epsilon = \kappa\epsilon_0$ being the dielectric constant of the liquid. At this distance from the positive ion, the energy of thermal motion of the electron is equal to the potential energy of attraction to the ion. Roughly, one can consider that at $r < r_c$ the electron has a large probability to recombine, while at larger distances it is more likely to escape the recombination.

The external field increases the escape probability and in the low field limit the theory provides a simple expression for the charge yield (Schmidt, 1997):

$$Q(E) = Q_0 e^{-r_c/l_{th}} \left(1 + \frac{E}{E_0}\right), \qquad (2.21)$$

with $E_0 = 2k_B T/er_c$, which can be considered as a measure of the upper limit for applicability of the Onsager theory, estimated to be about 1.3 kV/cm in LAr (Scalettar *et al.*, 1982). It is also assumed that the initial separation distance between the electron and the ion is equal to the mean thermalization length l_{th}.

The necessary condition for using the Smoluchowski equation is that the electron scattering length is negligibly small with respect to the scale of the problem, which in this case is r_c. In the rare gas liquids, however, the scattering length is rather long compared to the Onsager radius and therefore the Onsager theory is not applicable — while $r_c = 125$ nm in LAr and 49 nm in LXe, the momentum transfer scattering length is $\sim$20 nm and $\sim$100 nm, respectively. Indeed, poor agreement between the predictions of the Onsager theory and the measured charge yield for liquid argon has been pointed out by Scalettar *et al.* (1982) and analyzed theoretically by Tachiya (1988).

Another limitation of the Onsager theory is that it considers the electron as having thermal energy which is not the case of quasi-free electrons moving under the effect of electric field in heavy noble liquids for the fields used in radiation detectors.

The contribution of geminate recombination is not expected to be significant in rare gas liquids taking into account the long electron thermalization length of the order of some μm (as mentioned before) in comparison with the mean distance between the positive ions. The latter can be estimated from the value of LET that for minimum ionizing particles gives $\sim$130 nm in LAr and $\sim$40 nm in LXe. This was confirmed by diverse observations (see Dahl, 2009; Szydagis *et al.*, 2011 and references therein).

The Jaffé theory of columnar ionization (Jaffé, 1913) provides a more realistic framework representing a particle track as a cylinder within which the positive and the negative ions are distributed equally with radial density described by the Gaussian function. The recombination is considered as a volume process; the Coulomb attraction between the charges of the opposite signs is not explicitly present in the theory but incorporated, to some extent, into the recombination coefficient α. Diffusion and mobility of ions in an external electric field are also taken into account in the equations which in the final form are

$$\frac{\partial n_\pm(r,t)}{\partial t} = \mp \mu_\pm \vec{E} \cdot \vec{\nabla} n_\pm + D_\pm \Delta n_\pm - \alpha n_+ n_-. \qquad (2.22)$$

Here, $n_\pm$ denote volume concentration of ions of each sign, $\mu_\pm$ and $D_\pm$ are mobilities and diffusion coefficients, respectively.

Jaffé solved the equations for α-particles in gases, under an assumption that the mobility and the initial spatial distribution of both charge carriers are equal, first neglecting the recombination term and then introducing recombination as a perturbation. The general solution has a rather complex form but if the electric field is sufficiently high (above a few kV/cm) the charge yield can be expressed in a simple analytical form as

$$Q(E) = \frac{Q_0}{1 + K/E},\qquad(2.23)$$

where K is a constant characterizing the strength of recombination.

This equation can be easily fitted to the experimental data and has been (and still is) used with success to account for recombination in many gases and liquids including also LAr, LKr and LXe (see an example in Fig. 2.20(a)), in spite of the fact that the basic assumptions of the theory on the equality of the mobilities, diffusion coefficients and initial concentrations of the positive and negative charge carriers are not valid for pure rare gases, where electrons remain free (i.e., not forming negative ions) during the time significantly longer than the recombination time. Another drawback of the Jaffé theory is that it does not incorporate a mechanism of electron escape at zero field. In the absence of an electric field the general solution predicts recombination of all charge carriers at infinite time. This also does not correspond to the observations which show a rather high electron escape probability from the tracks of minimum ionizing particles in liquefied rare gases.

The inconsistency of Jaffé's method in the case of a condensed medium has been noticed by Kramers (1952). He took into account that ion mobilities and diffusion are smaller in the condensed medium and showed that the recombination term is, in fact, dominant over the two other terms. Then he solved the equations neglecting the diffusion term and introducing it afterwards as a perturbation. At high field values, the solution approaches the following form:

$$Q(E) = Q_0 \left(1 - \frac{K}{E}\right),\qquad(2.24)$$

which coincides with the Jaffé's solution for $\frac{K}{E} \ll 1$.

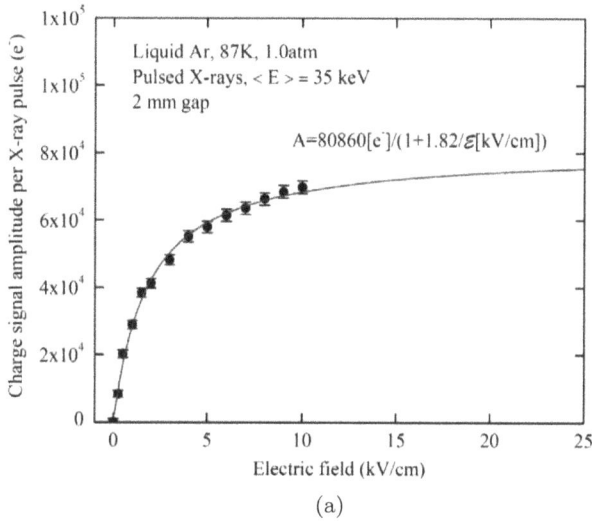

(a)

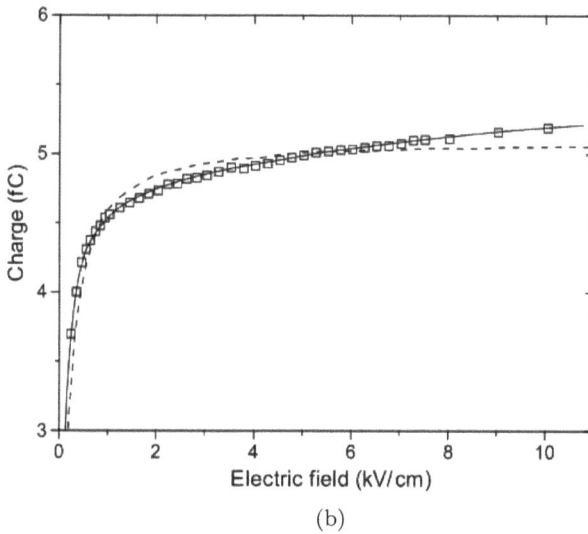

(b)

Figure 2.20: (a) Charge yield as a function of electric field for 35 keV X-rays in LAr. Solid line is the fit with Eq. (2.23). Redrawn from Bondar *et al.* (2016). (b) Charge yield as a function of electric field for 570 keV γ-rays in LXe. Dashed and solid lines — best fits with Eqs. (2.23) and (2.25), respectively. Redrawn from Aprile *et al.* (1991).

The Jaffé–Kramers theory does not take into account the structure of the particle track. In fact, recombination is different in the track parts with different ionization density. Systematic failures to adjust Eq. (2.23) to the data on charge yield have been noticed for relativistic electrons in liquid argon and xenon. An example of that is shown in Fig. 2.20(b) (dashed line). A modification to Eq. (2.23) has been proposed to separate regions with different ionization density (Egorov *et al.*, 1982; Miroshnichenko *et al.*, 1982).[15] In this model, the track of a minimum ionizing particle is considered as being composed by two distinct regions — the main track with relatively low ionization density and weak recombination, and a small number of short tracks of low energy δ-electrons forming, due to strong scattering, a kind of small ionization blobs with high density of ionizations inside them (see Figs. 2.6, 2.7 and 2.14). This can be formalized by adding a second term to Eq. (2.23) with a different coefficient responsible for recombination in those regions K_δ:

$$Q(E) = \frac{Q_0 - Q_\delta}{1 + K_p/E} + \frac{Q_\delta}{1 + K_\delta/E}. \tag{2.25}$$

Here, K_p characterizes recombination in the main part of the particle track. With this equation, a better fit to the experimental data could be achieved (solid line in Fig. 2.20(b)).

The dependence of the recombination coefficient K in Eq. (2.23) on the electron energy in liquid xenon has been studied by Voronova *et al.* (1989) using 662 keV γ-rays and low energy X-rays. The result is presented in Fig. 2.21(a) showing that the recombination coefficient increases with LET or, in other terms, increases as the electron energy decreases (upper scale in the plot), thus supporting the idea of existence of regions with different recombination strength along

---

[15]The fact that the best achieved energy resolution in liquid xenon ionization detectors was significantly worse than one would expect on the basis of the Fano factor was the principal reason for searching for additional sources of fluctuations. In such interpretation, these are due to fluctuation of the small number of low-energy electrons ($\sim$4–6 keV) forming blobs with high-ionization density (Egorov *et al.*, 1982; Machulin *et al.*, 1983).

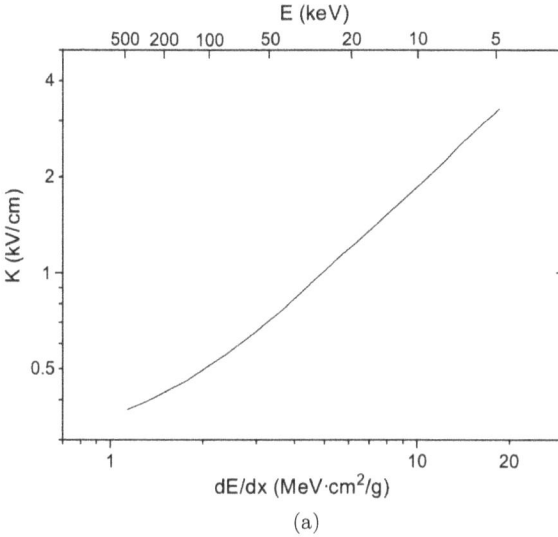

(a)

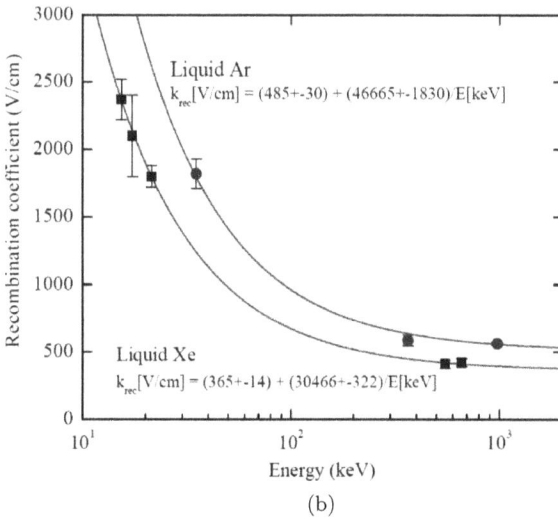

(b)

Figure 2.21: (a) Recombination coefficient K in the equation $Q(E) = Q_0 (1 + K/E)^{-1}$ as a function of LET for electrons in liquid xenon according to Voronova *et al.* (1989). Electron energy is shown in the upper horizontal axis. Redrawn from Lopes and Chepel (2005). (b) Recombination coefficient K as a function of the energy of electrons or γ-rays in liquid xenon and argon. The solid lines are fits with equation $K = a + b/E$. Redrawn from Bondar *et al.* (2020c).

the track of a minimum ionizing particle ($\sim$1 MeV for electrons). The recombination coefficient K as a function of energy in liquid argon and xenon is also shown in Fig. 2.21(b) (compiled by Bondar *et al.*, 2016, 2020c).

Thomas and Imel (1987) adapted the Jaffé equation to liquefied rare gases by setting to zero mobility of positive ions and neglecting diffusion of both charge carriers. The columnar boundary conditions were replaced by an assumption that the electrons and the ions are uniformly and equally distributed within some box of unknown dimensions which are adjusted to fit the experimental data. This resulted in a simple analytical solution for the charge extracted from the particle track

$$Q(E) = Q_0 \frac{1}{\xi} \ln(1 + \xi) \quad \text{with} \quad \xi = \frac{K'}{E}. \tag{2.26}$$

Here, $K' = \frac{N_0\alpha}{4a^2\mu_e}$ is a constant characterizing recombination strength, N_0 is the initial number of charge carriers of each sign populating a box of size a, α is the recombination coefficient in Eq. (2.22), and μ_e is electron mobility. For a sufficiently strong electric field, such that $K/E \ll 1$ and $K'/E \ll 1$, the two recombination constants in Eqs. (2.23) and (2.26) are related as $K' \approx 2K$.

The function in Eq. (2.26) was found to fit the experimental data for electrons in LXe somewhat better than the Jaffé solution in Eq. (2.23) up to the fields of $\sim$10 kV/cm (e.g., Aprile *et al.*, 1991). The coefficient K' was assumed to be independent on the field in those fits — an assumption in tension with the fact that the electron mobility in liquid xenon can be considered constant only for fields below $\approx$50 V/cm (see Section 3.1). For the fields of practical interest, the product $\mu_e E$ should be replaced by drift velocity $v(E)$ so that $\xi(E) = \frac{N_0\alpha}{4a^2 v(E)}$. The assumption on equal initial distributions of the electrons and the ions is questionable, too, given the long thermalization distance in the liquefied rare gases. In spite of all these contradictions, a function in the form of Eq. (2.26) is frequently chosen to describe field dependence of the charge yield in LXe from the electron tracks but also for xenon recoils (see Section 2.4.3).

Delta-electrons can be included into the box model of Thomas and Imel in a similar manner as in Eq. (2.25) (Thomas *et al.*, 1988):

$$Q(E) = (Q_0 - Q_\delta)\frac{1}{\xi_p}\ln(1 + \xi_p) + Q_\delta\frac{1}{\xi_\delta}\ln(1 + \xi_\delta) \qquad (2.27)$$

with $\xi_p = K'_p/E$ and $\xi_\delta = K'_\delta/E$. Here, K'_p and K'_δ are the coefficients characterizing recombination in the main track and in the δ-electron tracks, respectively.

The recombination coefficients K, K_p, K_δ and K', K'_p, K'_δ obtained by fitting Eqs. (2.23) and (2.25)–(2.27) to the charge yield data in LXe and LAr by different groups reveal a very significant dispersion (see compilations by Lopes and Chepel, 2005; Obodovski, 2005) thus indicating little sensitivity of the charge yield as a function of field (the observable) to the model parameters (underlying physics). This can be useful for practical needs but little adds to our understanding of the fundamental processes in the track. For the Thomas–Imel model, the fitting parameters were found to depend on the field region: Aprile *et al.* (1991) observed that the recombination coefficients K'_p and K'_δ for 570 keV γ-rays in LXe are significantly smaller (by factors of about 2 and 8, respectively) if only low field data from $\approx$50 V/cm to 1.8 kV/cm are considered instead of the whole data set up to 10 kV/cm.

A different approach to the recombination dynamics — a model of successive re-encounters and run-aways — has been developed by Mozumder (1995a, 1995b). According to his model, the track is considered as a line of positive ions with electrons at a distance equal to the thermalization length from it, initially at rest. The electrons move under the effect of the cylindrical field of the ions with velocity defined by their mobility and local field strength, having a certain probability of recombining when they come across the line of ions, κ. The recombination probability as well as the thermalization length are used as free parameters to adjust the model to the measured charge yield curves. Those electrons that did not recombine, move away from the track to the classical return point and turn back. The ion density is thus gradually reduced and the distance from the track to the classical return point increases as well. An external electric field is applied parallel to the track, in this model, resulting in additional motion of the oscillating electrons along the line of positive ions, finally escaping recombination when they reach the track end. The encounter recombination probability was found to be $\kappa \approx 0.01$ in LAr and ~ 0.01–0.03 in LXe for low-LET tracks. The model provided a simple microscopic insight to the recombination process in the liquefied rare gases although the agreement with the experiment is not as good as that of the parametric models. The thermal motion is not included.

In search for a simple parametrization of the charge yield as a function of electric field but also the ionization density in the track, the Birks' saturation function was found to be useful. Cennini *et al.*, 1994 have found that the charge yield from muon tracks per unit track length in the ICARUS liquid argon TPC, plotted against stopping power of the particle in that track segment, is well fitted by the Birks' formula $\frac{dQ}{dx} = A\frac{dE/dx}{1+B(dE/dx)}$, with constant B depending on the electric field strength (Fig. 2.22). Constant A in

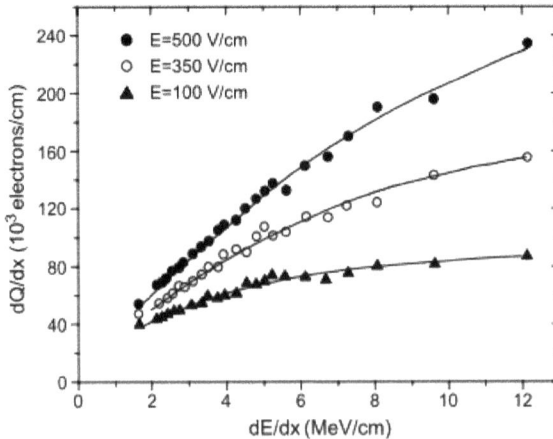

Figure 2.22: Charge yield per unit track length as a function of stopping power of relativistic muons in liquid argon (Cennini *et al.*, 1994). The data were obtained by measuring the charge extracted from different parts of muon tracks in the ICARUS time projection chamber. Solid lines correspond to fits of the Birks saturation function to the data (see text). Redrawn from Cennini *et al.* (1994).

the above equation represents the asymptotic value of the collected charge per unit of energy, which in that work was found to be about 36 electrons/keV (while from $W = 23.6$ eV one should be expecting 42 electrons/keV).

The Birks formula can be rewritten for the charge extracted from a particle track with a given $\frac{dE}{dx}$ as $Q(\frac{dE}{dx}) = Q_0 \frac{1}{1+B(\varepsilon)(dE/dx)}$ (here we use symbol ε for electric field and E for particle energy). On the other hand, the Jaffé equation states that $Q(\varepsilon) = Q_0 \frac{1}{1+K(E)/\varepsilon}$. By combining the two equations, one gets the relationship between the two constants $B(\varepsilon)\frac{dE}{dx} = \frac{K(E)}{\varepsilon}$ so that the charge yield from a part of the track with a given LET value $\frac{dE}{dx}$ is expressed in the form (Amoruso *et al.*, 2004)

$$Q\left(\varepsilon, \frac{dE}{dx}\right) = A\frac{Q_0}{1 + \frac{\eta}{\varepsilon}\frac{dE}{dx}}, \tag{2.28}$$

where A is a correction constant. The combination $\eta = K(E)(\frac{dE}{dx})^{-1}$ was found to be constant for different parts of muon tracks in LAr with LET values from 1.5 MeV/(g/cm^2) to 24 MeV/(g/cm^2), although it has been verified only for a limited range of fields between 0.2 and 0.5 kV/cm. The regions with high ionization densities in the track (δ-electrons) were suggested to be taken into account by adding more terms to Eq. (2.28) with different dE/dx values in the way similar to Eqs. (2.25) and (2.27).

Good proportionality of the charge signal to the electron/γ-ray energy was measured in ionization chambers for energies $\gtrsim 100$ keV with all liquefied rare gases. At lower energies, however, a significant variation of the charge yield with electron energy was observed, as illustrated by Fig. 2.23 for LXe and Fig. 2.24 for LAr. The observed charge signal is the product of two factors — the number of ionizations produced by a particle $N_i = E_0/W$ (E_0 is the deposited energy) and the probability for electrons to escape recombination $(1 - r)$, where r is the recombination probability which depends on the track structure details and on the electric field strength. Although one cannot completely exclude that the W-value of LXe is not constant at low energies (see comments to Table 2.3), its variation

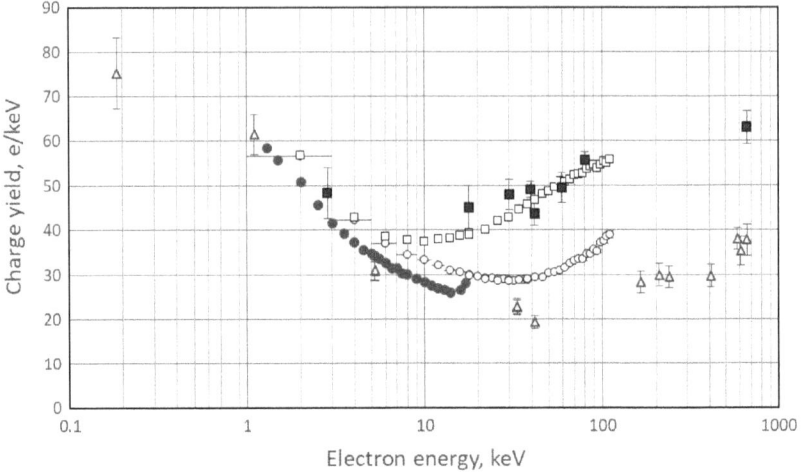

Figure 2.23: Charge yield from tracks of electrons and γ-rays in LXe as a function of electron energy for fields of 0.18 kV/cm — (●) from Akerib *et al.* (2016a) and (△) from Akerib *et al.* (2017a, 2017c); 0.19 kV/cm (○) and 2.32 kV/cm (□) — from Goetzke *et al.* (2017); 3.75 kV/cm (■) — from Akimov *et al.* (2014). Average vertical errors are approximately ±0.2 (Goetzke *et al.*, 2017), ±2 (Akerib *et al.*, 2016a); horizontal errors for Goetzke *et al.* (2017) are due to the geometry of the Compton scattering setup.

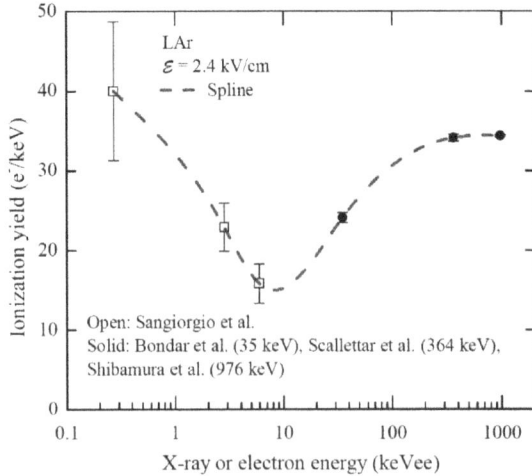

Figure 2.24: Charge yield from tracks of electrons and X-rays in LAr as a function of electron or X-ray energy for field of 2.4 kV/cm. The data points are those of Sangiorgio *et al.* (2013), Bondar *et al.* (2016), Scalettar *et al.* (1982) and Shibamura *et al.* (1975). Redrawn from Bondar *et al.* (2016).

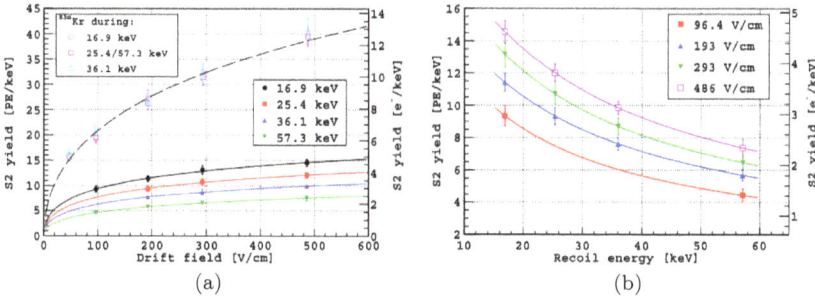

(a)   (b)

Figure 2.25:   Charge yield from tracks of nuclear recoils and electrons in LAr as a function of drift field (a) and the recoil energy (b) in a double phase TPC (Cao *et al.*, 2015). In the left plot, the charge yield for ^{83m}Kr γ-rays is also shown (41.5 keV total energy deposition resulting from a cascade of two photons of 32.1 keV and 9.4 keV). The charge is measured by extraction of the electrons to the gas phase where secondary scintillation develops. The authors present two vertical scales — in the number of photoelectrons per keV measured at the PMT photocathodes (a) and in the absolute number of electrons per keV of nuclear recoil energy (b). The calibration required for the latter adds another 10% of uncertainty. Redrawn from Cao *et al.* (2015).

is not expected to be significant and the recombination process is generally assumed to be the principal contributor.

The charge yield for low energy γ-rays in LAr as a function of electric field, according to Cao *et al.* (2015), is shown in Fig. 2.25(a) (together with nuclear recoils which will be discussed in Section 2.4.3). In these measurements, ^{83m}Kr diluted in the liquid was used as a source. This isotope emits a cascade of two photons with the energies of 32.1 keV and 9.4 keV seen in the detector as a single energy deposition of 41.5 keV. It should be noted that at low energies the charge signal is usually measured indirectly using double phase liquid/gas systems (dedicated setups or real dark matter detectors). In these systems, the electrons are extracted to the gas phase where proportional scintillation develops in the electric field. So that the number of photons emitted in the proportional scintillation is a measure of the number of extracted electrons (shown in the left vertical axis in units of photoelectrons/keV). The absolute calibration of the yield in the number of electrons in this work has been done indirectly as no signal from a single electron emitted to the gas could be well distinguished.

The dashed curve in the plot is a joint fit of a modified Eq. (2.26) to all the data (i.e., those for γ-rays and argon recoils). The parameter ξ is represented in the form $\xi = N_i C / \varepsilon^B$, where B and C are adjustable constants. The constant C characterizes the recombination in the tracks and is different for γ-rays and nuclear recoils while B is assumed to be the same for both particles and found to be $B = 0.61 \pm 0.03$. The fact that $B \neq 1$ is a consequence of the electron mobility being a function of field ε and not a constant as originally assumed by Thomas and Imel, 1987 (see Fig. 3.1 in Chapter 3).

2.4.3. *Charge yield and recombination (nuclear recoils)*

As for nuclear recoils, the charge yield has been intensively studied both in LAr and in LXe, boosted by the necessity of calibration of the energy scale in the dark matter search experiments. The charge yield from nuclear recoil tracks in LAr is shown in Fig. 2.25 as a function of the recoil energy and drift field as reported by Cao *et al.* (2015). These data however seem to be in tension with the measurements at higher energies by Bondar *et al.* (2014b). In LXe, quite a significant amount of data has been accumulated in the recoil energy range from several hundreds of keV down to $\sim$1 keV and even to a few hundreds of eV (Aprile *et al.*, 2006b, 2013, 2018b, 2019a; Dahl, 2009; Sorensen *et al.*, 2009, 2010; Angle *et al.*, 2011; Akerib *et al.*, 2016b, 2017a; Manzur *et al.*, 2010; Horn *et al.*, 2011, Lenardo *et al.*, 2019). Some of these data are represented in Figs. 2.26 and 2.27 as a function of the recoil energy and electric field, respectively. Although inconsistencies at low energies exist, for the energies above a few keV the picture is quite coherent in what general trend and also the absolute yields are concerned.

Along with the experimental data in Figs. 2.26 and 2.27, predictions of the data driven NEST v.2 model are also shown (Szydagis *et al.*, 2011, 2013; Lenardo *et al.*, 2015). We shall discuss the model in some details below in this section. Here, we only show a plot of computed charge and light yields for LXe as a function of the

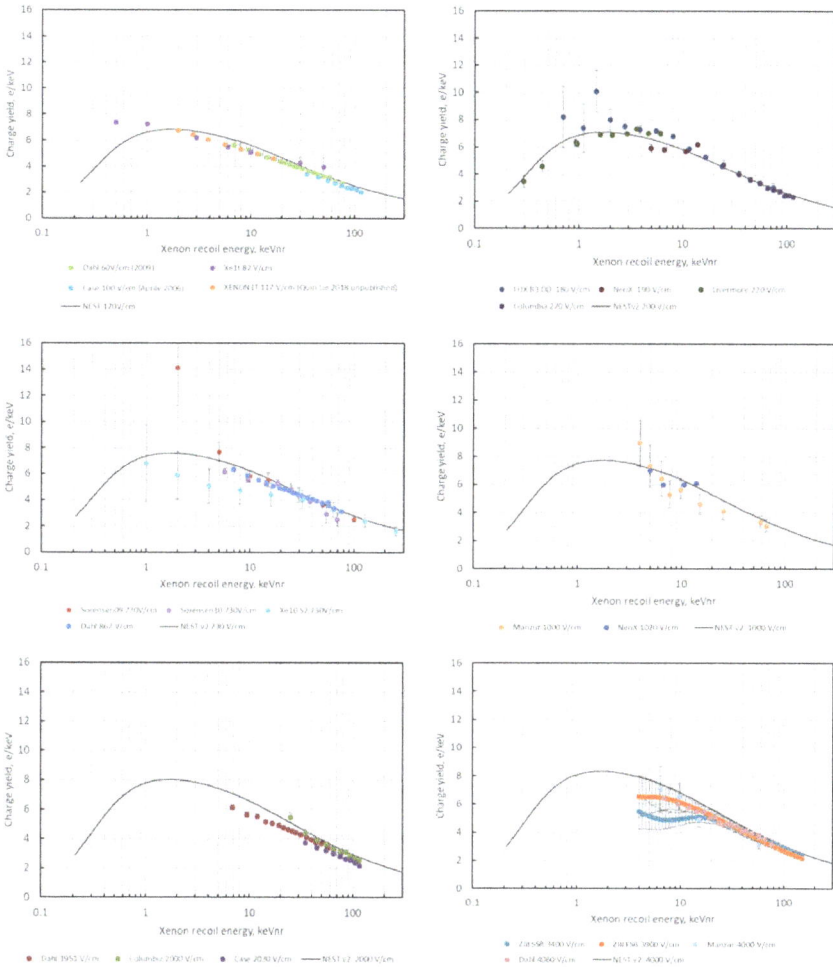

Figure 2.26: Charge yield from nuclear recoil tracks in LXe as a function of the recoil energy at different drift fields. Points — experimental values from Aprile *et al.* (2006b, 2018b, 2019a), Dahl (2009), Sorensen *et al.* (2009), Sorensen (2010), Manzur *et al.* (2010), Angle *et al.* (2011), Horn *et al.* (2011), Akerib *et al.* (2016b), Lenardo *et al.* (2019); solid line — NEST v2.0 prediction (help of M. Szydagis is greatly appreciated).

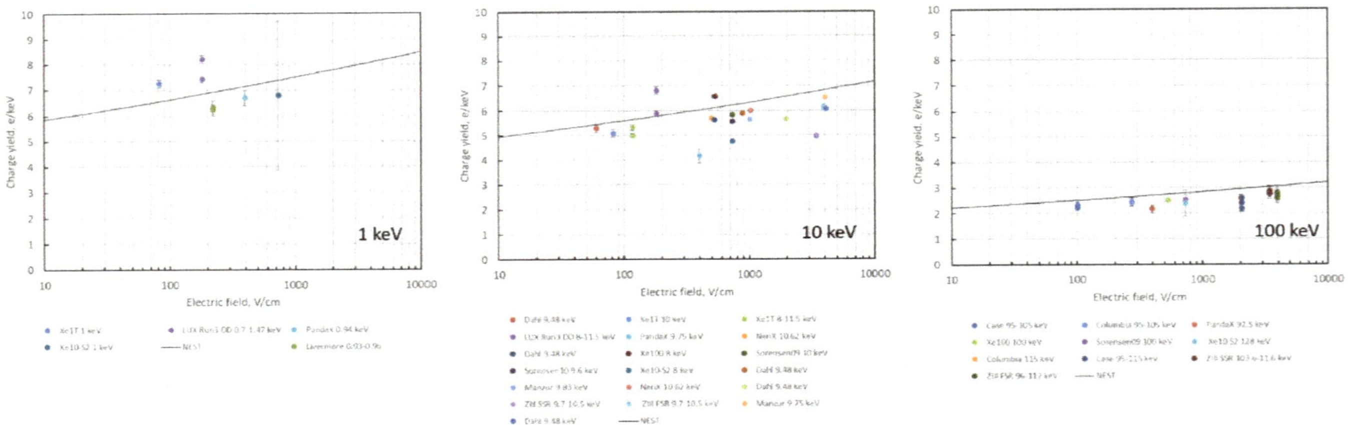

Figure 2.27: Charge yield from nuclear recoil tracks in LXe as a function of field for different recoil energies (indicated in the plot). Points — experimental values from Aprile *et al.* (2006b, 2013, 2018b, 2019a), Dahl (2009), Sorensen *et al.* (2009), Sorensen (2010), Manzur *et al.* (2010), Angle *et al.* (2011), Horn *et al.* (2011), Akerib *et al.* (2016b), Lenardo *et al.* (2019); solid line — NEST v2.0 prediction (help of M. Szydagis is greatly appreciated).

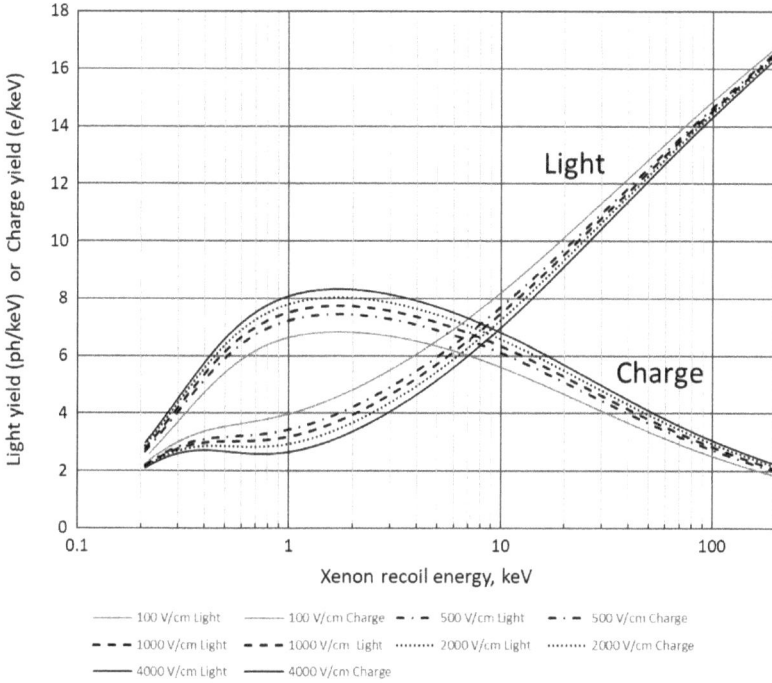

Figure 2.28: Light and charge yields from xenon recoil tracks in LXe as computed with NEST v2.0.

recoil energy for several field values in order to represent the whole set of data for liquid xenon in a single plot (Fig. 2.28).

The following observations can be made from these data. First of all, we notice that the charge extracted from the nuclear recoil tracks is several times smaller than that measured for electrons of the same energy (cf. Figs. 2.23 and 2.24). This is not surprising taking into account that only a small fraction of nuclear recoil energy goes to electronic excitations (ionizations and excitons) of the medium being most of the recoil energy spent in nuclear collisions. According to the LSS theory (Section 2.2.1, Eq. (2.5)), this fraction is a smoothly increasing function of the recoil energy, as can be seen in Fig. 2.25 for argon. In LXe this fraction is calculated to be $\approx$0.15 at 1 keV, $\approx$0.2 at 10 keV, and $\approx$0.3 at 100 keV. In LAr, the respective values are $\approx$0.2, $\approx$0.3 and $\approx$0.45. (Different theoretical models result in somewhat different values — see, e.g., Mangiarotti *et al.*, 2007; Bezrukov *et al.*, 2011 as well as references in Section 2.2.1).

The clearly observable trend of the charge yield to increase as the energy decreases from $\sim$100 keV to $\sim$1 keV is however opposite to that predicted by the LSS theory for the electronic component of energy transfer. Although some theoretical models do allow an upturn of the electronic stopping power as the recoil energy decreases, as pointed out by Bezrukov *et al.* (2011), it is not accompanied by the observed scintillation yield. This behavior can be interpreted as a consequence of an overwhelming reduction of the recombination probability due to a sparser distribution of the positive ions at low energies. Such increase is also observed for electrons (Fig. 2.23). The sparseness of the track structure may also explain a weak dependence of the yield on the electric field above $\sim$100 V/cm — a large distance between the ions and long electron thermalization length result in that quite a significant part of the electrons escape recombination even at zero field. On the other hand, if the sum of the number of the extracted electrons and the emitted photons $N_e + N_{ph}$ is taken as a measure of the recoil energy, the function $L(\epsilon)$ in Eq. (2.5) seems to reproduce correctly the experimental data for xenon recoils, as shown by Dahl (2009), and Sorensen and Dahl (2011).

As it has been observed long ago, there is no non-radiative de-excitation path from the lowest excited state of R_2^* to the ground state in pure liquefied rare gases (e.g., Doke *et al.*, 2002; Suzuki and Hitachi, 2011).[16] On the other hand, recombination of an electron and an ion/hole leads necessarily to formation of an exciton and hence emission of a photon. If we assume that there is no bi-excitonic quenching, which is significant only at high excitation densities, or other quenching mechanisms (e.g., energy transfer to impurities), this leads to a conclusion that while each of the two observable quantities — the number of collected electrons N_e and the number

---

[16]As experimental evidence for that fact, one can refer to the measurements of VUV absorption and luminescence spectra of Ar clusters in a molecular beam by Karnbach *et al.* (1993). The two spectra were found to be complementary to each other.

of emitted photons N_{ph} — is affected by the recombination process (in opposite ways), their sum is not. Using r for the probability of an electron to recombine with a hole, one can write for the number of electrons

$$N_e = N_i(1 - r),$$ (2.29)

and for the number of scintillation photons

$$N_{ph} = N_{ex} + N_i r,$$ (2.30)

that results in

$$N_{ph} + N_e = N_i + N_{ex}.$$ (2.31)

We also assume here that the electron attachment to impurities is negligible. The attachment does not break the balance but makes the attached charge practically unobservable in many detection systems due to the low ion mobility and significantly reduced probability of its extraction from the liquid to gas.

Therefore, the sum of the charge and scintillation signals should be the best estimator for the deposited energy, independent of r and, thus, the electric field. Taking into account that the sum $N_i + N_{ex}$ is also equal to the maximum possible number of primary scintillation photons that can be emitted from a particle track (corresponds to a complete recombination), one can write for the deposited energy $E_0 = (N_{ph} + N_e)W_s^{min}$. Here, W_s^{min} is the minimum value of $W_s = E_0/N_{ph}$ at the "flat-top" response in Fig. 2.14, which is equal to $W_s^{min} = E_0/(N_i + N_{ex})$ according to Eq. (2.15).

Another important aspect of using the sum of the two signals is that the fluctuations in the energy distribution between the ionization and the luminous channels can be substantially suppressed due to anti-correlation between them. This idea has been explored by several authors (e.g., Séguinot et al., 1992), being the accurate calculation of N_{ph} from the number of detected photons (photoelectrons, to be precise) the principal difficulty. It was later successfully implemented by Conti et al. (2003) and by Aprile et al. (2007) as illustrated here by Fig. 2.29, which shows a significant

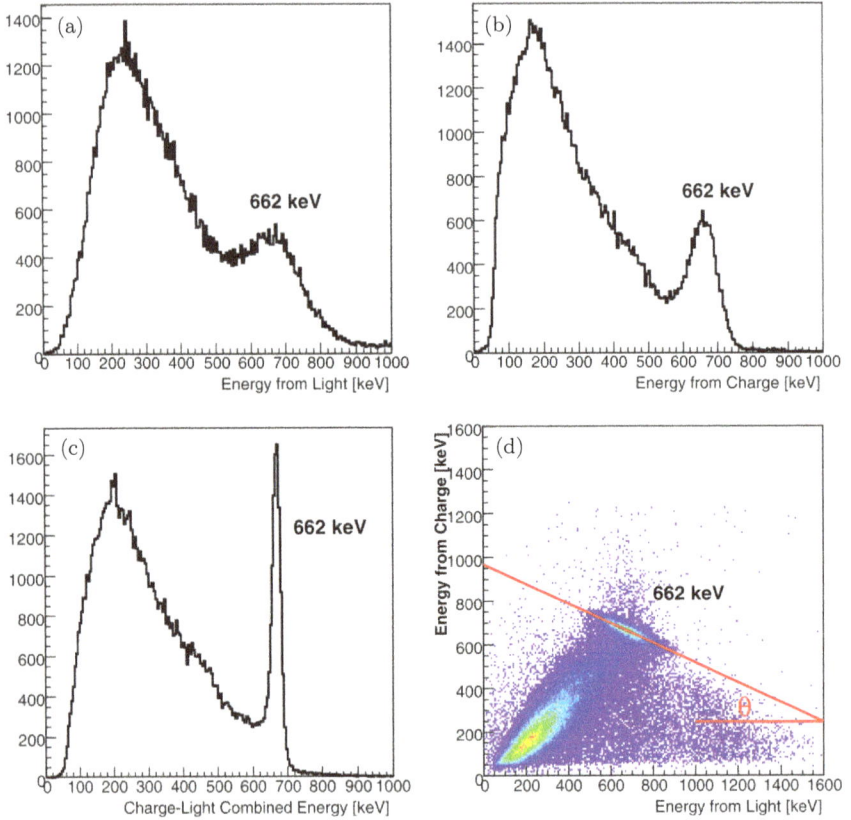

Figure 2.29: Gamma-ray spectrum in liquid xenon obtained at 1 kV/cm from the primary scintillation signal (a), from the ionization signal (b), and from their sum (c). The scatter plot in (d) shows the sum rule for the combined signal. Redrawn from Aprile *et al.* (2007).

improvement of energy resolution for 662 keV γ-rays in liquid xenon, reaching $\sigma/E \approx 1.7\%$ when the sum of the two signals is used as energy estimate. The missing coefficient between the number of emitted photons and the number of detected photoelectrons was determined from the correlation plot and using scales in energy units in both channels, calibrated with the respective photopeak positions. A similar energy resolution was measured for γ-rays of higher energy (2.458 MeV) by EXO Collaboration searching for neutrinoless double beta decay (Auger *et al.*, 2012).

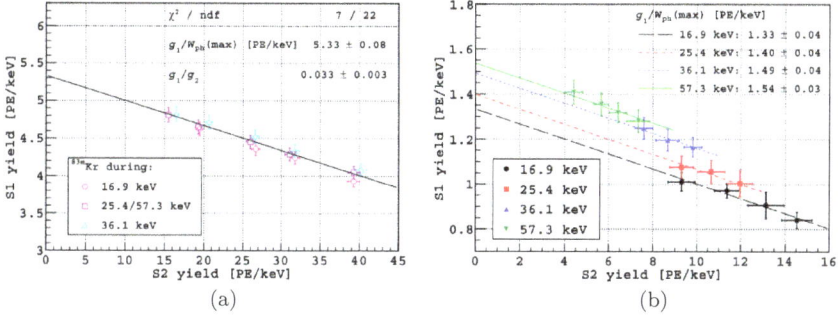

Figure 2.30: Anti-correlation between the ionization (S2) and primary scintillation (S1) signals in LAr double phase TPC: (a) — for 41.5 keV γ-rays from ^{83m}Kr at different values of drift field from 50 V/cm to 486 V/cm; (b) — for argon recoils of different energies (indicated in the plot) and at different drift fields from 96 V/cm to 486 V/cm. The ionization signal is measured by extracting the electrons to the gas phase and measuring the secondary scintillation from them in the gas. Both signals are, therefore, presented in units of photoelectrons/keV. Redrawn from Cao *et al.* (2015).

Linear anti-correlation of the primary scintillation and the ionization signals was also observed with electrons and γ-rays of lower energy in LXe (Aprile *et al.*, 2011, 2012a; Araújo *et al.*, 2012; Solovov *et al.*, 2012). Thus, $\sigma/E \approx 3.4\%$ was measured for 122 keV γ-rays by Solovov *et al.* (2012); at 40 keV, Aprile *et al.* (2011) observed an improvement from 16.2%, when only the primary scintillation signal was used, to 9% for the combined signal. Cao *et al.* (2015) reported on well linear anti-correlation of the primary scintillation and ionization signals in LAr both for 41.5 keV γ-rays from ^{83m}Kr and nuclear recoils in the energy range of 16.9–57.3 keV (Fig. 2.30).

Although the validity of the equality in Eq. (2.31) is, strictly speaking, limited to the situations when the exciton quenching can be neglected, Shutt *et al.* (2007) proposed using the sum $N_{ph} + N_e$ as the best energy estimator for both electrons and nuclear recoils in LXe in the energy range of 20–100 keV. They introduced a unified W-value as an average energy to create an excitation (either exciton or ionization) so that it can be written for the deposited energy

$$E_0 = (N_e + N_{ph})W. \tag{2.32}$$

From the calibration data with 122 keV γ-rays, they found that $W = 13.46 \pm 0.29$ eV. Using this value in the combined energy estimator both for electrons and for nuclear recoils, a significant improvement of particle identification was obtained. The combined signal is in fact used by all dark matter experiments with double phase detectors as the best estimator for the deposited energy.

For nuclear recoils, Eq. (2.32) has to be corrected for losses in nuclear collisions by the energy partition factor L, which in the framework of the LSS theory is described by Eq. (2.5):

$$E_0 L = (N_i + N_{ex})W \qquad (2.33)$$

(Dahl, 2009). The exciton quenching can be also taken into account by introducing a correction factor f_l in Eq. (2.30):

$$N_{ph} = (N_{ex} + r N_i) f_l. \qquad (2.34)$$

Keeping the ratio N_{ex}/N_i as a parameter and eliminating N_i from Eqs. (2.29), (2.33) and (2.34), one comes to the following expressions for the number of extracted electrons and emitted photons

$$N_e = \frac{E_0 L}{W}(1-r)\frac{1}{1+N_{ex}/N_i} \qquad (2.35)$$

$$N_{ph} = \frac{E_0 L}{W} f_l \left(1 - \frac{1-r}{1+N_{ex}/N_i}\right). \qquad (2.36)$$

Here, the field dependence is contained in the recombination coefficient r, while the particle energy is implicit in the Lindhard energy partition factor L, quenching factor f_l and also in r. For electrons, L and f_l are assumed to be equal to 1.

The above considerations on the recombination process and the energy partition for nuclear recoils constitute the theoretical model for the *Noble Element Simulation Technique* (NEST) (Szydagis et al., 2011, 2013; Lenardo et al., 2015),[17] — a computer program and database widely used by the dark matter community for prediction

---

[17]Given the limitations of the LSS model, in the most recent version of the program, v.2, the energy partition is described by an empirical function providing the best fit to the existing light and charge yield data.

of observable signals as a function of particle energy and drift field as well as for checking consistency of the calibration data. In this model, the "unified" W-value is assumed to be independent on particle type and energy, as well as the ratio N_{ex}/N_i. For LXe, the respective values are equal to 13.7 eV (Shutt *et al.*, 2007; Dahl, 2009) and 0.06. The recombination coefficient and its dependence on the electric field and particle type and energy are not known. Instead, an adjustable function is used to reproduce most of the known data on charge and light yield, simultaneously. The recombination coefficient is evaluated using the modified Thomas–Imel box recombination model, namely

$$r = 1 - \frac{\ln(1 + \xi)}{\xi}, \tag{2.37}$$

where $\xi = \gamma N_i/\varepsilon^\delta$. The ionization density effects are taken into account through N_i while the dependence of the electron drift velocity on the field strength ε are accounted through factor δ. A very good match with the existing data is obtained for all particles of interest in the energy range from GeVs down to a few hundreds of eVs — electrons, nuclear recoils, α-particles, γ-rays, fission fragments (which constitute the background in the rare event experiments).

Chapter 3

Electron Drift in and Emission from Noble Liquids

3.1. Electron Drift and Diffusion

Drift of the excess electrons in noble liquids is well studied experimentally in a wide range of electric fields and temperatures. The drift is usually characterized by drift velocity v_d and mobility μ related to each other as $v_d = \mu E$. Experimental data on electron drift velocities in liquid and solid argon, krypton and xenon are presented in Figs. 3.1–3.3. Some disagreement exists between several datasets, especially in the region of some tens of V/cm, but the general behavior is consistent. For xenon, this is illustrated by Fig. 3.4 where the data obtained by Miller *et al.* (1968), Huang and Freeman (1978), Yoshino *et al.* (1978), and Gushchin *et al.* (1982a) are presented in a single plot. One can notice three regions of different behavior of drift velocity as a function of the electric field strength. At low field values, $v_d \propto E$ meaning that the electron mobility is constant. At intermediate field values, $v_d \propto \sqrt{E}$ and so mobility is a decreasing function of the electric field. Above some field value, the drift velocity becomes practically independent on the field as one can see in the plots. Such behavior is similar both in the liquid and solid states although the drift velocity is higher in solids by a factor of $\sim$2. If we designate by E_1 and E_2 the approximate field values at which the electron drift velocity changes its behavior, we notice a tendency of both values to decrease from argon to xenon (Table 3.1).

Significant deviations from this behavior of drift velocity with field have been reported for $T \gtrsim 200$K in liquid xenon (Huang and

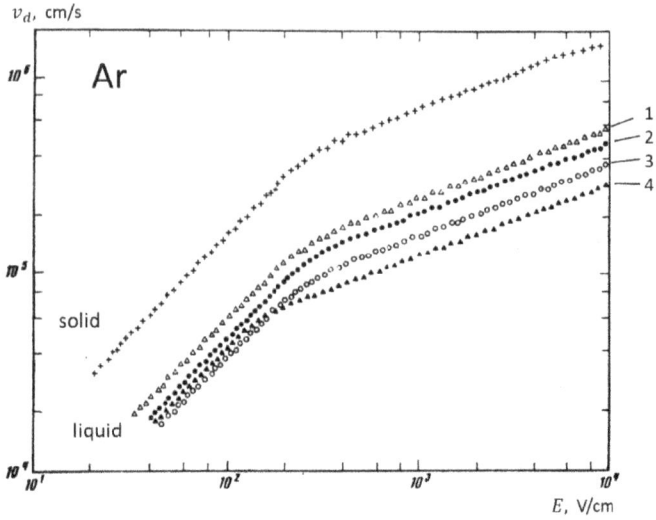

Figure 3.1: Drift velocity in liquid and solid argon at different temperatures:
solid — 80 K; liquid: 1 — 85 K, 2 — 100 K, 3 — 120 K, 4 — 130 K. Redrawn from
Gushchin *et al.* (1982a).

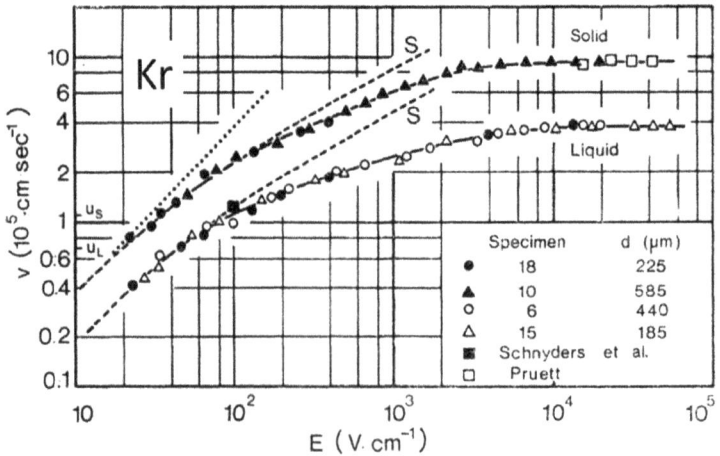

Figure 3.2: Drift velocity in liquid and solid krypton at different temperatures:
solid — 113 K; liquid — 117 K. Redrawn from Miller *et al.* (1968).

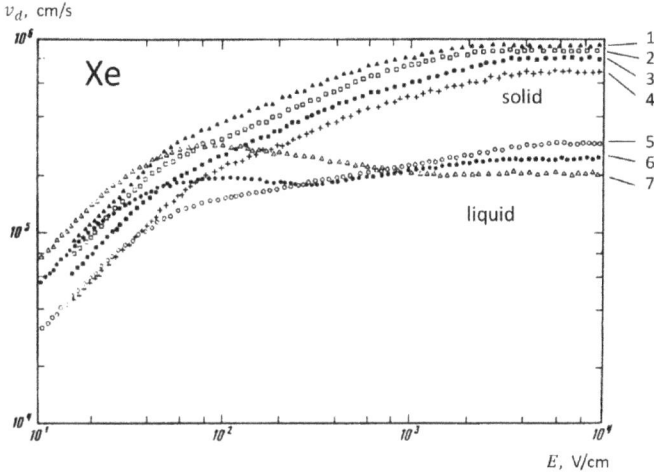

Figure 3.3: Drift velocity in solid and liquid xenon at different temperatures: solid 1 — 100 K, 2 — 120 K, 3 — 140 K, 4 — 155 K; liquid: 5 — 165 K, 6 — 200 K, 7 — 230 K. Redrawn from Gushchin *et al.* (1982a).

Freeman, 1978; Gushchin *et al.*, 1982a) revealing a broad maximum at $\sim$100 V/cm before the saturation is reached. Moreover, in the field region of some tens to several hundreds of Volts, a significant dependence of the electron drift velocity on temperature was found, for both liquid and saturated vapors, as illustrated in Fig. 3.5. In the liquid, the drift velocity varies non-monotonously, first increasing as the temperature increases, and dropping by almost two orders of magnitude close to the critical temperature. A similar trend, although apparently monotonous, is observed in the saturated vapor. The drop in the electron drift velocity from the triple point to the critical temperature can reach a factor as high as 300, for some fields. At present, there is no clear understanding of such complex behavior of the drift velocity in liquid xenon with temperature.

To get some initial insight on the electron transport in noble liquids under electric field, it is instructive to compare the distances between the atoms in different aggregation states with the thermal wavelength of the electrons and their scattering mean free path. In a gas at normal conditions, the distance between the atoms d_g is an order of magnitude larger than the interaction range of a few Å ($\sim r_m$,

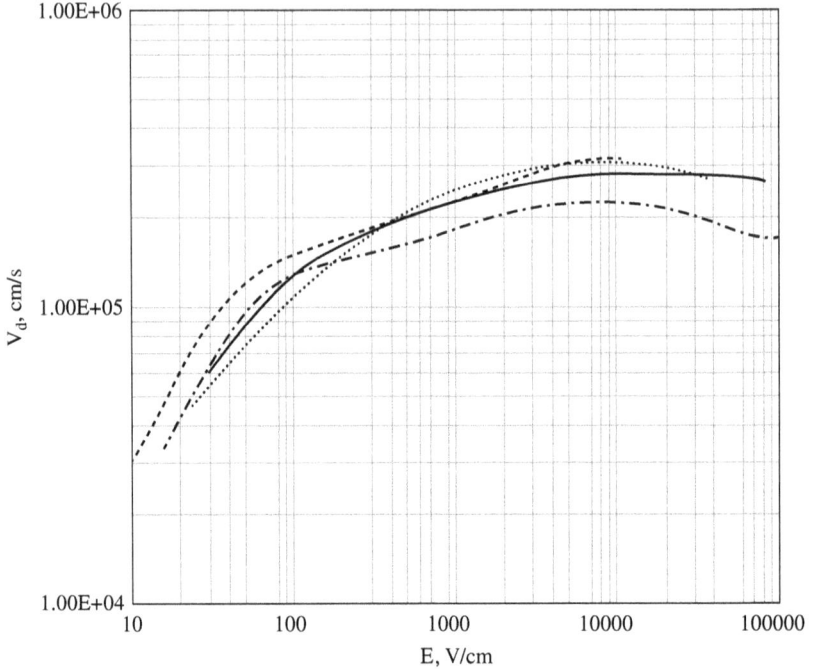

Figure 3.4: Drift velocity of electrons in liquid xenon at temperatures 163 K and 165 K according to different sources: Miller *et al.* (1968) (solid line), Gushchin *et al.* (1982a) (dashed line), Huang and Freeman (1978) (short dash grey line), and Yoshino *et al.* (1978) (dash-dotted grey line).

Table 3.1: Approximate values of the critical fields and saturation drift velocities for heavy noble liquids and solids near the triple point.

		T (K)	E_1, (V/cm)	E_2, (V/cm)	ν_{sat} (cm/s)
Ar	Liquid	85	250	$3 \cdot 10^4$	$8 \cdot 10^5$
	Solid	80	250	$1 \cdot 10^4$	$1.5 \cdot 10^6$
Kr	Liquid	117	80	$1 \cdot 10^4$	$4 \cdot 10^5$
	Solid	113	60	$5 \cdot 10^3$	$1 \cdot 10^6$
Xe	Liquid	163	50	$6 \cdot 10^3$	$3 \cdot 10^5$
	Solid	157	40	$4 \cdot 10^3$	$7 \cdot 10^5$

Note: Below E_1, $v_d \propto E$; in the interval between E_1 and E_2, $v_d \propto E^{1/2}$; above E_2, the drift velocity reaches saturation $v_d = v_{\text{sat}}$. Data from Miller *et al.* (1968), and Gushchin *et al.* (1982a).

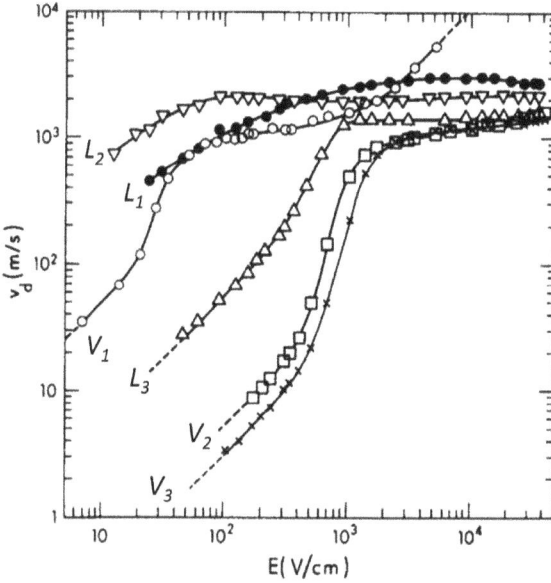

Figure 3.5:   Drift velocity of electrons in liquid xenon (marked with letter L) and its saturated vapours (marked with V) at temperatures from the triple to critical points. Liquid — 163 K (L_1), 216 K (L_2), and 288 K (L_3); vapor — 172 K (V_1), 273 K (V_2), and from 286 K to 289.73 K (V_3). The dash-line continuations of the curves at low field values have the slope equal to 1 thus confirming the proportionality $v_d \propto E$. Redrawn from Huang and Freeman (1978).

where $r_m \approx 3\text{--}4\,\text{Å}$ is the equilibrium distance of the pair potential), while in liquids it is close to it $d_l \approx r_m$ (see Table 2.2). This explains why the atoms can be treated, approximately, as independent non-interacting particles in gas but not in the liquid. The thermal electron wavelength is $\lambda_{\text{th}} = \frac{h}{\sqrt{3kTm_e}} \sim 60\,\text{Å}$ at room temperature and $\sim 120\,\text{Å}$ at the triple point of argon (84.8 K), that is, $\lambda_{\text{th}} \sim d_g$ in gas and $\lambda_{\text{th}} \gg d_l$ in the liquid. This fact results in that the contribution from the coherent scattering becomes significant in the liquid and, therefore, the electron behavior in liquids should be very different from their behavior in gas.

In what the electron scattering is concerned, λ_{th} should be compared with the mean free path between two successive collisions, which can be determined from the momentum transfer cross-section

σ_m and the number density n as $L = (n\sigma_m)^{-1}$. One can verify that $L \gg \lambda_{\text{th}}$ in gas but, again, not in liquid so that the meaning of mean free path is not so clearly defined in a dense medium. Translating this observation into time domain, one can picture the electron motion in gas as a free motion, most of the time, interrupted by short intervals of interaction with individual atoms, one at a time. At higher densities, this simple image becomes less plausible and at the limit of liquids turns out to be false — neither the atoms can be seen as non-interacting, nor the electron considered as interacting (from time to time) with a single atom. The concept of collision cross-section, so successful for a pair of particles, also loses its original clear meaning and has to be redefined.

Several theories and models have been developed to take into account the small distances between the atoms and their highly correlated positions in dense fluids, on one hand, and multiple scattering effects of the electron from many atoms simultaneously, on the other. A concise review of different approaches to the problem can be found, for example, in the introductory part of the paper by Boyle *et al.* (2015). A special attention should be paid to the theory developed by Lekner and Cohen (Lekner, 1967; Cohen and Lekner, 1967) which describes interaction of electrons with the atoms of fluid as scattering on an *effective potential* experienced by the electron throughout a single collision. The effective potential is built from the single atom potential as seen by an electron (the potential is assumed to be the same in the liquid), corrected for the screening of the long range polarization force by the presence of other atoms. The screening effect is accounted through the pair correlation function $g(r)$ which reflects the fluid structure. The electron motion through the fluid is therefore considered as a sequence of (effective) single collisions/scatterings on the effective potential — a picture reminding the electron motion through a dilute gas. Therefore, such convenient parameters as scattering cross-sections and mean free path can be used in the condensed phase.

Once the effective potential is defined, the transport parameters can be determined by solving the Boltzmann equation with the modified collisional term (Cohen and Lekner, 1967). As a sequence of the contribution from the coherent scattering in liquids, which leads to a significant anisotropy, the mean free path for transfer of energy is different from the mean free path of momentum transfer. Two cross-sections are therefore defined — the energy transfer cross-section $\sigma_0(\varepsilon)$ and the momentum transfer cross-section $\sigma_1(\varepsilon)$. The former is independent on the liquid structure while the latter depends on the structure factor $S(k)$ which can be measured by neutron diffraction or calculated.[1]

$$\sigma_0(\varepsilon) = 2\pi \int_0^\pi \sigma(\varepsilon, \theta)(1 - \cos\theta)\sin\theta \, d\theta, \tag{3.1}$$

$$\sigma_1(\varepsilon) = 2\pi \int_0^\pi \sigma(\varepsilon, \theta)S(q)(1 - \cos\theta)\sin\theta \, d\theta. \tag{3.2}$$

Here, $\sigma(\varepsilon, \theta)$ is the differential cross-section obtained from the effective scattering potential, $q = 2k\sin\frac{\theta}{2}$ is the momentum transfer and $\varepsilon = \hbar^2 k^2/2m_e$ is the electron energy. At the limit of low densities, $S(k) = 1$ (no correlation between the scattering centers) and the two cross-sections coincide giving the familiar momentum transfer cross-section σ_m in gases.

The original Lekner and Cohen theory was successfully used to obtain the electron drift velocities in liquid argon (Lekner, 1967) up to the fields of $\sim 10^4$ V/cm above which the experiment shows saturation of the drift velocity but the theory predicts its decrease (Miller *et al.*, 1968). For heavier noble gases, however, a significant discrepancy with the experiment was observed. A number of modifications have been proposed since then to improve the agreement for argon and also for other liquids (see Boyle *et al.*, 2015 for references), in particular in the high field region.

It has been thought that inelastic collisions do not play significant role in monatomic liquids until the electron energy exceeds the lowest electronic excitation energy (for liquid argon this would be at the fields of $\gtrsim 2 \cdot 10^5$ V/cm) and thus the electron behavior in these liquids

---

[1] The structure factor is related with the pair correlation function $g(r)$ through the equation $S(k) = \int e^{ikr}[g(r) - 1]d\boldsymbol{r}$.

is totally determined by the elastic scattering. On the other hand, the Raman scattering of visible light was observed in liquid argon resulting in frequency shifts of the scattered photons up to ≈ 12 meV (e.g., MacTaguie *et al.*, 1969), i.e., comparable with the energy of the vibrational level structure of diatomic argon molecule Ar_2 in its electronic ground state (see Fig. 2.1). This fact suggests that inelastic electron scattering may also occur on the fluctuations of the interatomic distances (meaning polarizability fluctuations). Basing on these observations, Sakai proposed to add low energy inelastic scattering into the scattering term of the Boltzmann equation and obtained a very good agreement with the experimental data for liquid argon, krypton and xenon (Sakai, 2005, 2007), although at the cost of introducing empirically defined inelastic cross-sections.

In a more recent work by Boyle *et al.* (2015), the adequacy of the interatomic potential used by Lekner and Cohen was questioned. The authors attempted to replace it by an *ab initio* potential obtained by solving the Dirac equation for argon atoms following the method of Chen *et al.* (2008), who tested their theory for krypton gas with success. In the case of liquid argon, a better agreement, than in the previous works, was obtained for the electron characteristic energy (although the data exists only in a very limited range of the electric field), but for drift velocities the agreement was not that good, in particular at high fields, where a decrease of the velocity was predicted as in the original work of Lekner and Cohen. This, again, may indicate the importance of inelastic collisions at low electron energies, specific for the condensed phase.

The region of constant mobility corresponds to thermal equilibrium of the drifting electrons with the medium so that their average kinetic energy is not altered by the electric field and remains equal to $\frac{3}{2}kT$. Taking into account that the average velocity of an electron between the subsequent collisions is equal to $\langle v \rangle = \frac{eE}{m}\tau$ (where τ is mean time between the collisions) and $v_d = L/\tau$ with $L = (n\sigma_m)^{-1}$, one gets for the electron mobility $\mu = e/(n\sigma_m\sqrt{3kTm_e})$, consistent with the measurements. In turn, the deviation from the linear behavior of the electron drift velocity with field is associated with the electron "heating".

At zero electric field, the electron velocities are distributed isotropically and in accordance with the Maxwell–Boltzmann distribution function. Applied electric field introduces anisotropy in the field direction and also distorts the shape of the energy distribution. At low field values, these effects are small but become more and more significant as the field strength increases. The electron energy distributions in liquid argon and xenon, calculated by Bakale *et al.* (1976) for two field values of 10 V/cm and 100 kV/cm, are shown in Fig. 3.6. More recent computations for liquid xenon in strong fields are shown in Fig. 3.7, where one can clearly see deviation from the Maxwell–Boltzmann distribution function at higher fields (Simonović *et al.*, 2019).

Compilation of some data on the mean electron energy as a function of the electric field is shown in Fig. 3.8. One should point out that the electron energy cannot be measured directly but can be estimated using the Einstein formula $\varepsilon_T = 3eD_T/2\mu$ which relates

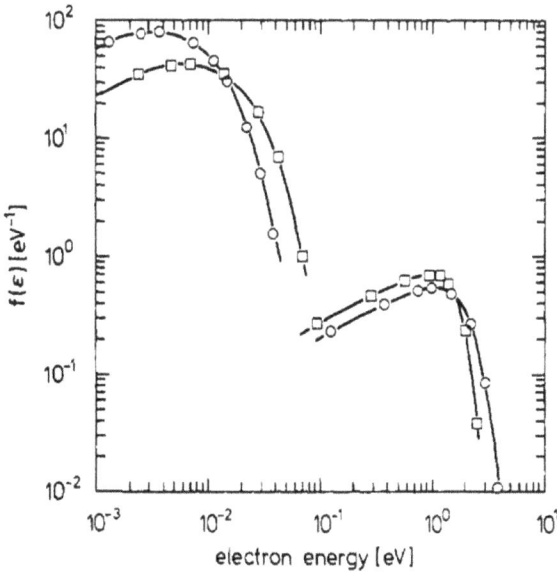

Figure 3.6: Energy distribution function for electrons at two values of the field strength, 10 V/cm (left pair of curves) and 100 kV/cm (right pair), in liquid argon (circles) and liquid xenon (squares). Redrawn from Bakale *et al.* (1976).

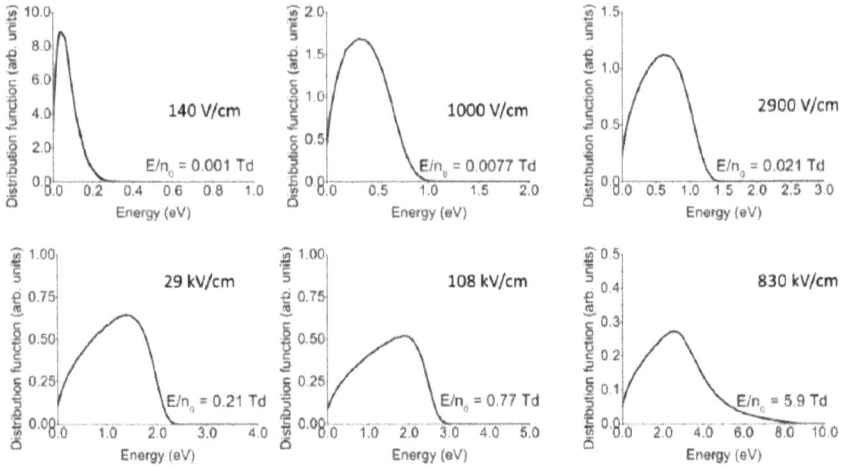

Figure 3.7: Energy distribution function for electrons in liquid xenon at several values of the electric field strength. Redrawn from Simonović *et al.* (2019).

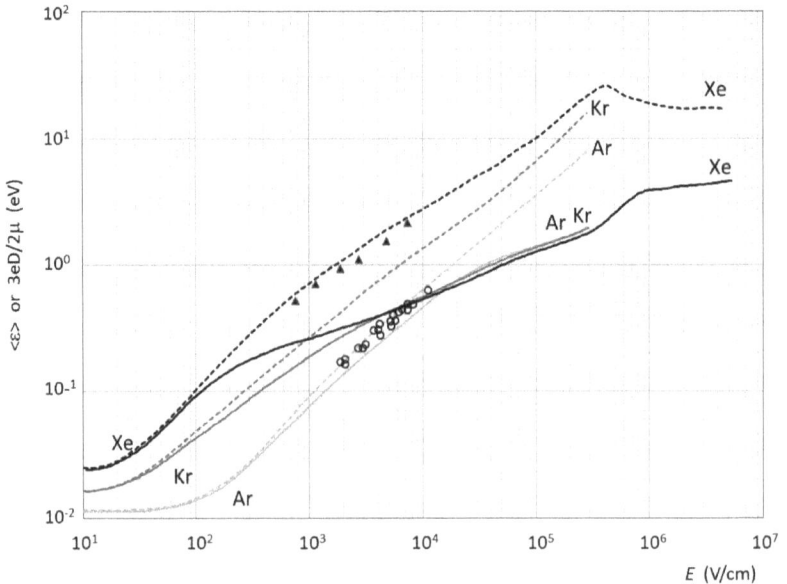

Figure 3.8: Characteristic energy $(3eD)/(2\mu)$ (dashed lines) and mean electron energy $\langle \varepsilon \rangle$ (solid lines) in liquid argon, krypton and xenon according to Sakai *et al.* (2006). Lines — theory, solid triangles — experimental data for xenon from Kubota *et al.* (1982) (measured by E. Shibamura), open circles — for argon from Shibamura *et al.* (1979), both for characteristic energy.

the energy (called characteristic energy) with the experimentally determined transverse diffusion coefficient and electron mobility. This value is normally used for testing theoretical models which among other parameters provide the electron average energy $\langle \varepsilon \rangle$ as a function of the field. Both values, the mean electron energy $\langle \varepsilon \rangle$ and the characteristic energy ε_T, for liquid Ar, Kr and Xe are plotted as a function of electric field in Fig. 3.8 from the work of Sakai *et al.* (2006). The curves represent theoretical calculations while the symbols are the experimental values for ε_T. The experimental results exist only in a limited range of fields to which theoretical models are adjusted. The mean electron energy in liquid xenon in comparison with that in the gas as a function of the reduced field is shown in Fig. 3.9 (Simonović *et al.*, 2019). If plotted as a function of the field strength as in Fig. 3.8, one can notice that the electron energy calculated in the two works differs by a factor of up to 1.7, depending on the field. Other calculations also result in somewhat different values (Atrazhev *et al.*, 2005) thus revealing the difficulties of theoretical assessment of the energy of drifting electrons in the liquid state.

In what the diffusion of electrons in noble liquids is concerned, it is known to be much smaller than in the respective gas, which is an

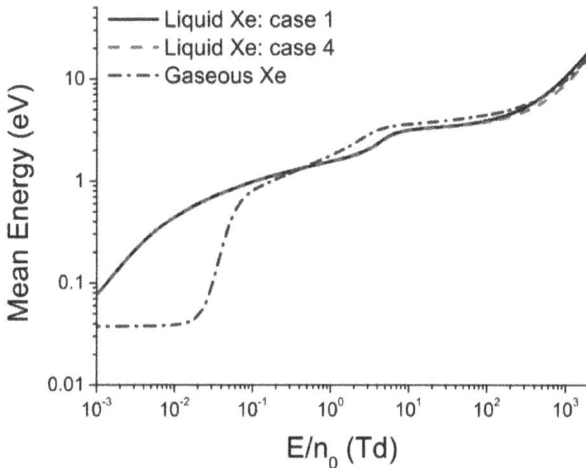

Figure 3.9: Mean electron energy in liquid and gaseous xenon as function of reduced electric field. Redrawn from Simonović *et al.* (2019).

advantage when a precise particle localization is required. Just to give an example, at the end of a 10 cm drift in liquid xenon in a 1 kV/cm field the transversal size of the electron cloud will be about 0.7 mm (sigma) while the longitudinal spread about half of that. The low electron diffusion is a great advantage of the condensed noble gases but it also makes its measurement a challenge. The experimental data on the electron diffusion coefficients in noble liquids were rather scarce until recently. Several large experiments, using detectors with liquid xenon and argon, made study of diffusion a part of their R&D program and carried out systematic measurements of the diffusion coefficients although in a limited range of field of interest for those experiments. The most studied liquid in this aspect is liquid xenon. Measurements of longitudinal and transverse diffusion require different techniques and are done with different setups. Recent papers of EXO Collaboration provided new data as well useful compilations of the existing information (Albert *et al.*, 2017; Njoya *et al.*, 2019), represented in Figs. 3.10 and 3.11, with theoretical predictions from other authors also added. In general, a reasonable agreement exists although the discrepancies can be as large as a factor of 2 in some field regions and even larger in the case of the longitudinal diffusion coefficient. At low fields, even the tendency observed experimentally is opposite to that predicted by calculations.

Diffusion coefficients for liquid argon are presented in Fig. 3.12. For transversal diffusion, the experimental data are from Shibamura *et al.* (1979) and Derenzo (1974). Recent theoretical results by Boyle *et al.* (2015) are also presented showing a good agreement with the data. Cennini *et al.* (1994) measured the longitudinal diffusion coefficient in the range of field between 100 V/cm and 350 V/cm and presented an average over all measurements which is reflected in the plot as the horizontal error bar. More recently, Li *et al.* (2016) measured the signal rise time as a function of field and drift distance and derived electron mobility and the effective energy with respect to the longitudinal diffusion, defined as $\varepsilon_L = eD_L/\mu$ in similarity with the Einstein formula for the characteristic energy. The authors do not present directly the diffusion coefficient but only ε_L. However they

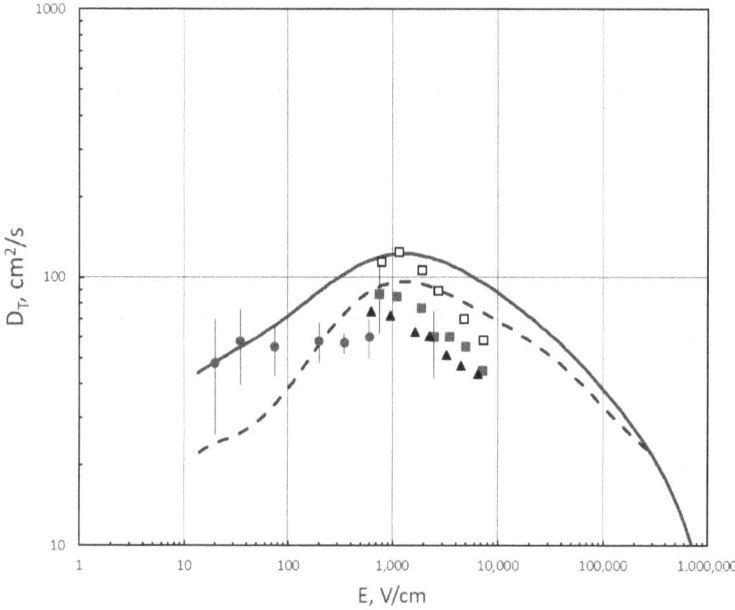

Figure 3.10: Transverse diffusion coefficient in liquid xenon. Experimental data: circles — Albert *et al.* (2017), triangles — Doke (1982), squares — Doke (1981), open squares — Shibamura *et al.* (1986) (unpublished, data retrieved from Atrazhev *et al.*, 2005). Calculations: solid line — from Atrazhev *et al.* (2005) (retrieved from the characteristic energy and electron mobility), dashed line — Boyle *et al.* (2016).

do offer a useful parameterization for D_L in the range of fields from $0.1\,\mathrm{kV/cm}$ up to $1.5\,\mathrm{kV/cm}$ including the temperature dependence. This parameterization is shown in the plot as a dashed line. They also compare their results on ε_L with those derived from the coefficient of transverse diffusion, measured by Shibamura *et al.* (1979), using a relationship between the two coefficients

$$\frac{D_L}{D_T} = 1 + \frac{E}{\mu}\frac{\partial \mu}{\partial E}, \tag{3.3}$$

which results from the equations relating diffusion coefficients, electron mobility and their temperature (e.g., Huxley and Crompton, 1974)

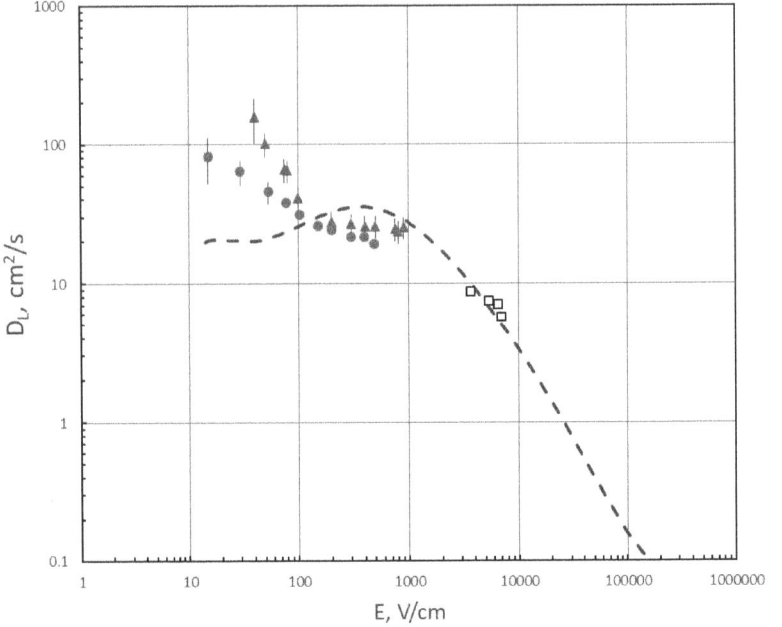

Figure 3.11: Longitudinal diffusion coefficient in liquid xenon. Experimental data: triangles — Njoya *et al.* (2020), circles — Hogenbirk *et al.* (2018), open squares — Shibamura *et al.* (1986) (unpublished, data retrieved from Njoya *et al.* 2020). Calculations: dashed line — Boyle *et al.* (2016).

$$D_T = \frac{kT}{e}\mu, \tag{3.4}$$

$$D_L = \frac{kT}{e}\left(\mu + E\frac{\partial\mu}{\partial E}\right). \tag{3.5}$$

The result of these calculations is shown in Fig. 3.12 as a dotted line.

3.2. Drift of Positive and Negative Ions

Drift velocity of positive ions in noble liquids is much lower than that of electrons resulting in a several orders of magnitude smaller current. In detectors with analogue readout, the integration time in the μs range is usually used effectively removing the ionic component of the current from the measured signal. In contrast, advanced time

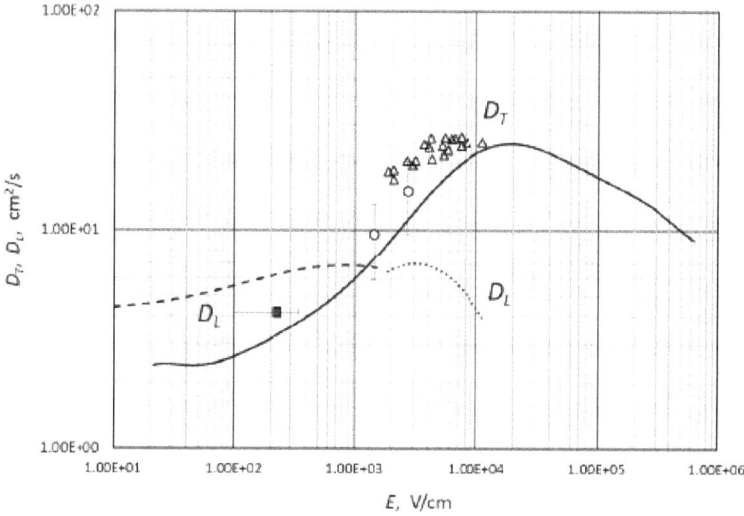

Figure 3.12: Coefficients of transversal and longitudinal diffusion in liquid argon. Transversal diffusion: triangles — measurements by Shibamura *et al.* (1979); circles — measurements by Derenzo (1974) (unpublished, the data are retrieved from Shibamura *et al.*, 1979); solid line — computations by Boyle *et al.* (2015). Longitudinal diffusion: filled square — measurements by Cennini *et al.* (1994); dashed line — measurements by Li *et al.* (2016) (the line represents a parametrization for D_L proposed in that work); dotted line — D_L calculated from D_T of Shibamura *et al.* (1979) using D_L/D_T relation (see text) following Li *et al.* (2016).

projection chambers, including the double phase detectors, digitize the current signal with a sampling rate typically in the nanosecond range and practically have no limitation on the length of the recorded timelines, except for technical constraints on processing and storage. The current dark matter and neutrino scattering experiments do not consider the ionic component of the signal sufficiently important for obtaining physically relevant information and therefore the ion drift is not measured. Experiments studying double beta decay, on the other hand, can significantly benefit from that as is the case of nEXO intending to detect $^{136}\mathrm{Ba}^+$ ions resulting from the $0\nu 2\beta$ decay of $^{136}\mathrm{Xe}$ (e.g., Miyajima *et al.*, 1994; nEXO, 2019). Ions (both positive and negative) may also contribute to some side effects such as, for example, "spontaneous" single electron emission observed in double phase detectors (see Section 3.4 for more details).

The ions in noble liquids remain in thermal equilibrium with the medium at all practical fields. Their mobility is thus constant and the drift velocity is directly proportional to the field strength $v_d = \mu E$. The available values for positive rare gas ions in their parent liquids are presented in Table 3.2. For liquid argon, a significant dispersion of the data exists. Here, only the range of the available values is presented; for more details and references the reader is referred to a compilation by Schmidt and Yoshino (2005).

The correct interpretation of the velocity measurements for positive charges in liquids is not straightforward as it is difficult (if not impossible) to identify which ion(s) contributes to the measured signal. In gaseous argon for example, three different mobility values were measured: $1.54 \, \text{cm}^2/\text{V} \cdot \text{s}$, $1.83 \, \text{cm}^2/\text{V} \cdot \text{s}$, and $2.6 \, \text{cm}^2/\text{V} \cdot \text{s}$, attributed to Ar^+, Ar_2^+ and Ar^{++}, respectively. The authors used mass spectroscopy to identify the ion species (Madson and Oskam, 1967) that is difficult to do in the case of liquids. The exact nature of the drifting positive ions in noble liquids remains therefore unclear. It is known however that the ionized rare gas atoms are prone to form single charged dimer molecules R_2^+ even in gas so it is highly likely that something similar happens in the liquid state.

Another factor which affects significantly the measured mobility is the presence of impurities in the liquid since other types of positive ions can be formed through chemical reactions of halogen-like R^+ ions and the impurity molecule. It follows from this the importance of gas purification and impurity control which was not sufficiently acknowledged in the earlier experiments. This might be the reason for the data in liquid xenon being more consistent — the measurements with liquid xenon are known to be more sensitive to the presence of impurities and therefore care is taken to purify xenon in most of the experiments.

We also notice in Table 3.2 that the mobility of positive rare gas ions in their parent liquid are in general an order of magnitude higher than that of impurity cations. This favours the hypothesis of

Table 3.2: Energy of electron state in noble liquids and solids V_0, zero field electron mobility μ_0 and ion mobilities μ_i near the triple point.

	Ne	Ar	Kr	Xe
Electron zero field mobility, μ_0 (cm^2/V·s)[a]				
liquid	1.6×10^{-3}[b]	475	1800	1900
solid	600	1000	3600	4000
Ion mobility, μ_i (cm^2/V·s)				
positive ion in its parent liquid				
TMSi$^+$	—	$(0.8 - 12) \times 10^{-3}$[c]	0.65×10^{-3}[d]	$(3 - 5.5) \times 10^{-3}$[e]
Tl$^+$	—	—	—	0.2×10^{-3}[f]
				0.13×10^{-3}[g]
O$_2^-$	—	$\sim 0.8 \times 10^{-3}$[h]	$\sim 1 \times 10^{-3}$[h]	0.7×10^{-3}[f]
SF$_6^-$	—	—	—	0.5×10^{-3}[f]
Hole mobility in solids, μ_h (cm^2/V·s)[i]	1.05×10^{-2}	2.3×10^{-2}	4.0×10^{-2}	1.8×10^{-2}
Potential energy of electron with respect to vacuum near the triple point, V_0 (eV)				
in liquid[j]	—	-0.165	-0.36	-0.64
in solid[k]	$+1.3$	$+0.3$	-0.3	-0.4

Source: Data from [a]Miller *et al.* (1968); [b]Loveland *et al.* (1972); [c]based on compilation by Schmidt and Yoshino (2005); [d]Palleschi *et al.* (1981); [e]Hilt and Schmidt (1994a); [f]Schmidt *et al.* (1994a); [g]Walters *et al.* (2003); [h]Davis *et al.* (1962); [i]Le Comber *et al.* (1975); [j]Steinberger (2005); [k]Schwentner *et al.* (1985).

hole formation in liquid xenon (Hilt and Schmidt, 1994a) — as there is no mass transfer in the hole conductivity mechanism (except for electrons jumping from one vacancy to another), the holes should have higher mobility than ions. It is however puzzling that the hole mobility in liquid xenon is five to six orders of magnitude lower than that of electrons — for example in semiconductors like silicon or germanium the difference between the two is just a factor of 3 or 4. Compared with rare gas solids, we find the hole mobility higher in solids than in liquids but also much smaller than the mobility of electrons.

In order to explain this fact, it was suggested that holes in rare gas solids can be seen as self-trapped molecular R_2^+ centres and that the small polaron theory can be applied (Druger and Knox, 1969; Le Comber *et al.*, 1975). This theory has also been successfully used to describe mobility of holes in liquid xenon (Hilt and Schmidt, 1994a). Classically, a polaron can be pictured as follows. Due to a high polarizability and weak interactions between the rare gas atoms, a charge placed in the liquid causes around itself a local deformation of the medium, which accompanies the charge (hole in this case) motion in the electric field. In terms of solid state physics, one can describe this situation as interaction with phonons and formation of a bound state. In liquid xenon, the binding energy of such state was estimated to be $\sim$60 meV. We refer to Khrapak *et al.* (2005) for more detailed interpretation and references on ion mobilities in rare gas liquids.

In recent years, a great bulk of high accuracy data has been obtained for mobilities of positive ions, other than xenon, in liquid xenon. This effort is motivated by the interest in searching for neutrinoless double beta decay of ^{136}Xe through detection of Barium daughter nucleus in the form of a positive ion ^{136}Ba$^+$. The obtained data are shown in Table 3.3. The mobility values are in good agreement with the theoretical estimates considering motion of a particle through a viscous liquid and using the Stock's law. The exact nature of the positively charged species whose drift is measured can, however, be questioned —

Table 3.3:   Mobility of alkaline and some other positive ions in liquid xenon.

Ion	T (K)	μ_i (cm^2/V $\cdot$ s)	Reference
Mg$^+$	167–173	2.42×10^{-4}	Jeng *et al.* (2009)
Ca$^+$	168–173	2.8×10^{-4}	*Ibid*
Sr$^+$	167	2.73×10^{-4}	*Ibid*
Ba$^+$	167–170	2.11×10^{-4}	*Ibid*
Tl$^+$	166–167	1.67×10^{-4}	*Ibid*
^{208}Tl$^+$	163.0	1.33×10^{-4}	Walters and Mitchell (2003)
^{226}Th$^+$	163.0	2.40×10^{-4}	Wamba *et al.* (2005)

a hypothesis of forming an alkaline–xenon molecular ion $M\mathrm{Xe}^+$ ($M = \mathrm{Mg, Ca, Sr, Ba}$) is discussed by Jeng *et al.* (2009) as well as the possibility of xenon "condensation" on the alkaline ion to form a $M^+\mathrm{Xe}_n$ cluster with the radius of 6.5–8.6 Å that requires between 1 and 2 layers of xenon atoms to surround the ion.

Rare gas atoms do not form negative ions but if electronegative molecules are present in the liquid, such as oxygen, water, and some other, electrons can attach to them forming a relatively stable negative ion (electron attachment is considered in some details in the next section). Mobilities of some negative ions in noble liquids are also shown in Table 3.2.

3.3. Electron Attachment

Some molecules, if present in the liquid, can capture a drifting electron and form a negative ion with very low mobility, several orders of magnitude lower than that of the electrons, thus leading to substantial reduction of the measured signal. This is why the liquid purity is of great concern in the detector operation.

In the most general form, the attachment process to a molecule X can be represented by the reaction

$$e^- + X_g \rightarrow X_g^- + \text{energy}, \tag{3.6}$$

where index g indicates that both the neutral and the ion are in the ground state. Energy can be released or absorbed in this process depending on the energy difference between the two states of the molecule (and in some sense on the kinetic energy of the electron as different reactions can take place at different energies — see below).

The necessary condition for a stable negative ion to be formed is $E_g^- < E_g^0$, i.e., the total energy of the ground state of the ion should be lower than that of the neutral molecule. The difference $E_g^0 - E_g^- = E_A$ is called *electron affinity*. Positive electron affinity means that it is energetically advantageous for the system electron–molecule to form an ion than to stay apart (zero kinetic energy of the electron is assumed). A negative value of electron affinity indicates that the negative ion is unstable. It is clear that atoms with complete electronic shells (rare gases) or highly symmetric molecules (e.g., CH_4 or CF_4) would not attach an extra electron while those with an open vacancy (halogens) are strongly attaching. The values of electron affinity for selected atoms and molecules, that may constitute interest for the condensed noble gas detectors, are presented in Table 3.4. One can notice that the electron affinity of homonuclear molecules with different numbers of atoms is different and differs also from that of a single atom (compare, for example, O, O_2, and O_3, among which O_2 is the less attaching). Molecules capable of attaching an electron are collectively called *electronegative*.

The following most common attachment processes can be distinguished.

1. Radiative attachment: the energy in excess is removed through emission of photons.

$$e^- + AB \rightarrow AB^- + h\nu \qquad (3.7)$$

2. Dissociative attachment: usually requires additional energy which comes from the kinetic energy of the electron. The threshold is typically in the range of 1 to $10\,\text{eV}$.

$$e^- + AB \rightarrow (AB)^- \rightarrow A + B^- \qquad (3.8)$$

Table 3.4: Electron affinity of some atoms and molecules (Radzig and Smirnov, 1985, if not stated otherwise). [a]Barabash and Bolozdynya, 1993; [b]Miller, 2019; [c]NIST Chemistry WebBook; [d]Raju, 2006; [e]Gutsev and Adamowicz (1995).

Molecule	Affinity, eV	Molecule	Affinity, eV	Molecule	Affinity, eV	Molecule	Affinity, eV
Ne	−1.2	H_2	<0[a]	O_3	2.103	NO_3	3.9[c]
Ar	−1.0	O_2	0.450	H_2O	<0[a]	CF_4	−1.22[e]
Kr	−1.0	F_2	3.08	CO_2	−(0.6–1.6)[c]	SF_6	1.03
Xe	−0.8	Cl_2	2.35	NH_3	1.21[d]	HNO_3	0.57[c]
H	0.75[c]	Br_2	2.53	NO_2	2.273	Fe_2O_3	3.06[c]
N	0.05[a]	I_2	2.524	SO_2	1.107	Acetone	0.0[c]
O	1.461[b]	CO	1.33[c]	N_2O	0.25[c]	p-Quaterphenyl	0.66[c]
F	3.401[b]	N_2	<0[a]	CS_2	0.5–0.6[c]		
Cl	3.613[b]	NO	0.026				
Br	3.36[c]	HCl	<0[a]				
I	3.06[c]	CsCl	0.455[b]				
		CsI	0.630[c]				

3. Direct three-body process: the excess of energy is transferred in collision to a third particle in the form of kinetic energy.

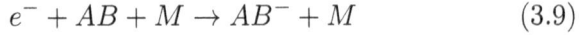

$$e^- + AB + M \rightarrow AB^- + M \qquad (3.9)$$

4. Two-step three-body process: the electron is captured by a molecule forming an unstable ion in an excited vibrational state which would normally release the captured electron returning to the neutral state (detachment) unless the excess of energy is transferred in collision to a third particle in the form of kinetic and potential energy.

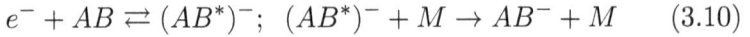

$$e^- + AB \rightleftarrows (AB^*)^-; \quad (AB^*)^- + M \rightarrow AB^- + M \qquad (3.10)$$

For the molecules with positive value of electron affinity, energy is released in the radiative and the three-body attachments. These reactions can take place even at zero electron energy and therefore are relevant at low electric field values where the electron energies are $\lesssim 1\,\text{eV}$. The cross-section of the radiative attachment is however small and the three-body processes are considered to be the most important for the detector operation. On the other hand, dissociative attachment has a threshold of a few eV above which it competes with the three-body processes and can become dominant. We also note that molecules with negative electron affinity, such as H_2, or CO_2, can attach electrons only through the dissociation mechanism.

In noble liquids, the energy of drifting electrons is below $1\,\text{eV}$ in all practical drift fields (see Fig. 3.8) so that attachment through the dissociative mechanism is unlikely. The exception are the detectors in which secondary scintillation or electron multiplication in the liquid are pretended to be used for signal amplification. These methods require fields of the order of 10^5 to $10^6\,\text{V/cm}$ (only possible in a very small region of space) and are not used in practical detectors, yet. In the double phase emission detectors, the amplification can be more conveniently done in the gas.

The variation of the number of electrons in the presence of electronegative molecules, attaching through a two-body process, is described by the differential equation

$$\frac{dN_e}{dt} = -k_2 n_s N_e. \tag{3.11}$$

Similarly, for a three-body process,

$$\frac{dN_e}{dt} = -k_3 n_s n N_e. \tag{3.12}$$

Here N_e is the number (or concentration) of drifting electrons, n_s is concentration of the impurity molecules (scavengers), and the number density of the buffer gas is n. The reaction rate coefficients k_2 and k_3 are specific for each type of attaching molecule; k_3 depends also on the buffer gas. Moreover, both coefficients depend on the electron energy. The values of k_3 for attachment of thermal electrons to diatomic oxygen in some buffer gases are shown in Table 3.5. One can observe from the table that more complex molecules are more efficient in the stabilization of the oxygen ion as they can absorb the energy in excess not only in the form of kinetic energy but also as internal energy (excitation and eventually ionization).

Table 3.5: Three-body attachment rate coefficients to O_2 for a selection of the buffer gases (after Christophorou, 1978a).

Buffer gas	k_3 $(10^{-30}$ $cm^6/s)$
He	0.033
Ne	0.023
Ar	0.05
Kr	0.05
Xe	0.085
H_2	0.48
N_2	0.06–0.26
CH_4	0.34
CO_2	~3
NH_3	~7
H_2O	14

The solution of Eqs. (3.11) and (3.12) for the number of drifting electrons is written as

$$N_e(t) = N_0 e^{-t/\tau_i}, \tag{3.13}$$

where index $i = 2$ or 3 corresponds to a two- or a three-body reaction, respectively. The reaction time scale is characterized by $\tau_2 = 1/(k_2 n_s)$ and $\tau_3 = 1/(k_3 n_s n)$, respectively. The solution can also be written as a function of drift distance $N_e(x) = N_0 e^{-x/\lambda_i}$ with $\lambda_i = v_d \tau_i$. If two or more processes are present, the inverse sums

$$\left(\sum_i \frac{1}{\tau_i} \right)^{-1} = \tau_a \quad \text{and} \quad \left(\sum_i \frac{1}{\lambda_i} \right)^{-1} = \lambda_a \tag{3.14}$$

represent the observable electron lifetime and attenuation length, respectively. These parameters are used in practice to characterize the purity of the detector medium.

If we substitute in the above equations the scavenger concentration by its relative content in the buffer gas $n_s = r_s n$ (r_s being expressed in parts per billion, ppb, for example) it becomes clear that the attachment probability increases with gas density linearly for a two-body process (like radiative or dissociative attachments), and as n^2 in the case of a three-body attachment. In fact, this dependence is used in the experimental studies of the attachment to distinguish the contributions from different reactions. Another important observation is that the relevance of the three-body attachment increases with the medium density. For low energy electrons, with the energy below the threshold for dissociative attachment (a few eV), the three-body attachment is well confirmed to be the dominant electron attachment process in many gases and their combinations.

The physics of the attachment process can be illustrated in the example of oxygen which is the most frequently found impurity in the noble gas detectors. The potential energy curves of the ground electronic states of O_2 and O_2^- are shown in Fig. 3.13. The minimum of energy of the ion is lower than that of the neutral, so it is energetically favorable for the system electron-O_2 molecule to form a negative ion. If the energy of electrons is $\lesssim 1\,\mathrm{eV}$, the negative ion is initially formed in the ground electronic state vibrationally excited

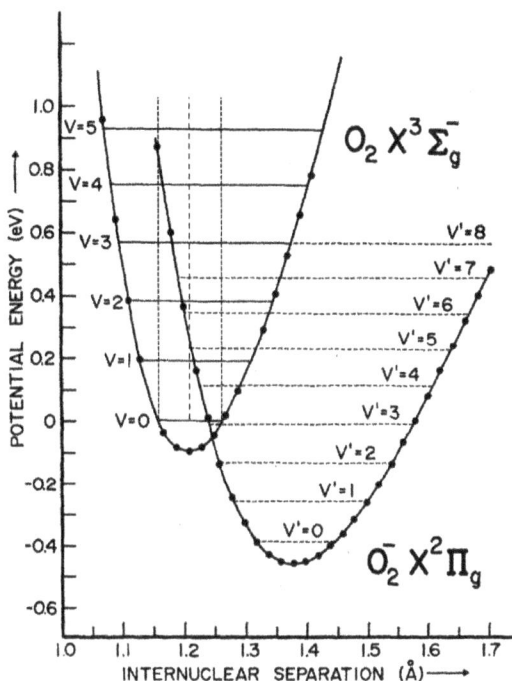

Figure 3.13: Potential energy curves and vibrational levels for the ground states of O_2 and O_2^-: the vertical dotted lines indicate the region limited by the classical turning points for the neutral molecule in the lowest vibrational state; according to the Frank–Condon principle, transitions occur within these limits. Redrawn from Bonnes and Schulz (1970).

Note: The equilibrium internuclear distance for O_2- is now established to be 1.35 Å (Radzig and Smirnov, 1985; NIST Chemical Webbook).

with $\nu' > 3$ (electronic excitation would require energies >4.3 eV — see, for example, Ewig and Tellinghuisen, 1991). The vibrational states of the ion with $\nu' > 3$ are unstable and the ion either decays to the neutral plus free electron (autodetachment) or stabilizes through the removal of the energy in excess by emission of an infrared photon or in collision with another molecule:

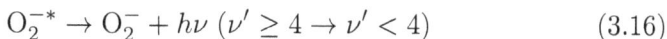

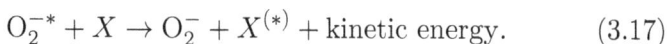

$$e^- + O_2 \rightleftarrows O_2^{-*} \, (\nu' \geq 4) \tag{3.15}$$

$$O_2^{-*} \rightarrow O_2^- + h\nu \, (\nu' \geq 4 \rightarrow \nu' < 4) \tag{3.16}$$

$$O_2^{-*} + X \rightarrow O_2^- + X^{(*)} + \text{kinetic energy.} \tag{3.17}$$

The lifetime of the $(O_2^-)^*$ in the $\nu' = 4$ state with respect to autoionization was found to be $\sim 2 \cdot 10^{-12}$ s and even shorter for higher vibrational states (Christophorou, 1978a,b). The probability of the radiative stabilization (the second process above) is low compared with the three-body collisional process. For example, one can estimate the ratio of the time constants for the two-body radiative attachment $e^- + O_2 \overset{k_2}{\rightarrow} O_2^- + h\nu$ to that of the three-body attachment $e^- + O_2 + X \overset{k_3}{\rightarrow} O_2^- + X$ as $\tau_2/\tau_3 = k_3 n/k_2$. Assuming that the buffer gas is argon at 1 bar and using the known reaction rates $k_2 \sim 10^{-19}$ cm^3/s and $k_3 \sim 5 \times 10^{-32}$ cm^6/s (Christophorou, 1978a), we get $\tau_2/\tau_3 \sim 10^7$ thus making evident that the three-body electron attachment to O_2 dominates at all practical gas densities.[2] The two-step three-body attachment mechanism (Eqs. 3.15 and 3.17) has been proposed by Bloch and Bradbury (1935) and further developed by Herzenberg (1969) and is now considered to be well established.

The frequency of collisions of an excited oxygen ion with the buffer gas molecules (some of which may result in the ion stabilization) can be estimated as $\sim n v \pi (r_1 + r_2)^2$, where v is the O_2 thermal velocity, n number density of the buffer gas, r_1 and r_2 are collisional radii of the O_2 and the buffer gas molecule, respectively. Taking argon at 1 bar as an example and $r_1 + r_2 \approx 0.7$ nm, one gets for the frequency $\sim 2 \times 10^9$ s^{-1}. The time between two successive collisions is then $\sim 5 \times 10^{-10}$ s, i.e., two orders of magnitude larger than the lifetime of the $(O_2^-)^*$ with respect to autodetachment. Thus, most of the collisions of the electrons with oxygen molecules in a gas do not result in the formation of a stable negative ion and do not affect the observable electron lifetime in the detector. At the liquid densities, however, the fraction of stabilized ions increases substantially.

In liquids, the excess of energy, needed to be removed for stabilization of the negative ion, can be absorbed by more than one atom due to interatomic interactions so that the attachment rate

---

[2]At very high densities, four-body reactions may become also relevant.

$(1/\tau)$ is not necessarily proportional to the medium density, which in practical applications does not vary significantly. Equations (3.11) and (3.12) are then rewritten as

$$\frac{dN_e}{dt} = -k_s n_s N_e,\tag{3.18}$$

corresponding to the process

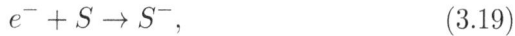

$$e^- + S \rightarrow S^-,\tag{3.19}$$

with the rate constant k_s depending on the electric field (electron energy) and possibly on the liquid density. The rate constant is linked to the electron capture cross-section σ_s through the equation

$$k_s(E) = \int_0^\infty v\sigma_s(v)F(v,E)\,dv,\tag{3.20}$$

where v is the electron velocity, and $F(v,E)$ is the electron velocity distribution as a function of the electric field strength E or, in terms of the electron energy ε,

$$k_s(E) = \int_0^\infty \sigma_s(\varepsilon)\,F(\varepsilon,E)\,d\varepsilon,\tag{3.21}$$

with $F(\varepsilon,E)$ as in Figs. 3.6 and 3.7.

There are very few studies of electron attachment to specific molecules in noble liquids. Figure 3.14 shows the attachment coefficient k_s for O_2, N_2O and SF_6 measured as functions of electric field in liquid argon and xenon (Bakale *et al.*, 1976).

The field dependence of the attachment rate to O_2 is consistent with the three-body mechanism of Bloch and Bradbury (1935), as in gas, although the role of autodetachment is diminished due to much higher collision rate (the estimated mean time between the successive collisions is $\sim 10^{-12}$ s, i.e., of the same order of magnitude as the lifetime of an isolated ion $(O_2^-)^*$ with respect to autodetachment). As for N_2O, both three-body and two-body mechanisms seem to contribute comparably, the latter being dissociative $e^- + N_2O \rightarrow N_2 + O^-$. An estimate of the time constant ratio for the two processes

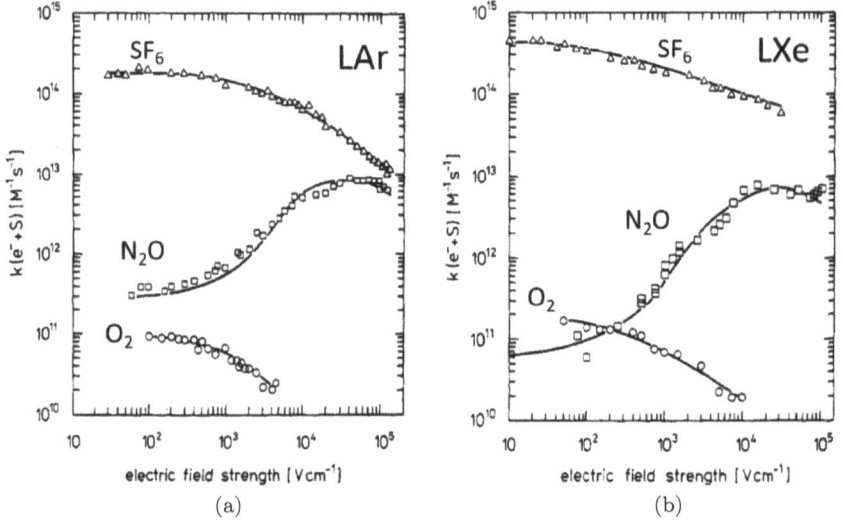

Figure 3.14: Rate constants for the attachment of electrons in liquid argon (a) and liquid xenon (b). Redrawn from Bakale *et al.* (1976).

$\tau_2/\tau_3 = k_3 n/k_2$ with k_2 and k_3 from Raju (2011) in gas (for the three-body process the buffer gas is nitrogen) gives $\tau_2/\tau_3 \sim 10$ for thermal electrons and ~ 0.1 for electrons with the energy of ~ 1 eV. For higher densities, the three-body attachment gains terrain proportionally to the density. The dependence on the electron energy observed in liquid argon and xenon is qualitatively consistent with that measured in gas (see data compilations in Raju, 2006 and 2011, as well as Fig. 3.15).

As already mentioned, oxygen is the most frequently found electronegative impurity in liquefied noble gases, its origin being usually attributed to outgassing and micro leaks in the system. In spite of the fact that highly sensitive and affordable residual gas analyzers (RGAs) became recently available and are indeed routinely used by all large experiments as well as in small laboratory setups, the specific content of impurities and contribution of each of them to the observed electron attachment continue to be unclear. In part, it is due to the difficulty of unambiguous interpretation of the measured mass spectra but also because of insufficient data on the attachment coefficients for a particular impurity in a particular

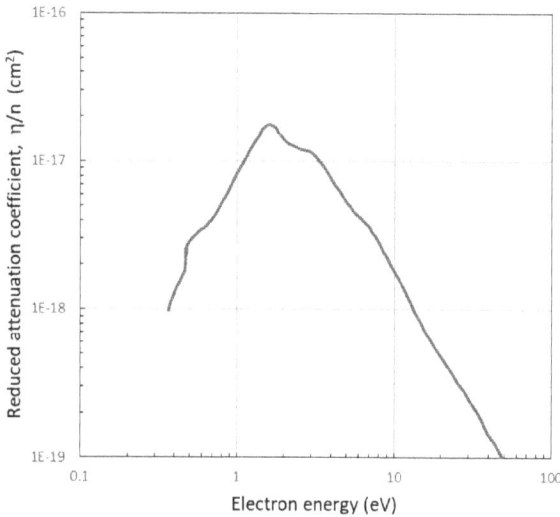

Figure 3.15: Reduced attenuation coefficient η/n in N_2O gas as a function of electron energy. Data from Raju (2011).

liquid. In practice, liquid purity in radiation detectors is determined by measuring the exponential decay of the ionization signal in time or with the distance travelled by electrons in the liquid. The electron lifetime τ is usually used as a universal characteristic of the detector performance. The concentration of impurities is then roughly estimated in terms of the oxygen equivalent (i.e., assuming that the attachment is totally due to the O_2 molecules), using the attachment coefficient from Bakale *et al.*, 1976 and the relation $\tau = 1/(k_s n_s)$. For example, for 1 ms lifetime in a field of $\sim$100 to 1000 V/cm the equivalent concentration of oxygen would be of the order of a few ppb (parts per billion, 10^{-9}) in liquid xenon and twice that in liquid argon. This partly explains why purification of liquid xenon is a more demanding task than purification of argon.[3]

---

[3]Another important factor favoring liquid argon in this aspect is its much lower operation temperature and also smaller polarizability of argon atoms that results in a lower dissolving capacity for some substances. Both factors favor deposition of impurities on the surfaces of the detector components.

The presence of other unidentified impurities with the attachment rate increasing with the electric field strength (contrary to what is observed for oxygen) has been reported in some setups.

3.4. Electron Emission

The energy state of an electron in condensed matter is different from that in vacuum or dilute gas. The potential energy of an electron in the condensed state, referred to that in vacuum, is called V_0. If the energy bands are formed, then V_0 is the energy at the bottom of the conduction band with respect to that in vacuum assumed to be zero. V_0 is found to be negative in liquid Ar, Kr, and Xe and in solid Kr and Xe (Table 3.2) meaning that it is energetically advantageous for electrons to stay in the condensed phase than in the vapor, in these cases. Solid Ar and liquid and solid Ne and He have positive V_0 values so that a gas bubble is formed in these media around an electron that explains the low electron mobility values.

There is a clear correlation between the value of V_0 and zero field electron mobility μ_0 (see Table 3.2) — the deeper the energy level of electrons in a liquid or solid, the higher their mobility. This holds for all rare gases including He: while V_0 decreases in the order He $\rightarrow$ Ne $\rightarrow$ Ar $\rightarrow$ Kr $\rightarrow$ Xe, the opposite happens to electron mobility (in the same order). In general terms, this is explained by an interplay of the repulsion between the excess electrons and the atoms, dominant in He and Ne, and the attractive charge-induced dipole interaction in highly polarizable gases (*cf.* atomic polarizabilities in Table 2.1). As a result, the electrons move faster in heavier noble gases than in Ne and especially He, where repulsive forces keep away the closest to the electron atoms resulting in a microscopic bubble around it which accompanies the electron motion through the liquid.

The value V_0 was found to depend also on the fluid density as shown in Fig. 3.16 which represents collectively the experimental data for liquid Ar, Kr, and Xe, discussed in detail by Steinberger (2005). Noticeably, the dependence is not monotonous for all three liquids. The densities where the minima of V_0 occur are the same where μ_0 is at maximum for the respective liquid. The theoretical

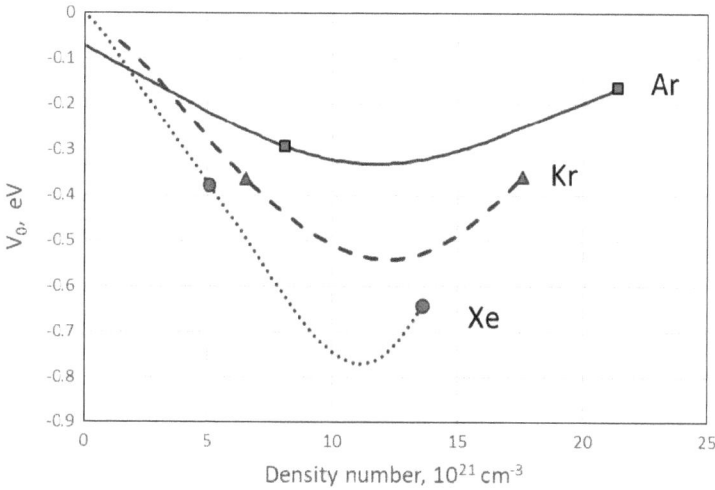

Figure 3.16:   Potential energy of electron V_0 in liquid Ar, Kr, and Xe as a function of fluid number density. The symbols on each curve correspond to the triple point (on the right) and to the critical density (on the left). Redrawn from Steinberger (2005).

explanations for the existence of the minimum value of V_0 meet difficulties — quite a number of different approaches have been suggested but no satisfactory and clear picture exists up to date. The reader is referred to Steinberger (2005) for a concise review of those theories as well as for experimental data and references.

A charge near the boundary between two dielectrics with different dielectric permittivities is subject to a force arising from different polarizations of the two media induced by the charge and resulting in an anisotropy in the direction perpendicular to the boundary. This force can be conveniently described using the method of image charges by replacing interaction of a point charge q with a dielectric by its interaction with an imaginary charge $q' = -q\frac{\epsilon_i - \epsilon_q}{\epsilon_i + \epsilon_q}$ placed on the opposite side of the boundary at the same distance from it as the original charge (e.g., Jackson, 1999). Here, index q is used to denote the medium where charge q is placed, and i denotes the medium where the image charge is located; $\epsilon_q = \kappa_q \epsilon_0$ and $\epsilon_i = \kappa_i \epsilon_0$ with κ_q and κ_i being relative dielectric constants of the respective medium. We notice that the image charge can be of the opposite sign, if $\epsilon_q < \epsilon_i$, or of the same sign (for $\epsilon_q > \epsilon_i$) as q. In the first case, the charge is attracted to the surface while in the second it is forced back. The force acting on the charge is then given by the equation

$$F_i(x) = -\frac{1}{4\pi\epsilon_q}\left(\frac{\epsilon_i - \epsilon_q}{\epsilon_i + \epsilon_q}\right)\frac{q^2}{(2x)^2}, \qquad (3.22)$$

and the image potential

$$\phi_i(x) = -\frac{1}{16\pi\epsilon_q}\left(\frac{\epsilon_i - \epsilon_q}{\epsilon_i + \epsilon_q}\right)\frac{q^2}{x}. \qquad (3.23)$$

Here, the axis Ox is perpendicular to the boundary; the charge q is placed at a distance x from the boundary (i.e., x is always positive).

The shape of the image potential for an electron at either side of the boundary separating the liquid phase (on the left) and the vapor (on the right) is illustrated by Fig. 3.17 by the dashed line. It

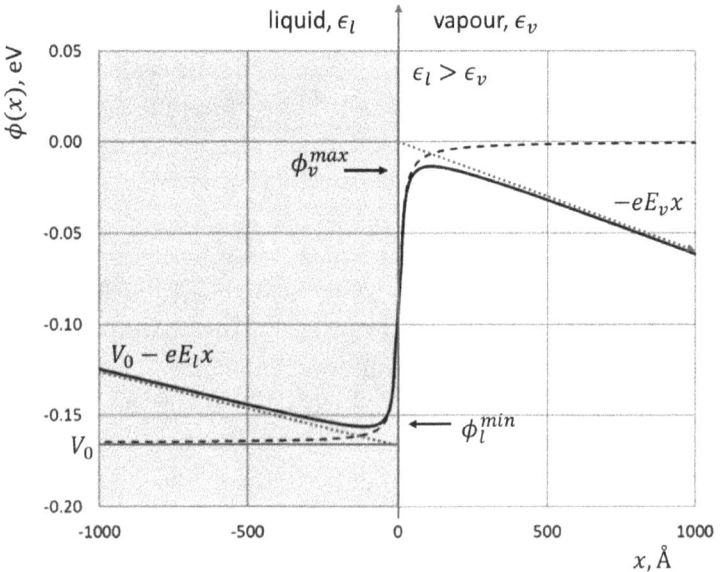

Figure 3.17:   Shape of the potential barrier at the liquid–vapor interface in argon (solid line). Dotted line shows the effect of the electric field of $E_l = 4\,\mathrm{kV/cm}$ perpendicular to the liquid surface; dashed line — the effect of the image potential. See text for explanation of the symbols.

is assumed that $\epsilon_l > \epsilon_g$ as is the case of Ar, Kr, and Xe (but not Ne or He; in these media the situation is opposite). One observes that an electron, approaching the boundary from the liquid side, is subject to a repulsive force from it while an electron placed at the vapor side is attracted to the surface. In other words, the dielectric force acts against the extraction of electrons to gas on both sides of the boundary.

If electric field E is applied, the potential energy of an electron near the liquid boundary is the sum of the image potential, the electric potential $-eEx$ and, in the liquid, V_0. Replacing ϵ_q and ϵ_i in Eq. (3.23) by $\epsilon_l = \kappa_l \epsilon_0$ for liquid and $\epsilon_v = \kappa_v \epsilon_0$ for vapour, depending on what side of the boundary the electron is placed, we obtain for the potential in the liquid $(x < 0)$

$$\phi_l(x) = -\frac{A_l}{x} - eE_l x + V_0, \qquad (3.24)$$

and in the gas/vapor $(x > 0)$

$$\phi_v(x) = -\frac{A_v}{x} - eE_v x, \qquad (3.25)$$

following the notation of (Borghesani *et al.*, 1990, 1991; Schmidt 1997), where A_l, A_v and δ_κ are defined as

$$A_l = \frac{e^2}{16\pi\epsilon_0\kappa_l}\delta_\kappa, \qquad (3.26)$$

$$A_v = \frac{e^2}{16\pi\epsilon_0\kappa_v}\delta_\kappa, \qquad (3.27)$$

$$\delta_\kappa = \frac{\kappa_l - \kappa_v}{\kappa_l + \kappa_v}. \qquad (3.28)$$

The resulting potential on both sides of the boundary at $x = 0$ for liquid argon is shown in Fig. 3.17 by solid line. In the liquid, a minimum of energy $\phi_l^{(\text{min})} = 2(eE_l A_l)^{1/2} + V_0$ is observed at $x_{\text{min}} = -\left(\frac{A_l}{eE_l}\right)^{1/2}$ while in the gas phase there is a maximum of $\phi_v^{(\text{max})} = -2(eE_v A_v)^{1/2}$ at $x_{\text{max}} = \left(\frac{A_v}{eE_v}\right)^{1/2}$. As $\kappa_v E_v = \kappa_l E_l$, the maximum and the minimum are located at the same distance from the boundary. For example, in liquid argon at $E_l = 1\,\text{kV/cm}$ this distance is about $200\,\text{Å}$ (we used here $\kappa_v = 1$ and $\kappa_l = 1.52$ from Table 2.1). The maximum in the potential energy curve near the boundary corresponds to the classical lower energy limit for electrons to cross the boundary. The existence of a minimum under the surface, on the other hand, means that the low energy electrons find themselves to be confined in a narrow gap of $\sim$10 nm under the liquid surface in the vertical direction but are free to move in the horizontal plane.

Under the electric field perpendicular to the liquid boundary and pointing to the liquid, the surface barrier is thus reduced, with respect to that at zero field $|V_0|$, by the value

$$\Delta\phi = 2e^{1/2}[(A_l E_l)^{1/2} + (A_v E_v)^{1/2}], \qquad (3.29)$$

or, in the explicit form,

$$\Delta\phi = \left[\frac{e^3}{4\pi\epsilon_0} \frac{\kappa_l^2 - \kappa_v^2}{\kappa_l^2 \kappa_v^2} (\kappa_l E_l)\right]^{1/2}. \qquad (3.30)$$

Assuming $\kappa_v \approx 1$, one comes to the following expression:

$$\Delta\phi = \left[\frac{e^3}{4\pi\epsilon_0} \left(1 - \frac{1}{\kappa_l^2}\right) E_v\right]^{1/2}. \qquad (3.31)$$

In liquid argon at $E_l = 1\,\text{kV/cm}$ (1.52 kV/cm in the gas phase), $\Delta\phi \sim$ 0.01 eV that should be compared with $|V_0| = 0.165$ eV (Table 3.2). For liquid xenon at $E_l = 3\,\text{kV/cm}$ (5.67 kV/cm in the gas) — this is the field at which an appreciable emission already exists from liquid xenon — $\Delta\phi \sim 0.024$ eV, to be compared with $|V_0| = 0.64$ eV.

The experimental extraction efficiency curves for liquid argon, krypton, and xenon are shown in Fig. 3.18 according to the data by Gushchin *et al.* (1982b) and Barabash and Bolozdynya (1993). More recent data for liquid xenon are shown in Fig. 3.19. It should be noted that absolute measurements of the fraction of extracted electrons are difficult and therefore the presented results should be regarded as relative, normalized to the maximum observed signal in each experiment. In liquid xenon, in fact, no saturation at high

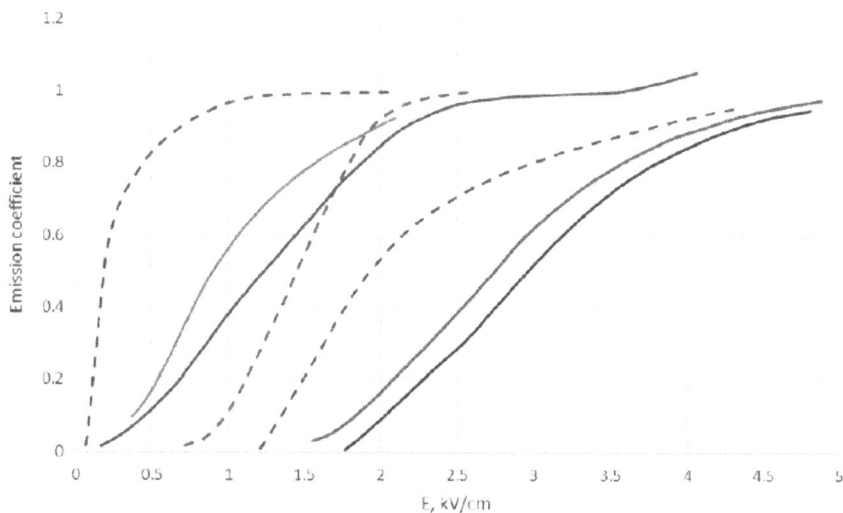

Figure 3.18: Electron extraction efficiency as a function of electric field strength in the condensed phases near the triple point: in liquids (solid lines), in solids (dashed lines), for argon (blue, left solid line — slow emission, right solid line — fast emission), for xenon (red), for krypton (brown). Redrawn from Gushchin *et al.* (1982b) (argon and xenon) and Barabash and Bolozdynya (1993) (krypton).

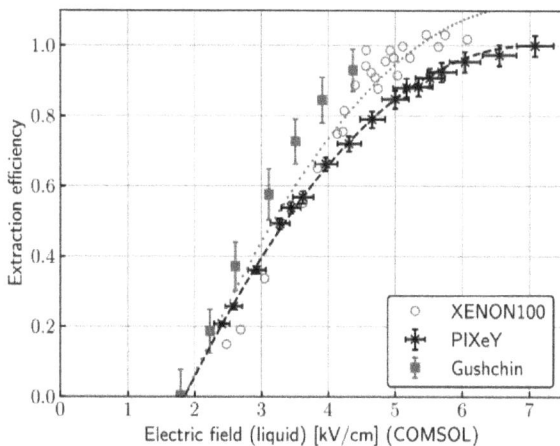

Figure 3.19: Electron extraction efficiency from liquid xenon as a function of electric field strength in the liquid as measured by Edwards *et al.* (2018) (marked as PIXeY), by Aprile *et al.* (2014a) (marked as XENON100), and by Gushchin *et al.* (1982b) (marked as Guschin). Redrawn from Edwards *et al.* (2018).

fields has been observed. Another significant source of uncertainty is the extraction field (in the condensed phase) which in practice is calculated from the voltage applied between the anode and the cathode and the phase boundary position between them. The field non-uniformity around the wires, if these are used for the electrodes as in Aprile *et al.* (2014a) and Edwards *et al.* (2018), also can affect the actual field strength in the extraction region by as much as ~10% (Edwards *et al.*, 2018).

As can be seen from the two plots, emission is a threshold phenomenon with the threshold field increasing significantly from argon to xenon. This is qualitatively consistent with the idea of surface barrier and with the V_0 values shown in Table 3.2 (although some discrepancy exists for Kr). Moreover, the threshold field in liquid argon was found to depend on the liquid temperature having a maximum at about the same temperature where V_0 reaches its minimum (Fig. 3.20 and Fig. 3.16). Temperature also affects the value by which the barrier is reduced by the electric field $\Delta\phi$ through density dependence of the dielectric constant of the liquid κ_l as can be seen in Eqs. (3.30) and (3.31).

Gushchin *et al.* (1982b) were also able to distinguish two emission components — a "fast" component and a "slow" one — in liquid

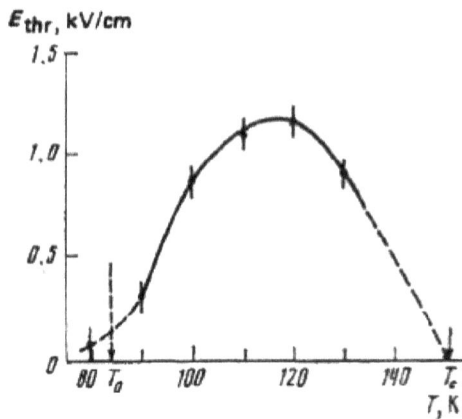

Figure 3.20:   Dependence of the threshold field for electron emission from liquid argon on temperature. Redrawn from Gushchin *et al.* (1982b).

argon but not in xenon (shown in Fig. 3.18). This phenomenon has been explored later in more detail by Borghesani *et al.* (1990) and Bondar *et al.* (2009b).

The distinct behavior of electron emission in liquid argon and liquid xenon seems to indicate the existence of different emission mechanisms. The theory of thermoionic emission from metals has been successfully applied to electron emission from room temperature liquid hydrocarbons, iso-octane in particular (Balakin *et al.*, 1977; Bolozdynya *et al.*, 1978). According to this theory, it is assumed that free electrons (quasi-free in dielectrics) are in thermal equilibrium with the medium and that their velocity distribution follows the Maxwell–Boltzmann distribution function. When electrons approach, under the effect of the electric field, the liquid boundary, some of them (in the high energy tail of the distribution) have enough energy to overcome the potential barrier and to cross the boundary immediately, while the others are reflected back to the liquid and become trapped in the potential well under the surface. Owing to collisions with the atoms, which are in average "hotter" then the unsuccessful electrons, these regain energy and their velocity vectors are rapidly randomized so that some of the electrons find themselves in the conditions to cross the surface, and so on. The necessary condition for emission to gas is that the momentum component perpendicular to the boundary $p_x > \sqrt{2m_e\phi_b}$ (here $\phi_b = |V_0| - \Delta\phi$ is the total height of the surface barrier for a given field). This is illustrated by Fig. 3.21.

The emission process in this picture should have some extension in time — although collisions of the trapped electrons with atoms are frequent, many are required for regaining energy from the atoms, as well as many attempts to cross the barrier. This indeed has been observed as exponentially decaying emission current with a time constant in the range from some tens of μs to seconds (see Table 2.1 in Bolozdynya, 2010 or Bolozdynya, 1999 for a summary).

The thermal emission mechanism can be important when the barrier height is not significantly larger than the thermal energy. For iso-octane for example, $k_B T \approx 0.026\,\text{eV}$ and $V_0 \approx -0.18\,\text{eV}$ (Bolozdynya, 1999). In liquid argon, V_0 is slightly higher ($-0.21\,\text{eV}$)

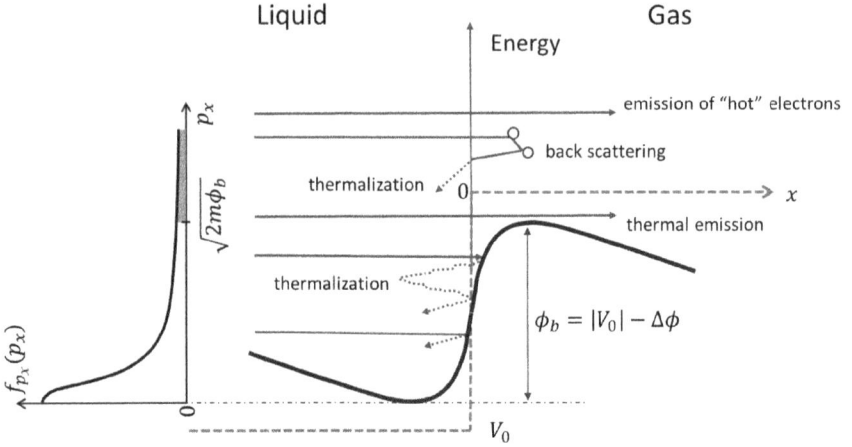

Figure 3.21:   Schematic diagram of the electron emission process. See text for explanation.

and the thermal energy is three times lower, $k_BT \approx 0.0075$ eV. On the other hand, the field in the liquid at which the mean electron energy starts to deviate from the thermal is <200 V/cm (which is within the field range where practical detectors operate) while tens of kVs are required to break thermal equilibrium in hydrocarbons. In spite of these facts, the thermal emission theory was successfully applied to describe the delayed emission from liquid argon by Borghesani *et al.* (1990) and Borghesani *et al.* (1991). The authors measured the characteristic emission time as a function of electric field (Fig. 3.22) that was found to be approximately inversely proportional to the electric field (through drift velocity)

$$\tau_e \propto \frac{\lambda_1}{\mu E}\exp\left(\frac{V_0 - \Delta\phi}{kT}\right), \qquad (3.32)$$

where λ_1 is the momentum transfer mean free path in the liquid ($\sim$200 Å in liquid argon, $\sim$700 Å in krypton and $\sim$1000 Å in xenon, according to Sakai, 2005). The effect of the field on the barrier height is small compared with V_0 ($\Delta\phi \propto E^{1/2}$ — Eq. (3.31)). Assuming the Maxwell–Boltzmann velocity distribution and neglecting the deviation of the mean electron energy from the thermal value,

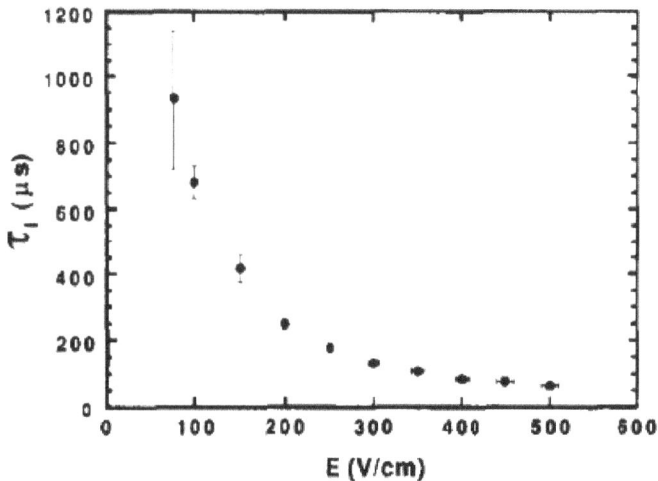

Figure 3.22: Experimental trapping time in liquid argon as function of electric field. Redrawn from Borghesani *et al.* (1990).

Borghesani *et al.* (1991) derived V_0 for liquid argon in good agreement with the values obtained in other experiments, where photoelectric emission from metal cathodes into the liquid was used. The same observation was earlier made for iso-octane (Balakin *et al.*, 1977; Bolozdynya *et al.*, 1978).

The effect of backscattering of emitted electrons due to elastic collisions with the gas atoms should also be taken into account. This effect is illustrated in Fig. 3.21 along with the other basic processes of electron emission in two-phase systems. It is more pronounced in pure noble gases, where elastic collisions are dominant, in contrast with molecular gases, where inelastic collisions prevail. This effect is well known for photoelectrons emitted from the photocathodes into the gas medium (Buzulutskov, 2008). The importance of the backscattering effect for electron emission from LAr was demonstrated in Bondar *et al.* (2009b), who analyzed the shape of the charge signal from the emitted electrons in pure Ar and in Ar doped with N_2. Figures 3.23 and 3.24 show that while the slow component dominates in pure argon (presumably defined by the backscattering effect, at higher electric fields), it disappears in the Ar/N_2 mixture, where

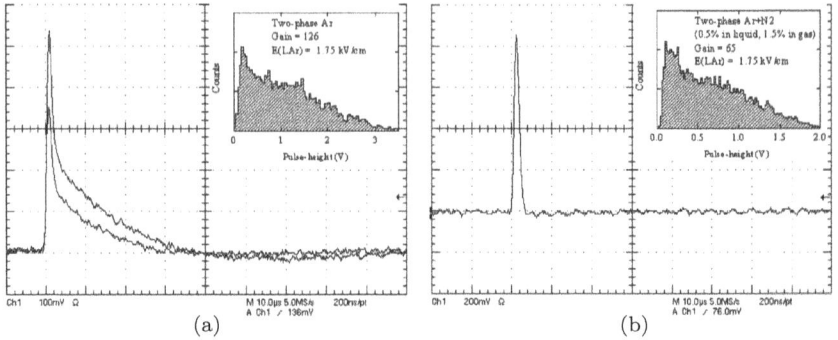

Figure 3.23:   Typical anode signals in a two-phase Ar detector (a) and in a two-phase Ar + N$_2$ (0.5% in the liquid, 1.5% in the gas) detector (b), operated with double-GEM multiplier at a gain of 126 and 65, respectively, induced by beta-particles from ^{90}Sr source at an electric field in the liquid of 1.75 kV/cm. In the insets, amplitude spectra are shown. Redrawn from Bondar *et al.* (2009b).

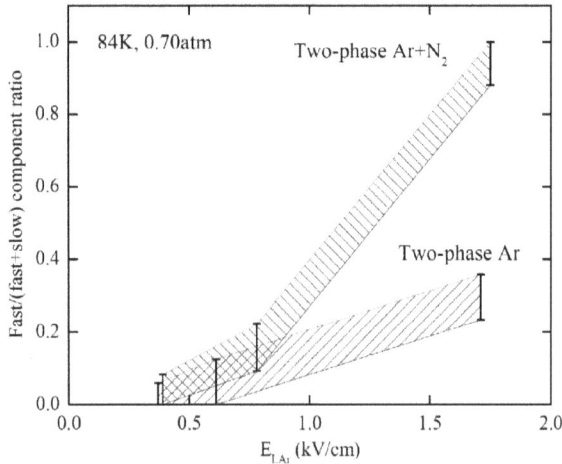

Figure 3.24:   Fast electron emission component fraction as a function of the electric field in the liquid in two-phase Ar and two-phase Ar + N$_2$ (0.5% in the liquid, 1.5% in the gas) at 84 K and 0.7 atm. Redrawn from Bondar *et al.* (2009b).

the contribution from elastic collisions is taken over by that from inelastic.

In contrast with liquid argon, in liquid krypton and xenon the height of the potential barrier at the boundary with gas is much larger than the thermal energy so that thermal emission can hardly be the principal mechanism. On the other hand, quasi-free electrons gain energy from the field much faster in those liquids than in argon (see Table 3.1 for critical field values).[4] The plots of extraction efficiency as a function of field in liquid krypton and xenon (Figs. 3.18 and 3.19) show practically no emission at low field values followed by a smooth step function with the threshold at about 1500 V/cm for krypton and 1700 V/cm for xenon. At these fields, many of the drifting electrons have kinetic energy above the surface barrier and therefore can cross the boundary at once (Figs. 3.6–3.9). The higher the field, the higher the proportion of such electrons. Thus, the emission from these liquids is frequently pictured as emission of "hot" electrons. The sub-barrier electrons are reflected back into the liquid and rapidly lose their energy in collisions with atoms ending up trapped in the potential well under the surface with nearly thermal kinetic energy and very little chance to be emitted to the gas through the thermoionic mechanism.

The ultimate fate of the under threshold electrons can be their attachment to electronegative impurities and formation of negative ions. These electrons can also diffuse in the plane parallel to the liquid surface eventually ending up at the detector walls. If the field lines are not exactly perpendicular to the surface, the tangential component of the field can drive the electrons in a certain direction as it has been observed in liquid krypton by Anisimov *et al.* (1984). The drift velocity of the trapped electrons was found to be equal to that in the bulk of the liquid thus indicating that, although confined to a 1D potential well, the electrons behave as quasi-free in two other dimensions. Akimov *et al.* (2016) observed electron drift in liquid xenon along the surface due to a small detector tilt with drift velocity and longitudinal diffusion coefficient consistent with the existing data for the bulk. As in liquid krypton, a massive emergence of electrons in the gas phase was detected in the periphery near the field shaping

---

[4]A comparison of the energy transfer mean free path for electrons in the three liquids supports this observation: $\lambda_0 = (n\sigma_0)^{-1} \sim 1\,\text{nm}$, $\sim 4\,\text{nm}$, and $\sim 6\,\text{nm}$ for argon, krypton, and xenon, respectively (cross-sections from Sakai, 2005).

electrodes (in both cases a parallel plate geometry was used although the liquid xenon detector was bigger and equipped with field shaping rings along ≈ 3 cm vertical drift). The localized electron emission at the end of the horizontal drift was tentatively attributed by the authors to non-uniformities of the electric field at the edges where the field is stronger thus resulting in a higher probability of emission.

In the absence of the tangential field component, the trapped electrons still can diffuse in the horizontal plane and eventually emerge in the gas at any point of the surface — although very small, there is still a finite probability of crossing the boundary through the thermal or tunnelling mechanisms. Remarkably, emission of single electrons has been observed in all large double phase xenon detectors for dark matter search, extending to many milliseconds after the main signal (which typically corresponds to emission of ~ 1000 electrons), so that even the correlation with it is difficult to establish for certain (Edwards *et al.*, 2008; Santos *et al.*, 2011; Aprile *et al.*, 2014a; Akimov *et al.*, 2016; Sorensen, 2017; Bondar *et al.*, 2020b). This phenomenon can be, in part at least, due to a sporadic emission of the trapped electrons, although taking into account the long time scale one can hypothesize that it is mostly negative ions that drift or diffuse under the liquid surface and not electrons. The emission is then due to detachment of the captured electron from the ion. Other hypotheses, such as photoelectric emission from the detector components and photoionization of impurity molecules in the liquid or formation of metastable excited negative ions in argon, have also been suggested. The exact nature of the single electron emission is at present unclear. It is however consensual that most likely several distinct processes are involved.

In spite of the fact that a number of physical questions are still to be answered, the emission process is sufficiently understood to make possible its use in particle detectors, some hitting the tonne mass scale, as it will be described in the following chapters.

Chapter 4

Light and Charge Amplification in Two-Phase Emission Detectors

4.1. Basic Concepts of Signal Amplification in Two-Phase Detectors

The ultimate goal for two-phase detectors is the development of large-volume detectors of superior sensitivity for rare-event experiments and medical applications. A typical deposited energy in such experiments might be rather low: of the order of 0.1 keV in coherent neutrino-nucleus scattering, 1–100 keV in dark matter search, ≥100 keV in astrophysical (solar and supernova) neutrino detection and 500 keV in positron emission tomography (PET) experiments. Accordingly, the primary ionization and/or scintillation signals have to be amplified in dense noble-gas media at cryogenic temperatures. This can be done by amplification of either light or charge signal or else a combination thereof. Many ways of light and charge signal amplification in two-phase detectors have been proposed: part of them is described in reviews (Buzulutskov, 2012; Chepel and Araujo, 2013; Gonzalez-Diaz *et al.*, 2018). Among this variety, two basic concepts can be distinguished, that most other concepts come from.

The basic concept of light signal amplification is that of the "classic" two-phase detector (Bolozdynya, 2010); it is provided by the effect of proportional electroluminescence (EL): see Fig. 4.1.

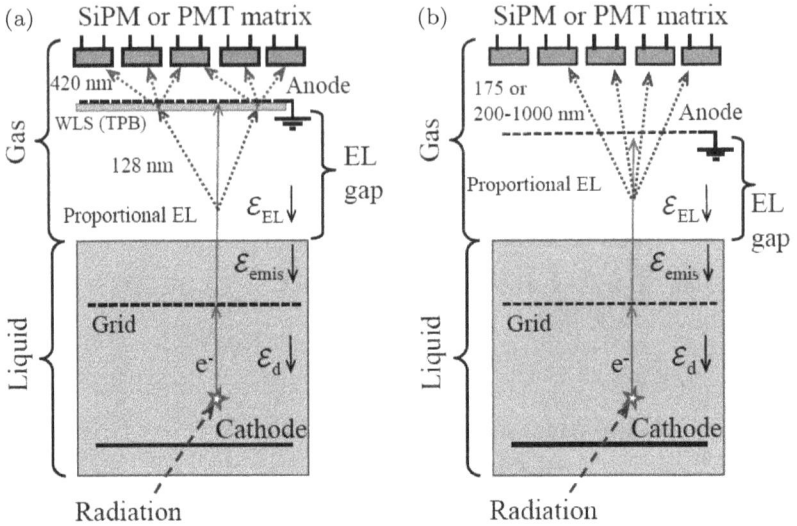

Figure 4.1:   Basic concepts of light signal amplification in two-phase detectors, using proportional EL in the EL gap. (a) With indirect optical readout via WLS, in two-phase Ar using excimer emission in the VUV at 128 nm. (b) With direct optical readout, either in two-phase Xe using excimer emission in the VUV at 175 nm, or in two-phase Ar using NBrS emission in the non-VUV range (at 200–1000 nm). Redrawn from Buzulutskov (2020).

Here each drifting electron of the ionization (S2) signal emits light in the EL gap under a high electric field, the light intensity being roughly proportional to the electric field. This concept splits into two sub-concepts, with indirect and direct optical readout. For the former, the optical readout goes via a wavelength shifter (WLS).

The basic concept of charge signal amplification in two-phase detectors is that using a GEM-like structure (gas electron multiplier, GEM) in the gas phase that provides electron avalanching within the GEM holes at cryogenic temperatures: see Fig. 4.2.

The purpose of this chapter is to describe in detail the physics and state-of-the-art of these basic concepts, as well as to consider other (most developed) concepts of signal amplification.

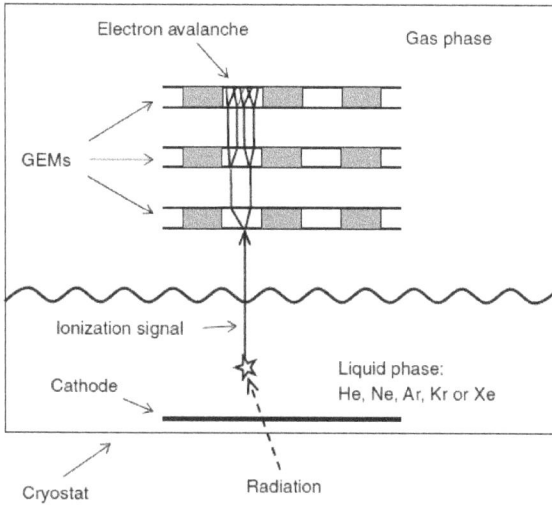

Figure 4.2: Basic concept of charge signal amplification in two-phase detectors, using GEM-like structures. Redrawn from Buzulutskov (2012).

4.2. Light Signal Amplification in the Gas Phase of Two-Phase Detector, Using Proportional Electroluminescence

4.2.1. *Three EL mechanisms*

According to modern concepts (Buzulutskov, 2020), there are three EL mechanisms responsible for proportional EL in gaseous Ar and Xe: that of excimer (Ar_2^* or Xe_2^*) emission in the VUV ("ordinary" EL), that of emission due to atomic transitions in the near infrared (NIR), and that of neutral bremsstrahlung (NBrS) emission in the UV, visible and NIR range. The energy level diagrams, appropriate reactions and their constants, relevant to the former two mechanisms, are presented in Fig. 4.3 and Table 4.1. Photon emission spectra for all three mechanisms and their absolute EL yields are shown in Figs. 4.4 and 4.5, respectively, on the Ar example. For other noble gases the emission spectra and EL yields look alike: see

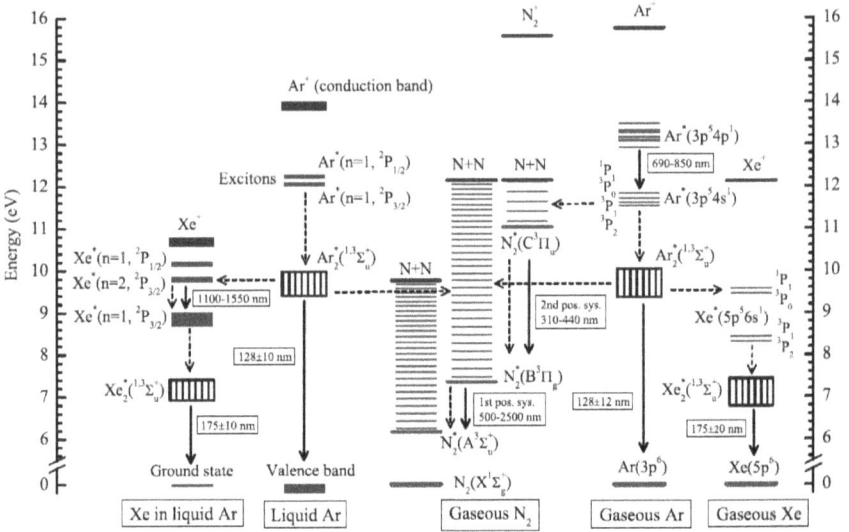

Figure 4.3: Energy levels of the lower excited and ionized states relevant to the ternary mixture of Ar doped with Xe and N_2 in the two-phase mode. These are shown for gaseous Ar, gaseous N_2, gaseous Xe, liquid Ar and Xe in liquid Ar. The solid arrows indicate the radiative transitions observed in experiments and relevant to the present review (i.e., when the excitation is induced by ionization or EL). The numbers next to each arrow show the photon emission band of the transition, defined by major emission lines or by full width of the emission continuum. The dashed arrows indicate the most probable non-radiative transitions induced by atomic collisions for Ar, Xe and N_2 species and their pair combinations in the gas and liquid phase. Redrawn from Buzulutskov (2017).

(Buzulutskov, 2012; Chepel and Araujo, 2013). For Ar, the EL mechanisms are described below in detail.

In the following, proportional EL is characterized by a reduced EL yield (Y/N), which by definition is the number of photons produced per drifting electron per atomic density and per unit drift path. It is typically plotted as a function of the reduced electric field (E/N), the latter being expressed in Townsends: 1 Td $= 10^{-17}$ V cm^2 atom^{-1}, corresponding to 0.87 kV/cm in gaseous Ar at 1.00 atm and 87.3 K.

Table 4.1: Basic reactions of excited species relevant to the performance in the two-phase mode, namely in Ar in the gas and liquid phase, doped with Xe (1000 ppm in the liquid and 40 ppb in the gas phase) and N_2 (50 ppm in the liquid and 135 ppm in the gas phase), their rate (k) or time (τ) constants reported in the literature and their time constants reduced to given atomic densities at 87 K (τ_{TP}), in particular for Ar to that of 8.63×10^{19} cm^{-3} and 2.11×10^{22} cm^{-3} in the gas and liquid phase, respectively (Buzulutskov, 2017).

No.	Reaction	k or τ	T	τ_{TP}
	Gaseous Ar + Xe(40 ppb) + N_2(135 ppm)			
(1)	$Ar^*(3p^5 4s^1) + 2Ar \rightarrow Ar_2^*(^{1,3}\Sigma_u^+) + Ar$	$k_1 \sim 1 \times 10^{-32}$ cm^6 s^{-1}	300 K	$\sim$13 ns
(2)	$Ar_2^*(^{1,3}\Sigma_u^+) \rightarrow 2Ar + h\nu(VUV)$	$\tau_2(^1\Sigma_u^+) = 4.2$ ns	300 K	4.2 ns
		$\tau_2(^3\Sigma_u^+) = 3.0 - 3.2\,\mu s$	300 K	3.1 μs
(3)	$Ar^*(3p^5 4p^1) \rightarrow Ar^*(3p^5 4s^1) + h\nu(NIR)$	$\tau_3 = 20 - 40$ ns	300 K	$<$100 ns
		$\tau_3 < 100$ ns	163 K	
(4)	$Ar^*(3p^5 4s^1) + Xe \rightarrow Ar + Xe^*$	$k_4 = (2-3) \times 10^{-10}$ cm^3 s^{-1}	300 K	$\sim$1 ms
(5)	$Ar_2^*(^3\Sigma_u^+) + Xe \rightarrow 2Ar + Xe^*(^1P_1, {}^3P_0)$	$k_5 \sim 5 \times 10^{-10}$ cm^3 s^{-1}	300 K	$\sim$0.6 ms
(6)	$Ar^*(3p^5 4s^1) + N_2 \rightarrow Ar + N_2^*(C)$	$k_6 \sim 1.5 \times 10^{-11}$ cm^3 s^{-1}	300 K	2.4 μs
		$k_6 = 3.6 \times 10^{-11}$ cm^3 s^{-1}	300 K	
		$k_6 \geq 6.5 \times 10^{-9}$ cm^3 s^{-1} (?)	87 K	$\leq$13 ns (?)
(7)	$Ar^*(3p^5 4s^1) + N_2 \rightarrow Ar + N_2^*(C, B, A)$	$k_7 \sim 3 \times 10^{-11}$ cm^3 s^{-1}	300 K	
		$k_7 = 3.6 \times 10^{-11}$ cm^3 s^{-1}	300 K	
(8)	$N_2^*(C) \rightarrow N_2^*(B) + h\nu(UV, 2nd\ pos.\ sys.)$	$\tau_8 = 30 - 40$ ns	300 K	35 ns
(9)	$N_2^*(B) \rightarrow N_2^*(A) + h\nu(NIR, 1st\ pos.\ sys.)$	$\tau_9 \sim 9\,\mu s$	119 K	$\sim$9 μs
(10)	$N_2^*(C) + Ar \rightarrow N_2^*(B) + Ar$	$k_{10} = 5.6 \times 10^{-13}$ cm^3 s^{-1}	300 K	21 ns
(11)	$N_2^*(B) + Ar \rightarrow N_2^*(A) + Ar$	$k_{11} = 1.4 \times 10^{-14}$ cm^3 s^{-1}	300 K	0.8 μs

(Continued)

Table 4.1: (*Continued*)

No.	Reaction	k or τ	T	τ_{TP}
(12)	$N_2^*(C) + N_2 \rightarrow N_2 + N_2^*(B)$	$k_{12} \sim 1 \times 10^{-11}\ \mathrm{cm^3\,s^{-1}}$	300 K	$\sim 8.6\ \mu s$
(13)	$N_2^*(B) + N_2 \rightarrow N_2 + N_2^*(A)$	$k_{13} \sim 1 \times 10^{-11}\ \mathrm{cm^3\,s^{-1}}$	300 K	$\sim 8.6\ \mu s$
(14)	$Ar_2^*(^3\Sigma_u^+) + N_2 \rightarrow 2Ar + N_2^*(B)$	$k_{14} \sim 3.3 \times 10^{-12}\ \mathrm{cm^3\,s^{-1}}$	300 K	$\sim 26\ \mu s$
	Liquid Ar + Xe(1000 ppm) + N_2(50 ppm)			
(15)	$Ar^*(n=1, {}^2P_{1/2,3/2}) + Ar \rightarrow Ar_2^*\left({}^{1,3}\Sigma_u^+\right)$	$\tau_{15} = 6$ps	87 K	6 ps
(16)	$Ar_2^*\left({}^{1,3}\Sigma_u^+\right) \rightarrow 2Ar + h\nu(\mathrm{VUV})$	$\tau_{16}\left({}^1\Sigma_u^+\right) = 7$ ns	87 K	7 ns
		$\tau_{16}\left({}^3\Sigma_u^+\right) = 1.6\ \mu s$		$1.6\ \mu s$
(17)	$Ar_2^*\left({}^{1,3}\Sigma_u^+\right) + Xe \rightarrow 2Ar + Xe^*(n=1,2, {}^2P_{3/2})$	$k_{17}\left({}^3\Sigma_u^+\right) \sim (0.8-1) \times 10^{-11}\ \mathrm{cm^3\,s^{-1}}$	87 K	~ 5.3 ns
		$\tau_{17}\left({}^3\Sigma_u^+\right) < 90$ ns	87 K	<90 ns
		$k_{17}\left({}^1\Sigma_u^+\right) \sim 3.3 \times 10^{-11}\ \mathrm{cm^3\,s^{-1}}$	87 K	~ 1.4 ns
(18)	$Xe^*(n=1,2, {}^2P_{3/2}) + Ar \rightarrow ArXe^*$	Immediate trapping	87 K	
(19)	$ArXe^* + Xe \rightarrow Ar + Xe_2^*\left({}^{1,3}\Sigma_u^+\right)$	$\tau_{19} \leq 20$ ns	87 K	≤ 20 ns
(20)	$Xe^*(n=1,2, {}^2P_{3/2}) + Xe \rightarrow Xe_2^*\left({}^{1,3}\Sigma_u^+\right)$	—	87 K	
(21)	$Xe_2^*\left({}^{1,3}\Sigma_u^+\right) \rightarrow 2Xe + h\nu(\mathrm{UV})$	$\tau_{21}\left({}^1\Sigma_u^+\right) = 4.3$ ns	165 K	4.3 ns
		$\tau_{21}\left({}^3\Sigma_u^+\right) = 22$ ns	165 K	22 ns
(22)	$Xe^*(n=2, {}^2P_{3/2}) \rightarrow Xe^*(n=1, {}^2P_{3/2}) + h\nu(\mathrm{NIR})$	$\tau_{22} < 170$ ns	87 K	<170 ns
	Reactions (17)–(21) in total ($\tau_{17} + \tau_{19}$):			≤ 110 ns
(23)	$Ar_2^*\left({}^3\Sigma_u^+\right) + N_2 \rightarrow 2Ar + N_2^*(B)$	$k_{23} = 3.8 \times 10^{-12}\ \mathrm{cm^3\,s^{-1}}$	87 K	250 ns
(24)	$ArXe^* + N_2 \rightarrow Ar + Xe + N_2^*(B,A)$	—	87 K	
(25)	$Xe_2^*\left({}^{1,3}\Sigma_u^+\right) + N_2 \rightarrow 2Xe + N_2^*(B,A)$	—	87 K	

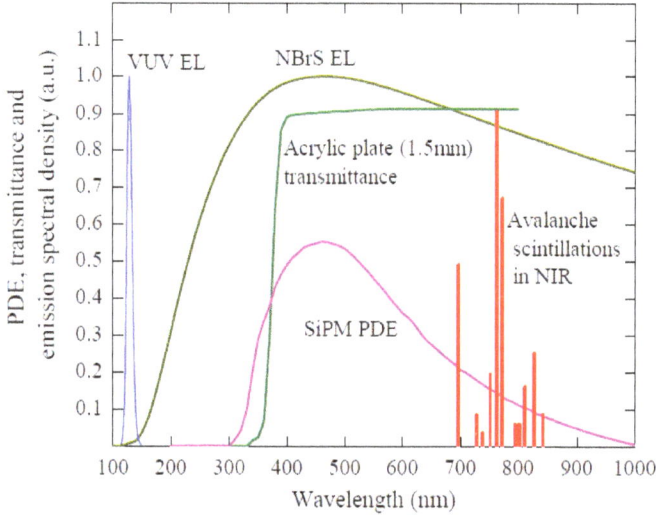

Figure 4.4: Photon emission spectra in gaseous Ar due to ordinary scintillations in the VUV range (redrawn from Morozov *et al.*, 2008), NBrS EL at 8.3 Td theoretically calculated in Buzulutskov *et al.* (2018) and avalanche scintillations in the NIR measured in (redrawn from Lindblom and Solin, 1988; Fraga *et al.*, 2000). Also shown are the Photon Detection Efficiency (PDE) of SiPM (MPPC 13,360–6050PE (Hamamatsu)) at overvoltage of 5.6 V and the transmittance of the acrylic plate (1.5 mm thick).

4.2.2. *Electroluminescence due to excimer emission*

In ordinary (excimer) EL, electrons accelerated by the electric field excite atoms, which in turn produce excimers (excited molecules) in three-body collisions. The excimers are produced in either a singlet $Ar_2^*(^1\sum_u^+)$ or a triplet $Ar_2^*(^3\sum_u^+)$ state. These states decay with photon emission in the VUV, peaked at 128 nm for Ar and 175 nm for Xe (see Figs. 4.3 and 4.4 and reactions (1) and (2) in Table 4.1):

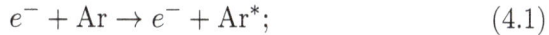

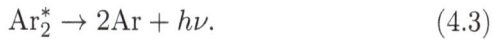

$$e^- + Ar \rightarrow e^- + Ar^*; \tag{4.1}$$

$$Ar^* + 2Ar \rightarrow Ar_2^* + Ar; \tag{4.2}$$

$$Ar_2^* \rightarrow 2Ar + h\nu. \tag{4.3}$$

The singlet and triplet excimer states are responsible, respectively, for the fast and slow emission components observed in proportional EL.

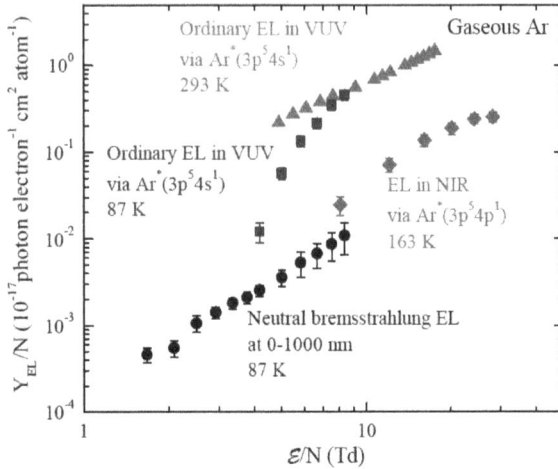

Figure 4.5:   Summary of experimental data on reduced EL yield in gaseous Ar for all known EL mechanisms: for NBrS EL below 1000 nm, measured in Buzulutskov *et al.* (2018) and Bondar *et al.* (2020a) at 87 K; for ordinary EL in the VUV, due to excimer emission going via $Ar^*(3p^54s^1)$ excited states, measured in Buzulutskov *et al.* (2018) and Bondar *et al.* (2020a) at 87 K and in Monteiro *et al.* (2008) at 293 K; for EL in the NIR due to atomic transitions going via $Ar^*(3p^54p^1)$ excited states measured in Buzulutskov *et al.* (2011) at 163 K. Redrawn from Buzulutskov (2020).

Their time constants are of 4.2 ns and 3.1 μs, respectively (Table 4.1). Ordinary EL has a threshold in the electric field, of about 4 Td (see Fig. 4.5), defined by the energy threshold of 11.55 eV of Ar atom excitation in reaction (4.1).

Ordinary proportional EL has two characteristic properties: a threshold and linear field dependence (far enough from the threshold). For Xe, this can be seen from Fig. 4.6 presenting compilation of the experimental data on EL yield by 2007: the results of different groups are close, with the exception of early works.

The absolute photon yields of ordinary (excimer) EL were theoretically calculated for Ne, Ar, Kr and Xe in Oliveira *et al.* (2011) using Monte Carlo simulation in a microscopic approach: see Fig. 4.7. One can see from the figure that well above the EL threshold, the theoretical predictions are in rather good agreement with the yields measured in experiment at room temperature, for

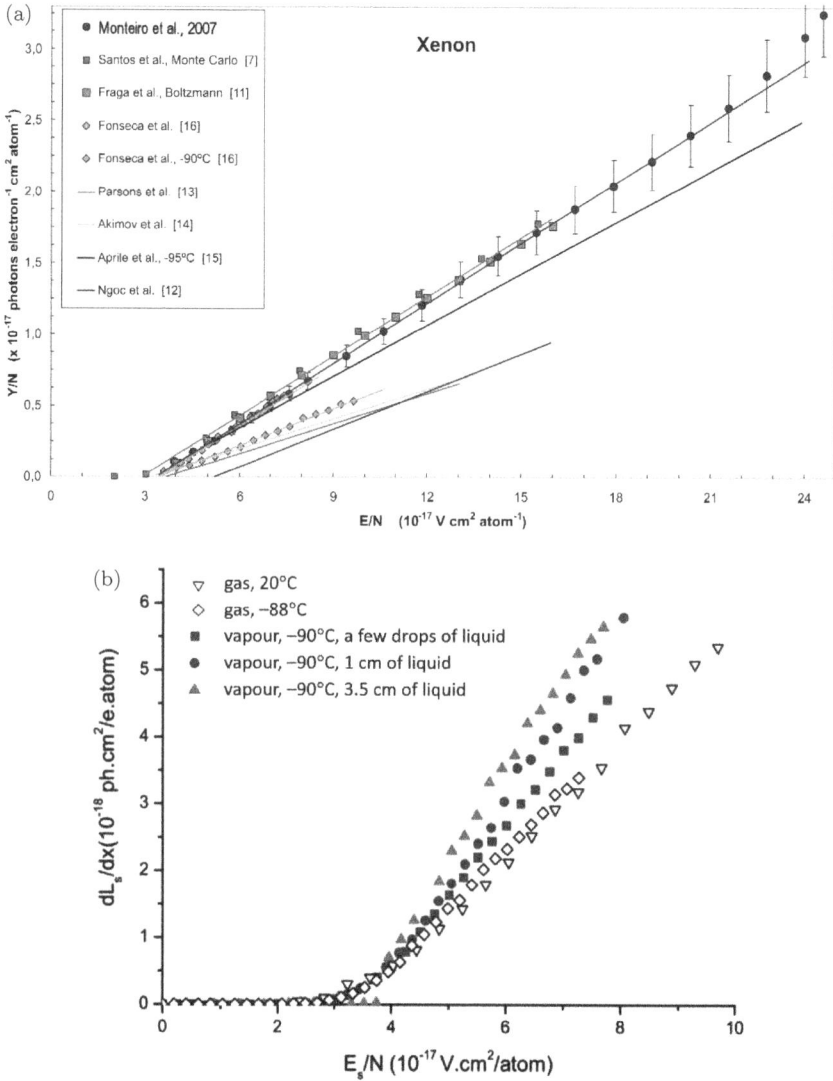

Figure 4.6: Reduced EL yields for ordinary (excimer) EL in gaseous Xe measured by different groups, as a function of the reduced electric field. Redrawn from Monteiro *et al.* (2007) (a) and from Fonseca *et al.* (2004) (b).

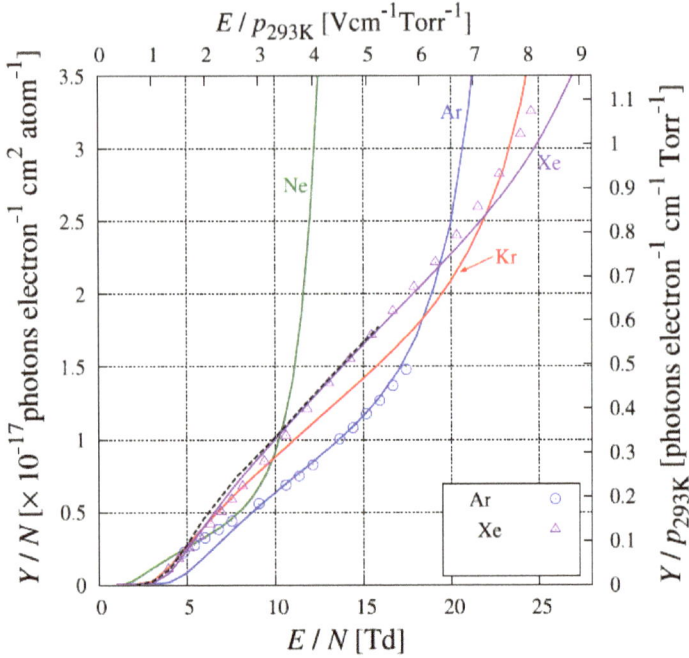

Figure 4.7: Measured reduced EL yield for ordinary (excimer) EL at room temperature (data points), as a function of the reduced electric field, compared with Monte Carlo simulation data (curves), for Ne, Ar, Kr and Xe. Redrawn from Oliveira *et al.* (2011).

Ar and Xe. Extrapolating the theoretical EL yield, the EL threshold was determined as 1.5, 4.1, 2.6 and 2.9 Td for Ne, Ar, Kr and Xe, respectively (Oliveira *et al.*, 2011).

Far enough from the threshold, the EL yield can be parametrized by a linear function as first observed in Conde *et al.* (1977), indicating the proportionality of the EL yield to the electric field:

$$Y/N [10^{-17} \text{photon electron}^{-1} \text{cm}^2 \text{atom}^{-1}] = a \cdot E/N - b, \quad (4.4)$$

where reduced electric field E/N is given in Td (10^{-17} V cm^2 atom^{-1}). This equation is universally valid for all temperatures and gas densities. Parameter b defines the electric field threshold for proportional EL, while parameter a describes the light amplification factor, namely the number of photons emitted by a drifting electron per applied voltage across the EL gap. For Ar, Kr and Xe this parameter was measured to be 81 (Monteiro *et al.*, 2008), 120 (Mano *et al.*, 2020) and 140 (Monteiro *et al.*, 2007) photon/kV, respectively. In the saturated vapours of xenon, the reduced light yield was found to be larger than in gas as can be seen in Fig. 4.6 (bottom). This is reflected in the respective increase of the slope a in Eq. (4.4) by a factor of 1.3 (Fonseca *et al.*, 2004). Direct excitation by the electron impact of xenon dimers (or higher polymers), concentration of which is higher in the saturated vapor, is a possible explanation for this effect.

The parametrization of the more recent experimental data by linear field dependence was presented in Monteiro *et al.* (2008) and Monteiro *et al.* (2007) for Ar and Xe, respectively:

$$Y/N = 0.081 \cdot E/N - 0.190, \text{for Ar};$$
$$Y/N = 0.140 \cdot E/N - 0.474, \text{for Xe}. \tag{4.4a}$$

However, near the EL threshold this parametrization gives overestimated EL yield values; in this case the experimental data should be used directly. In particular, in two-phase Ar at 87 K and 1.0 atm, the number of photons in the VUV emitted by drifting electron per 1 cm, deduced from experimental data of Fig. 4.5, is 345 and 43 at electric field of 8 and 4.6 Td, respectively, the latter corresponding to the nominal operation field of the DarkSide experiment.

4.2.3. *Electroluminescence due to neutral bremsstrahlung effect*

In addition to the ordinary EL mechanism, there is a concurrent EL mechanism, based on bremsstrahlung of drifting electrons scattered on neutral atoms, so-called NBrS (Butikov *et al.*, 1970; Buzulutskov *et al.*, 2018; Bondar *et al.*, 2020a; Tanaka *et al.*, 2020):

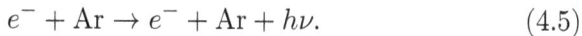

$$e^- + \text{Ar} \rightarrow e^- + \text{Ar} + h\nu. \tag{4.5}$$

The history of the issue is described in detail in Buzulutskov *et al.* (2018).

The NBrS differential cross-section is proportional to elastic cross-section $(\sigma_{el}(E))$ of electron-atom scattering (Buzulutskov *et al.*, 2018):

$$\frac{d\sigma}{d\nu} = \frac{8}{3}\frac{r_e}{c}\frac{1}{h\nu}\left(\frac{E-h\nu}{E}\right)^{1/2}\left[(E-h\nu)\cdot\sigma_{el}(E) + E\cdot\sigma_{el}(E-h\nu)\right].$$

(4.6)

Here, $r_e = e^2/m_e c^2$ is the classical electron radius, $c = \nu\lambda$ is the speed of light, E is the initial electron energy. For electron-atom scattering cross-sections in Ar, see Fig. 4.8.

Consequently, the NBrS EL yield is predicted to be maximal for electron energies of the order of 10 eV where the elastic cross-section has a maximum, thus corresponding to the electric fields of 4–5 Td. This is confirmed in Fig. 4.9 showing the absolute yield of NBrS EL theoretically calculated in Buzulutskov *et al.* (2018), in comparison with that of ordinary EL.

NBrS EL has a continuous emission spectrum, extending from the UV (200 nm) to the visible and NIR range (1000 nm) and even to the

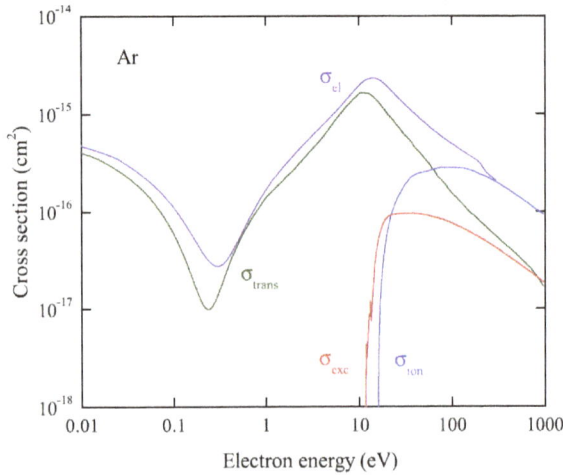

Figure 4.8: Electron scattering cross-sections in Ar: elastic (el), momentum-transfer (trans), excitation (exc) and ionization (ion). Redrawn from Buzulutskov *et al.* (2018).

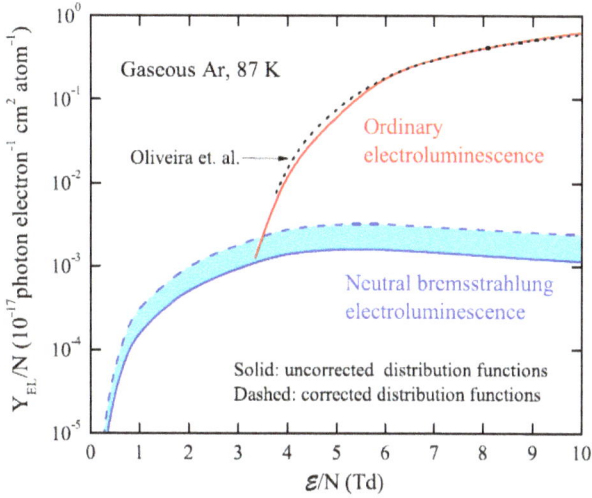

Figure 4.9: Reduced EL yield in gaseous Ar as a function of the reduced electric field, for NBrS EL (below 1000 nm) calculated in Buzulutskov *et al.* (2018) and compared to ordinary (excimer) EL calculated in Oliveira *et al.* (2011). Redrawn from Buzulutskov *et al.* (2018).

radio frequency range (Al Samarai *et al.*, 2016): see Fig. 4.4. From Fig. 4.4 and 4.9 one can see that NBrS EL, albeit being significantly weaker than ordinary EL above the Ar excitation threshold, has no threshold in the electric field, in contrast to ordinary EL, and thus dominates below the threshold. Accordingly, the NBrS effect can explain two remarkable properties of proportional EL observed in experiment (Buzulutskov *et al.*, 2018; Bondar *et al.*, 2020a): that of the photon emission below the Ar excitation threshold and that of the substantial contribution of the non-VUV spectral component above the threshold.

At lower electric fields, below 4 Td corresponding to the Ar excitation threshold, the NBrS theory developed in Buzulutskov *et al.* (2018) correctly predicts the absolute value of the EL yield. This is seen when comparing the experimental data on EL yields in Fig. 4.5 to those of the theory in Fig. 4.9 and when comparing the experimental and theoretical photon emission spectra in Fig. 4.10. On the other hand, at higher fields (above 5 Td), the experimental EL

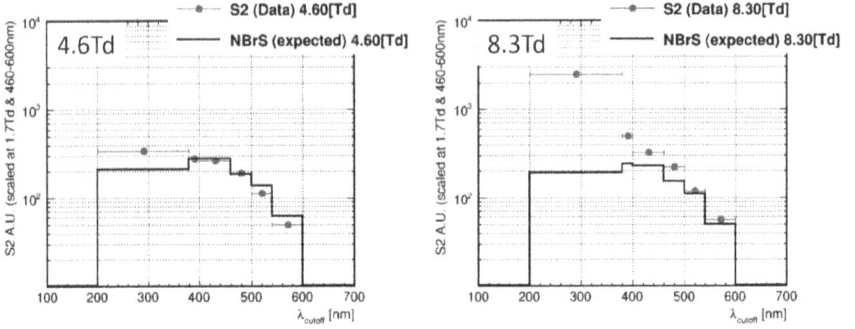

Figure 4.10:   Photon emission spectra of proportional EL in Ar in the UV and visible range measured at room temperature in Tanaka *et al.* (2020) (data points) in comparison with those of the NBrS EL theory (Buzulutskov *et al.*, 2018) (lines), at reduced electric field of 4.6 and 8.3 Td. Redrawn from Tanaka *et al.* (2020).

yields quickly diverge from those of the theory, exceeding those in the UV spectral range (below 400 nm). In Buzulutskov *et al.* (2018) this discrepancy was hypothesized to be explained by the contribution of electron scattering on sub-excitation Feshbach resonance (going via intermediate negative ion state $Ar^-(3p^54s^2)$), which might be accompanied by the enhanced photon emission (Dyachkov *et al.*, 1974).

Near the threshold of ordinary EL, namely in the field range of 4–6 Td, it becomes important to take into account the possible contribution of NBrS EL, to correctly measure the EL yield of ordinary EL. Note that just in this field region, namely nominally at 4.6 Td, the DarkSide dark matter search experiment operates (Agnes *et al.*, 2015). In Buzulutskov *et al.* (2018), Bondar *et al.* (2020a) the NBrS contribution was measured, using dedicated spectral devices insensitive to the VUV, and appropriately subtracted from the data obtained with VUV-sensitive devices. The resulting EL yield of ordinary EL, shown in Fig. 4.11, is in good agreement with the theory. On the other hand, the experimental data of Monteiro *et al.* (2008) in Fig. 4.11 has an excess over those of Buzulutskov *et al.* (2018)

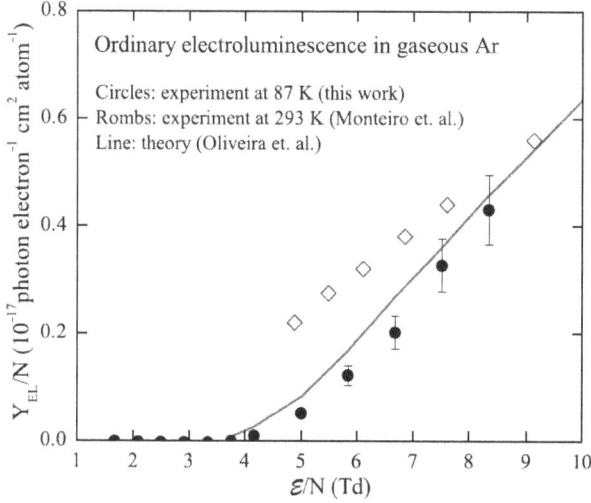

Figure 4.11: Reduced EL yield in gaseous Ar as a function of the reduced electric field, for ordinary EL (due to excimer emission in the VUV) near its threshold, measured at 87 K (Buzulutskov *et al.*, 2018) and at 293 K (Monteiro *et al.*, 2008). The theoretical calculation of Oliveira *et al.* (2011) is also shown. Redrawn from Buzulutskov *et al.* (2018).

and theory (Oliveira *et al.*, 2011), below 6 Td. It would be logical to explain this discrepancy by the NBrS contribution that was not taken into account in Monteiro *et al.* (2008).

4.2.4. *Electroluminescence due to atomic transitions in the NIR*

At higher electric fields, above 8 Td, another "non-standard" EL mechanism comes into force, namely that of EL in the NIR due to transitions between excited atomic states (Lindblom and Solin, 1988; Fraga *et al.*, 2000; Buzulutskov *et al.*, 2011, 2012; Oliveira *et al.*, 2013). Here, the higher atomic levels $Ar^*(3p^5 4p^1)$ are excited; the transitions between those and the lower excited levels are responsible for the fast emission in the NIR, at 700–850 nm (see Fig. 4.3 and reaction (3) in Table 4.1):

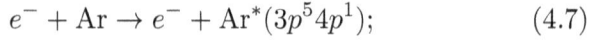

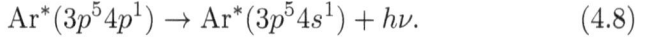

$$e^- + Ar \rightarrow e^- + Ar^*(3p^5 4p^1); \qquad (4.7)$$

$$Ar^*(3p^5 4p^1) \rightarrow Ar^*(3p^5 4s^1) + h\nu. \qquad (4.8)$$

Atomic EL in the NIR has a line emission spectrum: see Fig. 4.4. In addition, it has a higher threshold in the electric field than ordinary EL. The absolute yield of atomic EL in the NIR was measured in Buzulutskov *et al.* (2011) and Bondar *et al.* (2012); it was well described by the theory using a microscopic approach (Oliveira *et al.*, 2013): see Fig. 4.12. It should be remarked that atomic emission in the NIR is particularly noticeable at even higher fields, above 30 Td, where the avalanche multiplication of the electrons takes place, accompanied by corresponding secondary scintillations: by so-called "avalanche scintillations" (Fraga *et al.*, 2000; Bondar *et al.*, 2010).

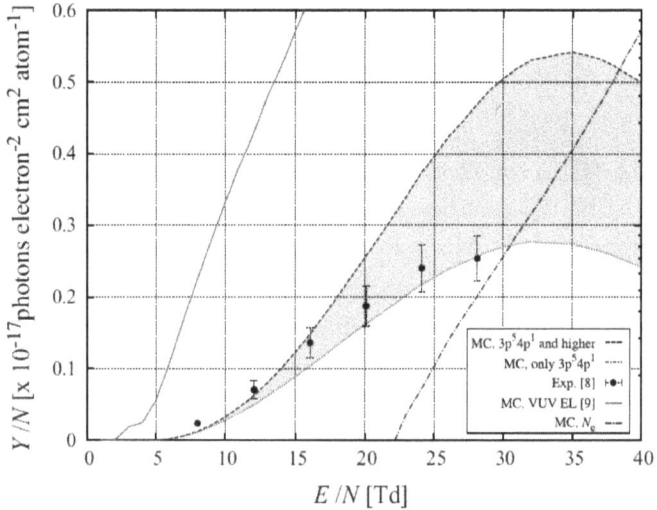

Figure 4.12: Reduced EL yield in gaseous Ar as a function of the reduced electric field, for atomic EL in the NIR due to atomic transitions going via $Ar^*(3p^5 4p^1)$ excited states measured in Buzulutskov *et al.* (2011) at 163 K (data points) and theoretically calculated in Oliveira *et al.* (2013) (hatched area between two curves). For comparison, that for ordinary EL in the VUV going via $Ar^*(3p^5 4s^1)$, theoretically calculated in Oliveira *et al.* (2013) (solid curve), is shown. Redrawn from Oliveira *et al.* (2013).

4.2.5. *Concepts of light signal amplification*

Ordinary EL in Ar and Xe results in two "standard" concepts of light signal amplification in two-phase detectors, depicted in Fig. 4.1. In the first concept in Ar, the EL gap is optically readout by the SiPM or PMT matrix indirectly, via a WLS, typically TPB. This is because the typical cryogenic PMTs and SiPMs are not sensitive to ordinary EL of Ar, at 128 nm.

In the second concept in Xe, the EL gap is optically read out directly by the PMT matrix, provided that the typical cryogenic PMTs has quartz windows and thus become sensitive to ordinary EL of Xe, at 175 nm.

The concept of direct optical readout (Fig. 4.1(b)), standard for Xe, becomes non-standard for Ar if to use NBrS EL in the non-VUV range, at 200–1000 nm. Such an alternative readout concept has the advantage of doing without WLS, in two-phase Ar detectors. This may lead to more stable operation due to avoiding the problems of WLS degradation and its dissolving in liquid (Sanguino *et al.*, 2006; Asaadi *et al.*, 2019), as well as that of WLS peeling off from the substrate. Such a concept was realized in Aalseth *et al.* (2020a). It might be considered as backup solutions for large-scale two-phase Ar detectors, in case the problem of WLS instability cannot be resolved.

4.3. Charge Signal Amplification in the Gas Phase of Two-Phase Detector, Using Electron Avalanching

4.3.1. *Charge signal amplification concepts at cryogenic temperatures*

Over the past two decades, there has been a growing interest in so-called "Cryogenic Avalanche Detectors": see review Buzulutskov (2012). In the wide sense these define a class of noble-gas detectors operated at cryogenic temperatures with electron avalanching performed directly in the detection medium, the latter being in a gaseous, liquid or two-phase state, provided by electron impact ionization reaction:

$$e^- + \mathrm{Ar} \rightarrow 2e^- + \mathrm{Ar}^+. \tag{4.9}$$

Earlier attempts to obtain high and stable electron avalanching directly in noble gases and liquids at cryogenic temperatures, using "open-geometry" multipliers, have not been very successful: rather low gains (≤ 10) were observed in liquid Xe (Derenzo *et al.*, 1974; Masuda *et al.*, 1979; Policarpo *et al.*, 1995; Aprile *et al.*, 2014b) and Ar (Kim *et al.*, 2004) and low gains (≤ 100) in gaseous Ar (Grebinnuk *et al.*, 1978) and He (Masaoka *et al.*, 2000) at cryogenic temperatures, using wire, needle or micro-strip proportional counters. Moreover, two-phase detectors with wire chamber readout, which initially seemed to solve the problem, turned out to have unstable operation in the avalanche mode due to vapor condensation on wire electrodes (Dolgoshein *et al.*, 1973).

The problem of electron avalanching in cryogenic noble-gas detectors was solved in 2003 (Buzulutskov *et al.*, 2003) after introduction of cryogenic gaseous and two-phase detectors with GEM readout. GEM (Sauli, 2016) and thick GEM (THGEM, Breskin *et al.*, 2009) belong to the class of micro-pattern gas detectors (MPGDs). GEM is a thin insulating film, metal clad on both sides, perforated by a matrix of micro-holes, in which gas amplification occurs under the voltage applied across the film. The typical GEM geometrical parameters are the following: dielectric (Kapton) thickness is 50 μm, hole pitch — 140 μm, hole diameter on metal — 70 μm. THGEM is a similar, though more robust structure with 10-fold expanded dimensions. In particular, the typical THGEM geometrical parameters are the following: $t/p/d/h = 0.4/0.9/0.5/0.1$ mm. Here, $t/p/d/h$ denotes "dielectric thickness/hole pitch/hole diameter/hole rim width".

Contrary to wire chambers and other "open geometry" gaseous multipliers, cascaded GEMs and THGEMs have a unique ability to operate in dense noble gases at high gains (Buzulutskov *et al.*, 2000), including at cryogenic temperatures and in the two-phase mode (Buzulutskov, 2007). Consequently, at present, the basic idea of cryogenic avalanche detectors (CRADs) in the narrow sense is that of the combination of MPGDs with cryogenic noble-gas detectors, operated in a gaseous, liquid or two-phase mode.

The original CRAD concept (Buzulutskov *et al.*, 2003) is the following (see Fig. 4.2): electron avalanching at cryogenic temperatures is performed in pure noble gases using GEM multiplier, either in a gaseous or two-phase mode. In the latter case the conventional two-phase electron-emission detector is provided with charge signal amplification using electron avalanching in the gas phase: the primary ionization electrons produced in the liquid are emitted into the gas phase by an electric field, where they are multiplied in saturated vapor using cascaded GEM multiplier. The proof of principle of this concept was demonstrated in 2003–2006 in two-phase Ar, Kr and Xe (Buzulutskov *et al.*, 2003; Bondar *et al.*, 2006) at appropriate cryogenic temperatures (at around 87, 120 and 165 K, respectively) and in gaseous He and Ne at lower temperatures, down to 2.6 K (Buzulutskov *et al.*, 2005; Galea *et al.*, 2006).

Later on, the original CRAD concept was elaborated: it was suggested to provide CRADs with new features. There are more than a dozen different concepts with charge and light amplification in two-phase and liquid detectors developed by different groups since 2003: their gallery by 2012 is shown in Fig. 4.13. For their detailed description, we refer to review Buzulutskov (2012) and appropriate references listed in the figure: (Buzulutskov *et al.*, 2003, 2011; Bondar *et al.*, 2006, 2008, 2010; Periale *et al.*, 2005; Rubbia, 2006; Buzulutskov and Bondar, 2006; Ju *et al.*, 2007; Gai *et al.*, 2007; Lightfoot *et al.*, 2009; McConkey *et al.*, 2010; Akimov *et al.*, 2009, 2011; Duval *et al.*, 2009, 2011;). The references to such concepts and their realization after 2012 are as follows: (Bondar *et al.*, 2012, 2013a; Akimov *et al.*, 2013b; Breskin, 2013; Arazi *et al.*, 2015; Erdal *et al.*, 2020; Mavrokoridis *et al.*, 2014; Hollywood *et al.*, 2020; Ye *et al.*, 2014).

In this and the following sections, we confine ourselves to the discussion of the most developed and successfully realized concepts, as well as of the physical effects and specific technology underlying them.

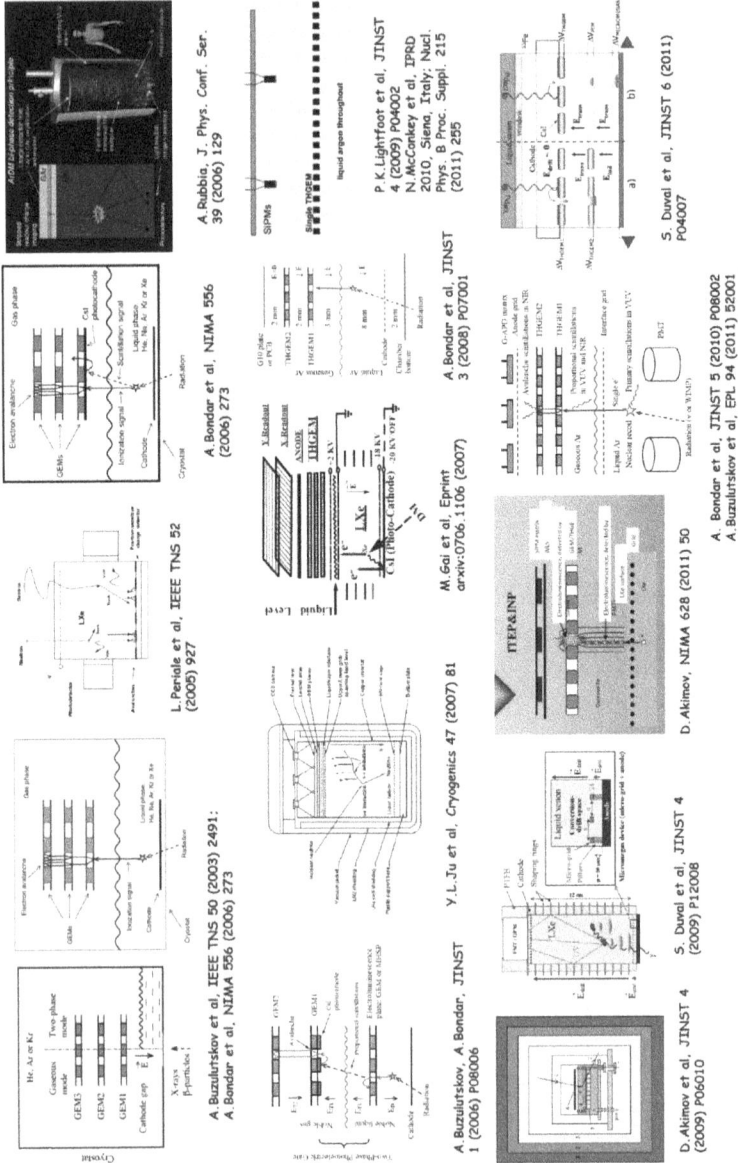

Figure 4.13: Gallery of concept pictures for charge signal amplification in two-phase detectors, using electron avalanching in the gas phase and charge or optical readout, by 2012 in order of introduction. Redrawn from Buzulutskov (2020).

4.3.2. *GEM operation in pure noble gases at cryogenic temperatures*

In most CRAD concepts, the MPGD multiplier should be able to operate at high gains in dense pure noble gases, in particular in saturated vapor and at cryogenic temperatures. Fortunately, unlike open-geometry gaseous multipliers (e.g., wire chambers), cascaded GEMs and THGEMs permit attaining high ($\geq 10^3$) charge gains in "pure" noble gases at atmospheric pressure, presumably due to considerably reduced photon-feedback effects. At room temperature, this remarkable property was discovered first for Ar (Bressan *et al.*, 1999; Buzulutskov *et al.*, 2000) and then for other noble gases (Bondar *et al.*, 2002a, 2002b; Buzulutskov, 2007, 2012), using GEM multipliers. For operation of other MPGD multipliers in "pure" noble gases at room temperature, we refer to review Buzulutskov (2012).

It should be remarked that the stable operation of GEMs at cryogenic temperatures is a non-trivial fact, since the electrical resistance of their dielectric materials considerably increases with the temperature decrease, i.e., by an order of magnitude per 35 degrees for Kapton GEMs (Bondar *et al.*, 2004), which might result in strong charging-up effects within the holes. Fortunately, the latter did not happen: the charging-up effects for GEMs have not been observed. The GEM performance in gases at cryogenic temperatures was found to be generally independent of temperature down to ~100 K, at a given gas density (Bondar *et al.*, 2004): stable and high-gain GEM operation was observed in all noble gases and in their mixtures with selected molecular additives that do not freeze in a wide temperature range (CH_4, N_2 and H_2). This is seen from Fig. 4.14(a): rather high triple-GEM gains were reached at cryogenic temperatures, exceeding 10^4 in Ar, Kr and Xe+CH_4.

It is amazing that GEMs were able to operate in electron avalanching mode at even lower temperatures, down to 2.6 K in gaseous He (Buzulutskov *et al.*, 2005; Galea *et al.*, 2006). On the other hand, high GEM gains observed in He and Ne above 77 K were reported to be due to the Penning effect in uncontrolled ($\geq 10^{-5}$) impurities (i.e., N_2) which froze out at lower temperatures,

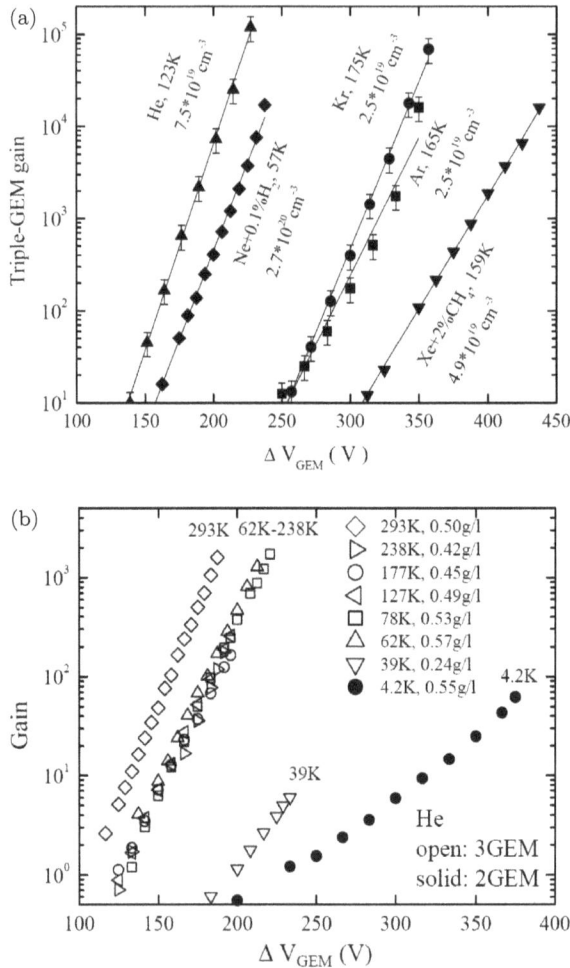

Figure 4.14: Gain characteristics of GEM multipliers at cryogenic temperatures. (a) In gaseous He, Ar and Kr, in the Penning mixture Ne+0.1%H_2 (its density corresponding to saturated Ne vapor in the two-phase mode) and in Xe+2%CH_4. (b) In gaseous He at low temperatures (down to 4.2 K). The appropriate temperatures and densities are indicated. In He at 39 K and 4.2 K, the maximum gains were limited by discharges, while at other temperatures and in other gases the discharge limit was not reached. The multiplier active area was 2.8×2.8 cm^2. Redrawn from Buzulutskov (2012).

resulting in the considerable gain drop at temperatures below 40 K:
see Fig. 4.14(b). A solution to the gain drop problem at lower
temperatures was found in Buzulutskov *et al.* (2005): Ne and He
can form high-gain Penning mixtures with H_2 at temperatures down
to $\sim$10 K. This is seen from Fig. 4.14(a): triple-GEM gains exceeding
10^4 were obtained at 57 K in the Penning mixture Ne+0.1%H_2, its
density corresponding to that of saturated Ne vapor in the two-phase
mode. Unfortunately, this does not work for two-phase He, due to
the very low H_2 vapor pressure at 4.2 K.

4.3.3. *Two-phase detectors with GEM multipliers*

Regarding GEM operation in two-phase detectors, most promising
results were obtained for two-phase Ar detectors (see Fig. 4.15): the
charge gain of 5000 with triple-GEM readout was routinely attained
(Bondar *et al.*, 2006) and the stable operation for tens of hours at this
gain was demonstrated (Bondar *et al.*, 2009a). This maximum gain

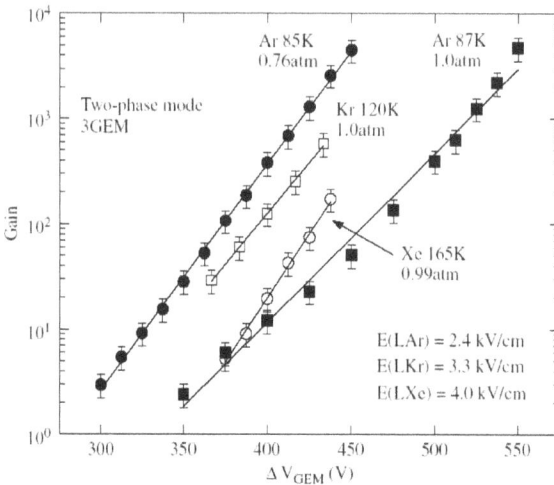

Figure 4.15: Gain characteristics in two-phase Ar, Kr and Xe detectors with
triple-GEM multiplier charge readout. The appropriate temperatures, pressures
and electric fields in the liquids are indicated. The multiplier active area was
2.8×2.8 cm^2; the maximum gains were limited by discharges. Redrawn from
Bondar *et al.* (2006).

value should be compared to that of 600 and 200 obtained in GEM-based two-phase Kr and Xe detectors, respectively: see Fig. 4.15.

In this review, the charge gain is defined as the ratio of the output anode charge of the GEM or THGEM multiplier incorporated into the two-phase detector to the input "primary" charge, i.e., to that of prior to multiplication measured in special calibration runs.

Stable operation of GEM in high purity xenon gas at low temperature of $-90°$C as well as in saturated vapor have been reported in Solovov *et al.* (2007) and Balau *et al.* (2009). For a constant gas density, voltage across the GEM and extraction field, higher gains were measured in the cold gas than at room temperature. The maximum achievable gain was also found to be higher at low temperature. In a two-phase system, the maximum gain as high as 150 was obtained with a single GEM foil at $-108°$C (Balau *et al.*, 2009). Xenon condensation in the GEM holes, observed in some measurements, could be avoided by keeping a slight temperature gradient in the chamber.

The successful performance of two-phase Ar detectors with triple-GEM charge readout is illustrated in Figs. 4.16 and 4.17. In particular, the high triple-GEM gain provided the operation of the two-phase detector in single-electron counting mode (Fig. 4.16) (Bondar *et al.*, 2007). It should be remarked that though the single-electron spectra are well separated from electronic noise at gains exceeding 5000, they are described by an exponential function (rather than by a peaked function). The latter is generally a rule for gaseous multipliers operated in proportional mode, due to intrinsically considerable fluctuations of the avalanche size. Consequently, the single- and double-electron events can hardly be distinguished in two-phase Ar detectors with GEM (or THGEM) charge readout. For that, the combination with PMT- or SiPM-based optical readout should be used, as suggested in Buzulutskov (2011) and Bondar *et al.* (2012). Accordingly, the function of GEM- or THGEM-based charge readout would be to provide superior spatial resolution.

Figure 4.17 demonstrates the unique ability of the two-phase Ar detector with GEM charge readout to observe directly and simultaneously the fast and slow components of electron emission

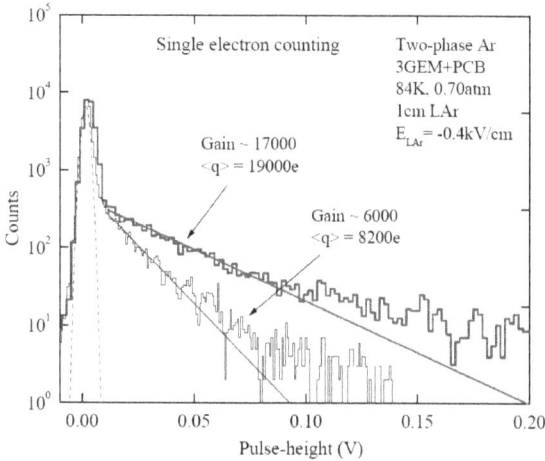

Figure 4.16: Pulse-height spectra in a two-phase Ar detector with triple-GEM multiplier charge readout, operated in single electron counting mode with external trigger, at charge gains of 6000 and 17,000. The electronic noise spectrum (dashed line), corresponding to the Equivalent Noise Charge of $\sigma = 940e^-$, is also shown. The multiplier active area was 2.8×2.8 cm^2. The results were obtained for specially selected GEM foils resistant to discharges, to reach the highest gain. Note that the stable maximum gain of typical triple-GEM multipliers in two-phase Ar was somewhat lower: of about 5000. Redrawn from Bondar *et al.* (2007).

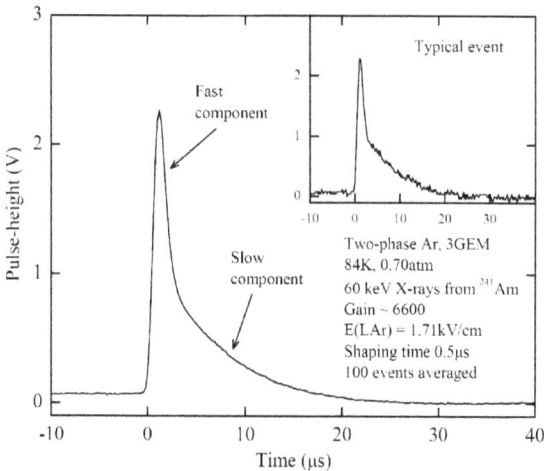

Figure 4.17: Typical anode signals in a two-phase Ar detector with triple-GEM multiplier charge readout at emission electric field in the liquid of 1.7 kV/cm. The fast and slow components of electron emission through the liquid–gas interface are distinctly seen. Redrawn from Bondar *et al.* (2009b).

through the liquid–gas interface (Bondar *et al.*, 2009b). The fast component is due to the hot electrons, heated by an electric field, while the slow component is due to the electrons reflected from the potential barrier and then thermalized (Buzulutskov, 2012). Note that the two-phase Ar detector with EL gap optical readout cannot provide such an ability, due interference with the slow component of excimer (VUV) emission. It should also be emphasized that the slow component has never been observed in two-phase Kr and Xe systems, presumably due to higher potential barrier at the interface as compared to Ar. And vice versa, the fast component has never been observed in two-phase He and Ne systems since the electrons in liquid He and Ne are localized in the bubbles and thus cannot be heated by the electric field. Consequently, the two-phase Ar system is unique in this sense.

It should be remarked that all the gain data of this section were obtained with GEMs with relatively small active area, of 2.8×2.8 cm^2. Unfortunately, as we will see later, the maximum gain decreases with the multiplier active area.

4.3.4. *Two-phase detectors with THGEM multipliers*

In two-phase detectors with THGEM charge readout, the maximum gains are comparable to those with GEM readout: see Figs. 4.18 and 4.19. Gains as high as 3000 (Bondar *et al.*, 2008) and 600 (Bondar *et al.*, 2011a) were obtained in two-phase detectors in Ar and Xe, respectively, with double-THGEM multipliers having an active area of 2.5×2.5 cm^2.

However, for larger active area, of 10×10 cm^2, the maximum gain in two-phase Ar detector decreased three-fold: down to about 1000 for double-THGEM multiplier (Bondar *et al.*, 2013b) and 100–200 for single-THGEM multiplier (Badertscher *et al.*, 2011; Bondar *et al.*, 2013b): see Figs. 4.19(a) and 4.20. This could result from the larger number of holes, implying a larger discharge probability on defects. Note that the nominal electric field in Fig. 4.20, defined as the THGEM voltage divided by its thickness, is higher than the real electric field in the THGEM hole.

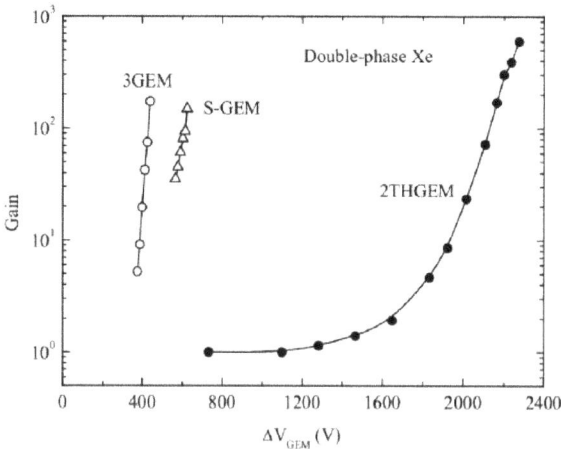

Figure 4.18: Gain characteristics of double-THGEM multipliers in a two-phase Xe detector for THGEM active area of 2.5×2.5 cm^2 (Bondar *et al.*, 2011a). Gain characteristics of triple-GEM (Bondar *et al.*, 2006) and single-GEM (Balau *et al.*, 2009) multipliers (of similar active area) are shown for comparison. Here the maximum gains were limited by discharges. Redrawn from Buzulutskov (2012).

The maximum gains of 100–1000 are obviously not sufficient for charge readout in a single-electron counting mode. Nevertheless, the gain of about 80, obtained in two-phase Ar detector with single-THGEM having 2D readout (Badertscher *et al.*, 2011), permitted to obtain track images, demonstrating the excellent imaging capability of two-phase Ar detectors with THGEM multiplier charge readout even at such a moderate gain. This technique is of particular importance for giant liquid Ar detectors for long-baseline neutrino experiments. At present, the R&D of this technique has continued as part of the ArDM and DUNE experiments (Badertscher *et al.*, 2013; Cantini *et al.*, 2015; Aimard *et al.*, 2018). It should be remarked that in that community another terminology is used, namely "dual-phase TPC" and "large electron multiplier" (LEM).

The further reduction of the maximum charge gain in a two-phase Ar detector when increasing the active area up to 40×40 and 50×50 cm^2, was reported in Badertscher *et al.* (2013) and Aimard *et al.* (2018), respectively. For the latter, the THGEM multiplier could not operate at nominal electric fields higher than 28 kV/cm,

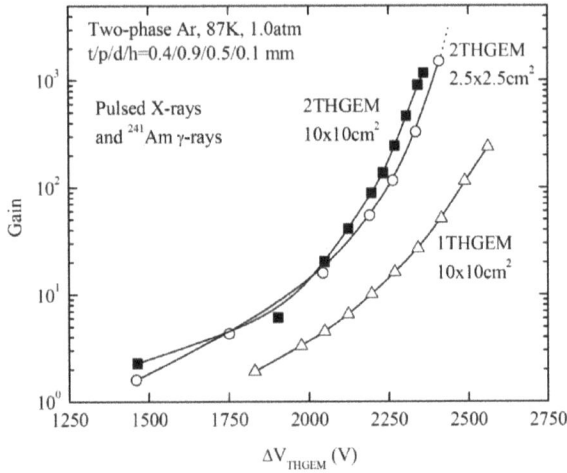

(a)

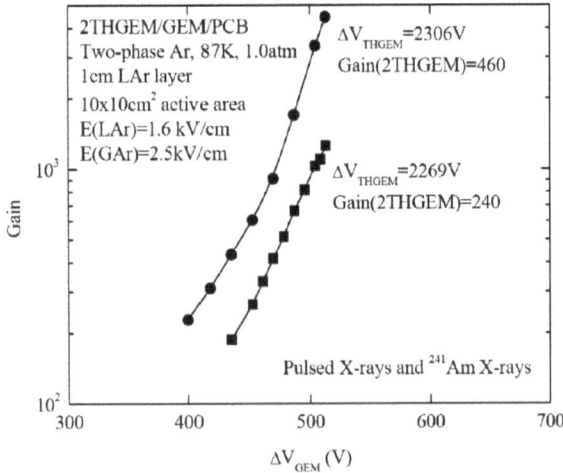

(b)

Figure 4.19: (a) Gain characteristic of double-THGEM multipliers in two-phase Ar detector for THGEM active area of 10×10 cm^2 (maximum gain was limited by discharges) compared to that of 2.5×2.5 cm^2 (maximum gain was not reached); also shown is the single-THGEM multiplier (maximum gain was limited by discharges). (b) Gain characteristics of hybrid three-stage 2THGEM/GEM multiplier in a two-phase Ar detector for an active area of 10×10 cm^2; shown is the overall multiplier gain as a function of the voltage across GEM, at two fixed voltages across each THGEM, i.e., at two values of 2THGEM gain indicated in the figure. Here the maximum gains were limited by discharges. Redrawn from Bondar *et al.* (2013b).

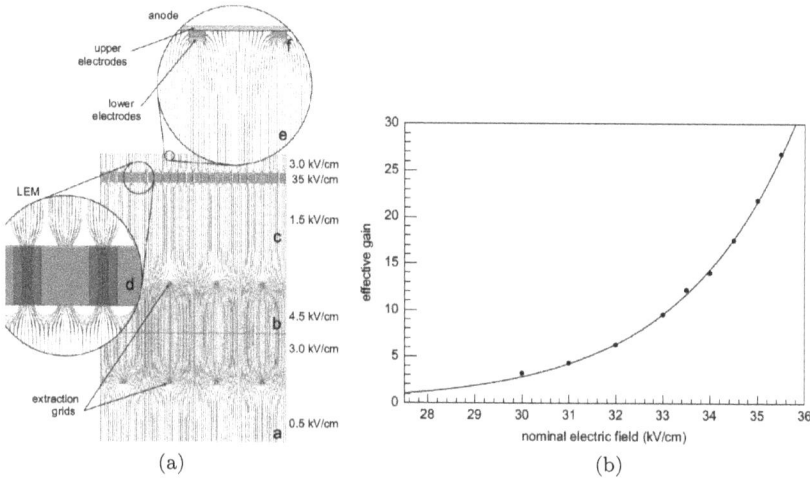

(a)   (b)

Figure 4.20: Electric field configuration in two-phase detector with THGEM (LEM) 2D readout. (a) "Effective" gain characteristics for single-THGEM 2D readout in a two-phase Ar detector, of an active area of 10×10 cm^2 (Badertscher *et al.*, 2011), as a function of the nominal electric field in THGEM. (b) The maximum gain was limited by discharges. The "effective" gain value should be multiplied by about a factor of 3, to be normalized to the gain definition of the present review (see details in Buzulutskov (2012)). Redrawn from Badertscher *et al.* (2011).

i.e., at charge gain higher than 3 according to Fig. 4.20(b). This is the real problem; it is discussed below.

An interesting way to increase the maximum charge gain in two-phase Ar CRADs was suggested in Bondar *et al.* (2013b), namely that using a hybrid three-stage 2THGEM/GEM multiplier: see Fig. 4.19(b). The idea was that the avalanche charge from a THGEM hole would be distributed between several GEM holes. This might reduce the overall avalanche charge density, believed to be responsible for the discharge development, and thus might increase the discharge voltage. The maximum gain as high as 5000 was obtained for 10×10 cm^2 active area, in practical configuration when the last multiplier stage was followed by a printed circuit board (PCB) anode, convenient for readout electronics coupling. This is almost enough for stable operation in single-electron counting mode with 2D readout, for which charge gains of 10^4 are needed.

It should be noted that the idea of hybrid multiplier, combining hole multipliers with different hole densities, can be realized without "fragile" GEMs: by cascading more robust THGEMs of different geometry, with sequential increasing hole densities and decreasing thicknesses.

4.3.5. *Gain limit, gain stability and discharge-resistance problems in two-phase detectors with GEM and THGEM multipliers*

Several problems of the performance of two-phase Ar detectors with GEM and THGEM charge readout remain unsolved. The first problem is that of the gain limit due to discharges. The obvious way to increase the gain would be adding an additional multiplication stage, i.e., up to overall four stages in the case of GEMs and three stages in the case of THGEMs. Also, it looks attractive to combine GEMs and THGEMs in a hybrid multiplier like that discussed above. Another way is an optical readout from THGEMs. Indeed, the THGEM charge gain of the order of 100 is still significant if combined with optical readout using SiPM matrices (Bondar *et al.*, 2013a) or CCD cameras (Mavrokoridis *et al.*, 2014), and might be sufficient for track imaging and even for single-electron counting mode. These will be discussed in the following section.

The second problem is that of the resistance to electrical discharges of "standard" (thin Kapton) GEMs. It was observed (Buzulutskov, 2020) that when operating two-phase Ar detectors with triple-GEM multiplier at maximum gains (approaching 10^4), the triple-GEM was not able to withstand electrical discharges on a long term: after several series of measurements, the maximum reachable gain decreased by several times. Obviously, this is due to the low resistance of thin GEMs to discharges because of metal or carbon evaporation and their deposition on the insulator in the GEM holes. The solution of the problem might be switching to thicker THGEMs that were found to behave in a more reliable way under discharges, or else combining THGEM and GEM multipliers or THGEM multipliers with increasing hole density, as discussed above.

There is another unsolved problem in two-phase CRAD performances. Namely, the performance of MPGD multipliers in two-phase Ar and Xe is not fully understood: not all multiplier types were able to operate with electron multiplication in saturated vapor. In two-phase Ar, while G10-based THGEM multipliers successfully operated for tens of hours with gains reaching a few thousands, others, namely RETHGEM and Kapton THGEMs, did not show stable multiplication in the two-phase mode (Buzulutskov, 2012; Bondar *et al.*, 2013b): they either did not multiply at all (with gain below 1) or showed unstable operation due to large gain variations. In addition, in Lighfoot *et al.* (2005) it was reported on the unsuccessful performance of the Micromegas multiplier in two-phase Xe: in half an hour the multiplication collapsed. The most rational explanation of these instabilities is the effect of vapor condensation within the THGEM holes or Micromegas mesh that prevents electron multiplication. The criteria for such a condensation are not yet clear. It could be that the specific properties of the holes and electrodes might play a role, i.e., the wetting capability depending on the electric-field non-uniformity and consequently on the MPGD geometry, as well as on the temperature gradients, the latter in turn depending on the electrode's heat conductivity.

4.3.6. *THGEM as interface grid in two-phase detectors*

In addition to the performance as charge multiplier, THGEM can be used as an interface grid immersed in the liquid (see Fig. 4.1), acting as an effective electron transmission electrode defining the electron emission and EL regions of the two-phase detector (Bondar *et al.*, 2019). Examples of such THGEM electrodes used in practice (Buzulutskov *et al.*, 2018; Aalseth *et al.*, 2020a) are shown in Fig. 4.21(a). The feature of such THGEM interface grid is that one has to apply a high enough voltage across it, to have the 100% electron transmittance. In particular, for THGEM electrode with 28% optical transparency, the nominal electric field should exceed 5 kV/cm to provide the 100% electron transmission

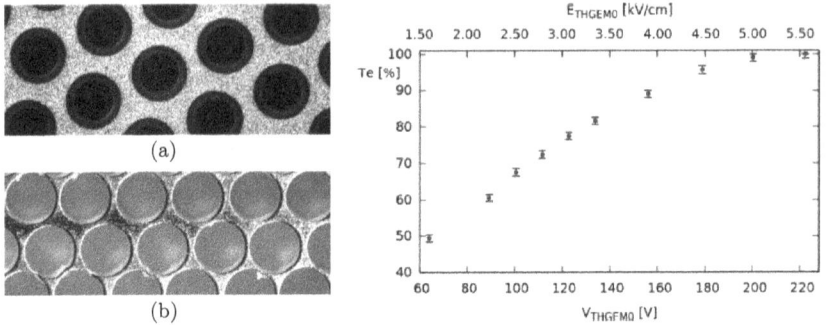

Figure 4.21:   (a) Microscope images of two THGEM electrodes used as interface grid in two-phase Ar detectors (Buzulutskov *et al.*, 2018; Aalseth *et al.*, 2020a): with 28% and 75% optical transparency (Bondar *et al.*, 2019). (b) Calculated electron transmission through 28% THGEM acting as an interface grid immersed in liquid Ar as a function of the voltage across THGEM (bottom axis) and nominal electric field in THGEM (top axis) (Bondar *et al.*, 2019). The electric fields below and above the THGEM were 0.56 and 4.3 kV/cm, respectively. Redrawn from Bondar *et al.* (2019).

(Bondar *et al.*, 2019): see Fig. 4.2(b). On the other hand, it was shown there that the THGEM with 75% optical transparency provides 100% electron transmission at already 2 kV/cm of nominal electric field. The advantage of such THGEM electrodes as compared to wire electrodes, is that they are quite rigid which avoids the problem of wire grid sagging under high electric field in large-area two-phase detectors.

4.4. Combined Charge/Light Signal Amplification in the Gas Phase of Two-Phase Detector, Using Avalanche Scintillations

4.4.1. *Two-phase Ar detector with combined THGEM/SiPM-matrix multiplier*

Figure 4.22(a) shows the concept of combined charge/light signal amplification in a two-phase detector with EL gap, using avalanche scintillations. In this concept, the combined THGEM/SiPM-matrix multiplier is coupled to the EL gap: the THGEM provides the charge signal amplification by operating in electron avalanching mode, while

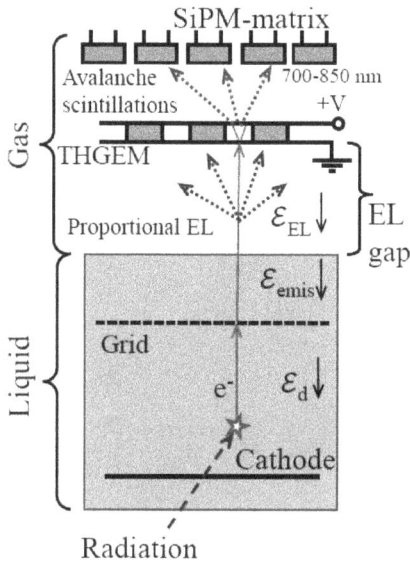

Figure 4.22: Concept of combined charge/light signal amplification in two-phase detector with EL gap, using avalanche scintillations and combined THGEM/SiPM-matrix multiplier. Redrawn from Buzulutskov (2020).

the SiPM matrix optically records avalanche scintillations produced in the THGEM holes thus providing the light signal amplification with position resolution. Since ordinary SiPMs produced by industry are typically sensitive to the NIR but not sensitive to the VUV, the avalanche scintillations can be recorded in two ways: either directly, using avalanche scintillations due to atomic transitions in the NIR (see Figs. 4.4 and 4.5), or indirectly via WLS film in front of the SiPM matrix, using excimer avalanche scintillations in the VUV. Earlier work on realizing this concept in two-phase Ar (Bondar *et al.*, 2013a) and Xe (Akimov *et al.*, 2013b) detectors should be mentioned, albeit with smaller active area and SiPM matrix size.

In the most elaborated way, the concept was realized in two-phase Ar detector with combined THGEM/SiPM-matrix multiplier of 10×10 cm^2 active area in Aalseth *et al.* (2020a), using a 11×11 SiPM matrix having 1 cm pitch of SiPM elements. The amplitude and position resolution properties of the detector are illustrated in Fig. 4.23. As large as about 500 photoelectrons were recorded by the

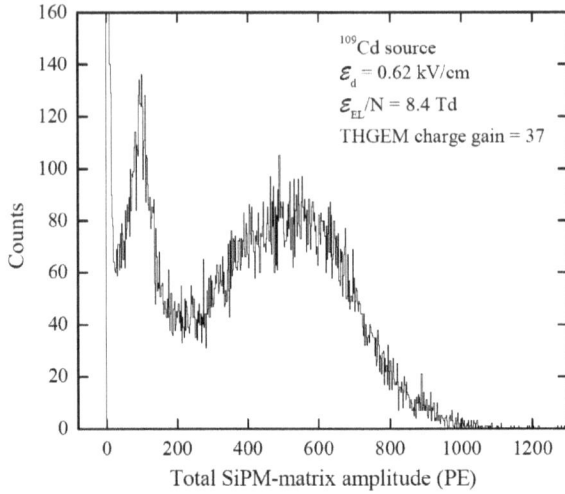

(a)

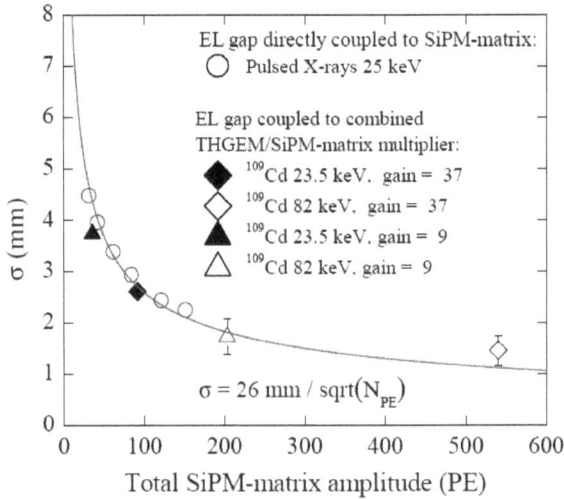

(b)

Figure 4.23:   Total SiPM-matrix amplitude spectrum for gamma-rays of ^{109}Cd source (a) and its position resolution (b), in a two-phase Ar detector with EL gap and combined THGEM/SiPM-matrix multiplier, shown in Fig. 4.21 (Aalseth *et al.*, 2020a). The THGEM charge gain was 37. The two characteristic peaks of low (22–25 keV) and high energy (60–70 and 88 keV) lines of ^{109}Cd source on W substrate are well separated in the amplitude spectrum. The position resolution (standard deviation) is shown as a function of the total number of photoelectrons recorded by the SiPM matrix. The curve shows the fit by inverse root function. Redrawn from Aalseth *et al.* (2020a).

SiPM matrix for 88 kev gamma-rays from ^{109}Cd source, at rather moderate THGEM charge gain, of 37. This would correspond to the detector yield of about 1–2 photoelectrons per drifting electron in the gas phase (Aalseth *et al.*, 2020a). Moreover, the highest position resolution was obtained in this detector than ever was measured for two-phase detectors with EL gap: $\sigma = 26\text{mm}/\sqrt{N}_{\text{PE}}$, where N_{PE} is the total number of photoelectrons recorded by the SiPM matrix. Such photoelectron yield and position resolution would be sufficient for applications in neutrino and dark matter search experiments.

4.4.2. *Two-phase Ar detector with combined THGEM/CCD-camera multiplier*

Figure 4.24(a) shows the concept of combined charge/light signal amplification in a two-phase Ar detector, using avalanche scintillations in combined THGEM/CCD-camera multiplier (Mavrokoridis *et al.*, 2014). It is similar to that considered above, with the difference that the SiPM matrix is replaced by the CCD cameras. Here, the avalanche scintillations were recorded indirectly via WLS film behind the THGEM plate, using excimer avalanche scintillations in the VUV. In the most elaborated way the concept was realized in Hollywood *et al.* (2020) in a two-phase Ar detector with 54×54 cm^2 active area and overall LAr mass of 1 t. The advanced CCD cameras, with additional electronic amplification of the signal, namely Electron Multiplying Charge Coupled Device (EMCCD) were used, which allowed for single photoelectron counting.

Using such a detector, 3D reconstruction of cosmic muon tracks and beamline interactions has been successfully performed. The detector was proposed to be employed in future large scale two-phase LAr neutrino experiments as an alternative readout method, replacing the highly segmented anode planes.

One can see from Fig. 4.24(b), showing the light signal amplitude dependence on the THGEM electric field, that two operation modes were observed: that of proportional EL (linear part) and that of electron avalanching (exponential rise part). The optical readout allows to measure the THGEM charge gain *ad hoc*: using the

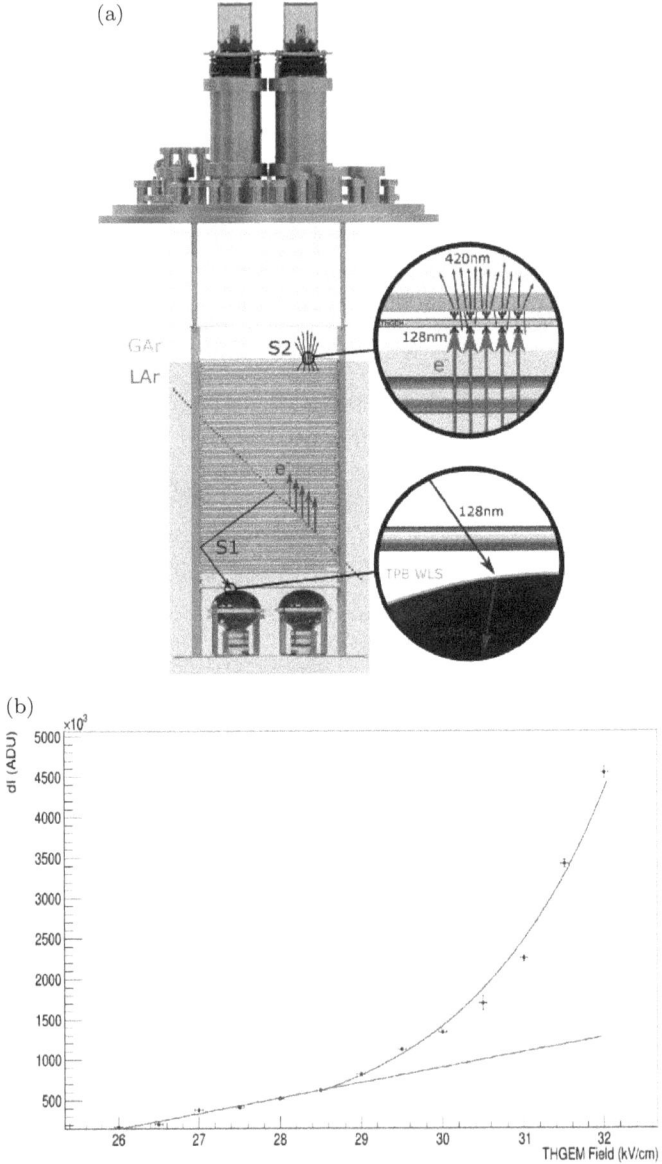

Figure 4.24: (a) Concept of combined charge/light signal amplification in two-phase detector, using avalanche scintillations in combined THGEM/CCD-camera multiplier (Mavrokoridis *et al.*, 2014), in the most elaborated way realized in Hollywood *et al.* (2020), with 54×54 cm^2 active area and overall LAr mass of 1 t. (b) Light signal amplitude on the CCD-cameras as a function of the nominal electric field in THGEM holes. Redrawn from Hollywood *et al.* (2020).

dependence of the light signal intensity on the electric field. In particular, at the maximum field of 32 kV/cm, the avalanche light gain, defined as the measured light intensity normalized to that of the linear extrapolation from the lower electric fields, is close to 4. This avalanche light gain corresponds to larger charge gain, because secondary (avalanche) electrons produce less EL light per drifting electron compared to the initial electron, since they drift over a smaller distance in the hole. If we approximate the THGEM hole by a parallel-plate gap, it can be shown that the light gain of 4 corresponds to the charge gain of 8.

4.5. Charge and Light Signal Amplification in the Liquid Phase

In the liquid, the excited, ground and ionized atomic states transform to the exciton, valence and conduction bands, respectively (see Fig. 4.3), in particular in Ar, to the $Ar^*(n = {}^{1,2}P_{3/2})$ and $Ar^*(n = {}^{1,2}P_{1/2})$ excitons, which reflect the $^3P_1$ and $^1P_1$ levels of the $Ar^*(3p^54s^1)$ atomic states. According to reaction (15) of Table 4.1, the excitons are immediately trapped in singlet or triplet excimer states $Ar_2^*(^{1,3}\Sigma_u^+)$. The singlet $^1\Sigma_u^+$ and triplet $^3\Sigma_u^+$ excimers provide the fast and slow emission components in the VUV for scintillations in liquid Ar (reaction (16) of Table 4.1)) with a lifetime of 7 ns and 1.6 μs, respectively (see Buzulutskov, 2017 and references therein). This mechanism is the basic one for scintillations in liquid Ar and is similar to that of ordinary EL in gaseous Ar described in Section 4.3.2. As for the other scintillation mechanism considered in that section, namely that of atomic transitions in the NIR in gaseous Ar, it does not work in liquid Ar due to the disappearance of analogues of the higher excited levels. On the other hand, there are some indications that the NBrS emission may still exist in liquid Ar: see Sections 4.5.1 and 4.5.2.

4.5.1. *Charge and light signal amplification in liquid Xe and liquid Ar using wires, strips and needles*

The attempts to obtain high and stable charge amplification directly in noble-gas liquids, using "open-geometry" multipliers, have not

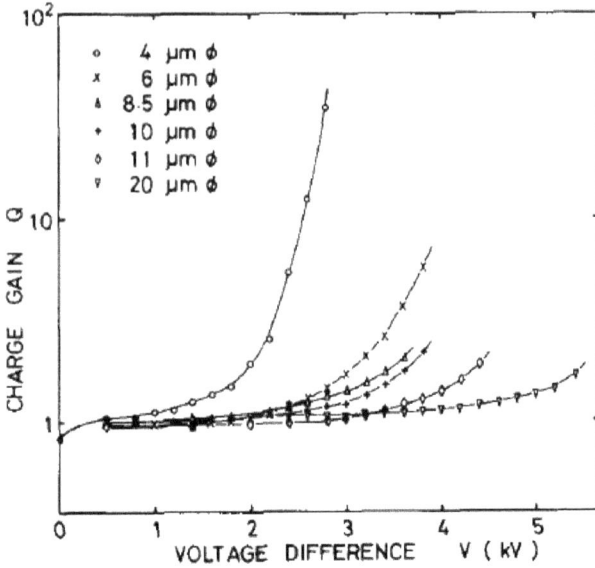

Figure 4.25:   Charge gain vs. applied voltage for different diameters of the anode wire in liquid Xe. The maximum gains are limited by discharges. Redrawn from Masuda *et al.* (1979).

been very successful: rather low charge gains (≤ 10) were observed in liquid Xe using wire proportional counters (Derenzo *et al.*, 1974; Masuda *et al.*, 1979; Aprile *et al.*, 2014b) and micro-strip counter (Policarpo *et al.*, 1995) and in liquid Ar using needle counter (Kim *et al.*, 2004). This is clearly seen from Fig. 4.25 showing charge gain in liquid Xe for different wire diameters as a function of the applied voltage.

On the other hand, proportional EL observed in liquid Xe using thin wires and VUV-sensitive PMTs (Masuda *et al.*, 1979; Aprile *et al.*, 2014b, Ye *et al.*, 2014) looks more promising for applications: see Fig. 4.26. In particular, the maximum proportional EL yield of about 300 photons per drifting electron was obtained using 10 μm diameter wire (Aprile *et al.*, 2014b). In a wide voltage range, proportional EL gain was well described by a linear function, defining a nominal EL threshold at a certain electric field. In particular, the nominal thresholds for proportional EL and electron multiplication

(a)

(b)

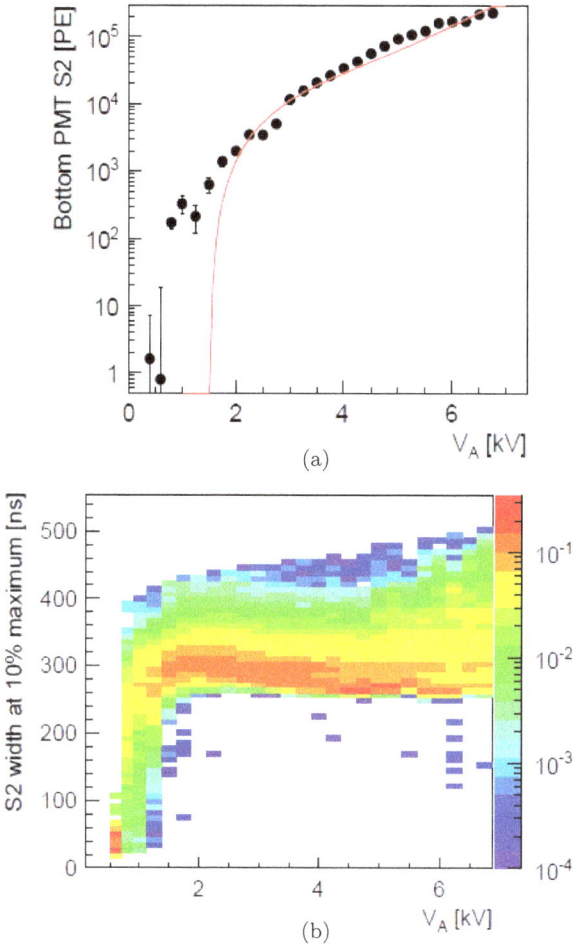

Figure 4.26: (a) proportional EL gain vs. applied voltage for 10 μm anode wire in liquid Xe. The curve is the linear fit to the data points. Note that the experimental data points diverge from this fit near and below the nominal EL threshold. (b) The EL pulse width at 10% of pulse height. Notice the EL pulse width decrease below the nominal EL threshold, presumably indicating an alternative EL mechanism below the threshold. Redrawn from Aprile *et al.* (2014b).

were measured as about 400 and 700 kV/cm, respectively. The linear dependence of proportional EL gain on the electric field and the existence of the threshold for EL in liquid Xe were confirmed also in a needle-type device (Schussler *et al.*, 2000): see Fig. 4.27.

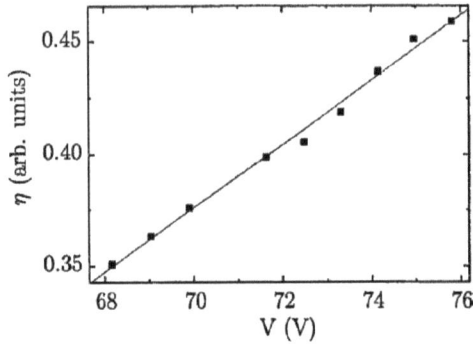

Figure 4.27: Light gain of needle-type device in liquid Xe as a function of the applied voltage. The electric field near the tip of the needle varies from 30×10^3 kV/cm to 2×10^3 kV/cm over a distance of 0.1 μm. If to extrapolate to zero, light emission has a threshold at 44 V. Redrawn from Schussler *et al.* (2000).

It is interesting however, that below the nominal EL threshold the EL signal was still observed in Aprile *et al.* (2014b): see Fig. 4.26(a). Moreover, the pulse width for such an "alternative" EL dramatically dropped down compared to "ordinary" EL (see Fig. 4.26(b)), indicating that alternative EL is fast. This under-threshold and fast EL in liquid Xe looks very similar to under-threshold and fast EL in gaseous Ar described in Section 4.2, where it was convincingly explained by the NBrS effect: compare Fig. 4.26(a) to Figs. 4.5 and 4.9. This similarity has led us to a conclusion that proportional EL in liquid Xe, observed under the nominal EL threshold, of about 400 kV/cm, with high probability is due to the NBrS effect as well. Note that a similar hypothesis has been proposed in Buzuluskov *et al.* (2018) to explain proportional EL in liquid Ar, observed in immersed THGEMs and GEMs at relatively low fields as follows.

4.5.2. *Light signal amplification in liquid Ar using THGEMs and GEMs*

In liquid Ar, proportional EL, with its characteristic properties of having a threshold and linear field dependence, was observed in immersed THGEM and GEM structures in Lightfoot *et al.* (2009)

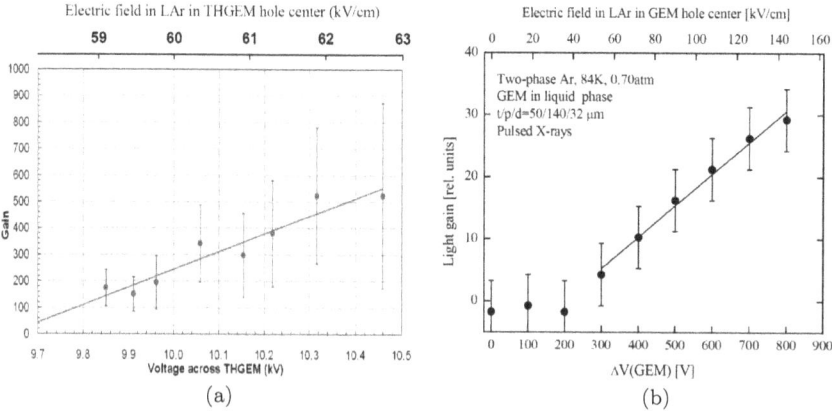

Figure 4.28: Relative amplitude of the EL signal generated in THGEM (a) and GEM (b) with the plate immersed in liquid Ar as a function of the voltage across the plate. The appropriate electric fields in the THGEM and GEM hole centers are shown on the top axes. Redrawn from Lightfoot *et al.* (2009) and Buzulutskov (2012).

and Buzulutskov (2012): see Fig. 4.28. The mystery of the results obtained is that liquid Ar EL started at much lower electric fields, at about 60 kV/cm, which is almost an order of magnitude lower than that of liquid Xe. In addition, these electric fields are far less than those of 3×10^3 kV/cm expected from theoretical calculations for EL in liquid Ar due to atomic excitation mechanisms (Stewart *et al.*, 2010).

It was supposed in Buzulutskov *et al.* (2018) that the NBrS effect could be responsible for proportional EL in liquid Ar observed in immersed GEM-like structures. Indeed, the electric fields in the center of GEM or THGEM holes used in liquid Ar, of 60–140 kV/cm, correspond to $E/N = 0.3$–0.7 Td that are not that small. For such reduced electric fields, the theory predicts that NBrS EL already exists: see Fig. 4.9. It also predicts the linear dependence of the EL yield observed in the experiment. Thus, one cannot exclude the NBrS origin of proportional EL in noble-gas liquids at lower electric fields.

Another possible explanation might be the presence for some reason of gaseous bubbles associated to the THGEM and GEM holes, within which proportional EL in the gas phase could develop.

This effect will be discussed below. This hypothesis, of bubble-assisted EL, could be confidently tested by measuring the EL emission spectrum in the NIR: it would be confirmed if the emission spectrum would consist of atomic lines corresponding to gaseous Ar. Otherwise, the continuous spectrum would indicate the real liquid Ar emission. This method has been realized in Auger *et al.* (2016) when studying electrical breakdown in liquid Ar; the results will be discussed in the following.

4.5.3. *Liquid-hole multipliers in liquid Ar and liquid Xe*

The liquid-hole multiplier (LHM) concept (Breskin, 2013; Arazi *et al.*, 2015) has emerged after observation of proportional EL in THGEMs immersed in liquid Xe: see Fig. 4.29 (Arazi *et al.*, 2013). Note the linear voltage dependence of EL yield and rather low voltage threshold for EL, corresponding to the electric field in THGEM holes of about 10 kV/cm. This field value is one and two orders of magnitude smaller than that for proportional EL in liquid Ar using THGEMs and in liquid Xe using wires, respectively. Therefore,

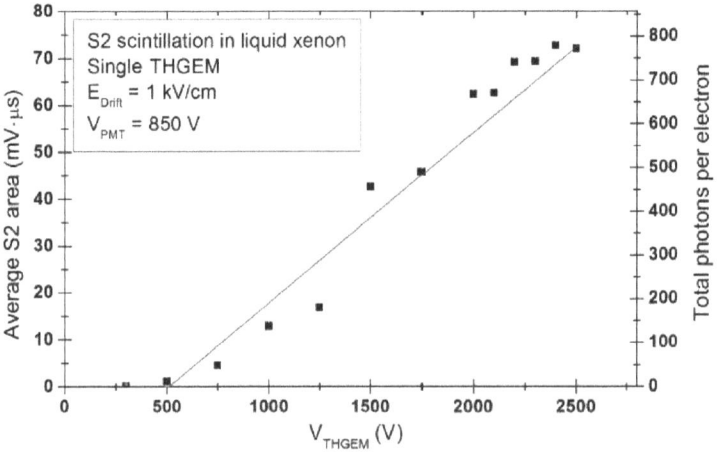

Figure 4.29: Amplitude of EL signal (S2 pulse-area) and its EL yield (number of photons per drifting electron) produced by THGEM in liquid Xe as a function of the THGEM voltage. Redrawn from Arazi *et al.* (2013).

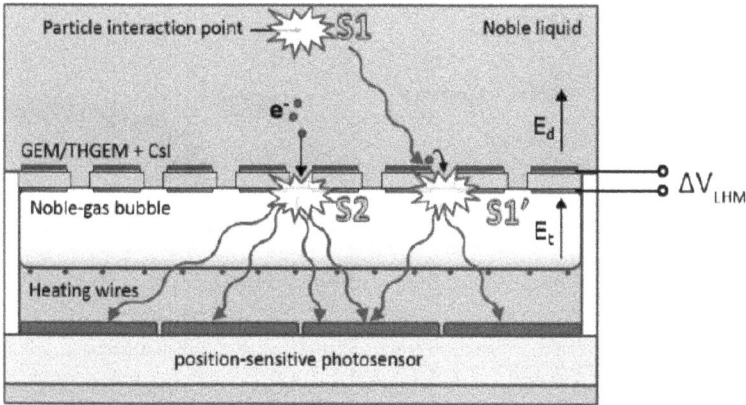

Figure 4.30: Concept of LHM detector (Breskin, 2013; Arazi *et al.*, 2015; Erdal *et al.*, 2020). Redrawn from Erdal *et al.* (2020).

such a low threshold EL can hardly be explained by ordinary EL or NBrS EL. This effect was proved to be due to proportional EL in the bubbles formed in the holes of the THGEM and under it (Arazi *et al.*, 2015; Erdal *et al.*, 2015).

In the LHM concept, shown in Fig. 4.30, the noble-gas bubble is formed in noble-gas liquid under the THGEM or GEM plate in a controlled way using heating wires. Radiation-induced ionization electrons and scintillation (S1)-induced photoelectrons from a CsI photocathode (optionally deposited on top of the THGEM) are focused into the THGHEM holes; both cross the liquid–vapor interface and create EL signals (S2 and S1, respectively) in the bubble.

In some sense, the LHM bubble resembles the DarkSide scheme (Aalseth *et al.*, 2018), where a gas pocket is formed using heating wires, confined from above by a metallized quartz-plate anode. However, there is a fundamental difference: the gas bubbles in LHM are kept from coming up through the THGEM holes due to surface tension. This allows ionization to enter the gas phase from both sides of the liquid: from below and above the bubble.

Figure 4.31 illustrates the LHM performance in liquid Xe and liquid Ar: the EL (S2) signals are distinctly seen along with the

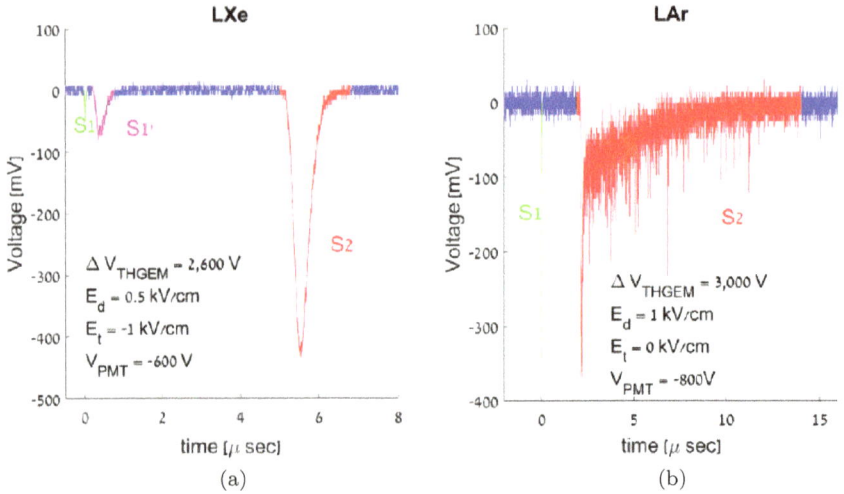

Figure 4.31:   Sample EL signals induced by alpha particles, recorded from a LHM detector: (a) in liquid Xe and bare-PMT, with a CsI-coated THGEM electrode and (b) in liquid Ar and TPB-coated PMT. Redrawn from Erdal *et al.* (2020).

primary scintillation (S1) signals. In liquid Xe, the EL yield of as high as 400 photons per drifting electron was reached, which is comparable with that obtained in a traditional two-phase Xe detector with a 1-cm thick EL gap. Nevertheless, further research is needed before this technique can be applied in real rare-event experiments.

4.5.4. *Breakdowns in noble-gas liquids*

Breakdowns in noble-gas liquids are relevant to the subject of this chapter because they are accompanied by both charge multiplication and EL. Their study is important for practical implementation of large liquid Ar TPCs for neutrino experiments, where the high electrical potentials are envisioned, from 100 kV to 1 MV.

The breakdown mechanism is not fully understood; it includes a number of physical effects like electron field emission, streamer propagation and EL. Nevertheless, the universal empirical rule was revealed in breakdown measurements at cm scale (Gerhold *et al.*, 1994; Acciarri *et al.*, 2014; Auger *et al.*, 2016; Tvrznikova *et al.*, 2019): the threshold for dielectric breakdown in any noble-gas liquid

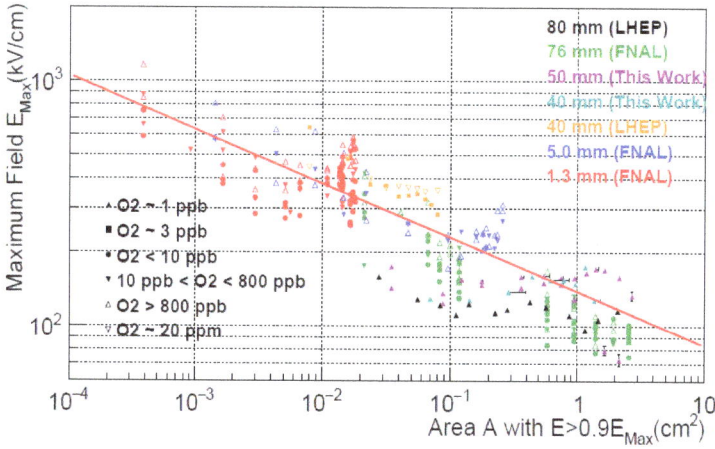

Figure 4.32: Breakdown field vs. stressed area of the cathode in liquid Ar. The stressed area is defined as the area with an electric field greater than 90% of the maximum electric field in the gap. The fit line represents the dependence $E_{\max} = C \cdot A^p$ with $C = 139$ and $p = -0.22$. Redrawn from Auger *et al.* (2016).

decreases with the increase of the surface area of the electrodes according to inverse power law:

$$E_{\max} = C \cdot A^p, \qquad (4.10)$$

where C is a material-dependent constant, A is the stressed area (defined as the area with an electric field intensity above 90% of its maximum), and $p \approx -0.25$. See also Fig. 4.32 showing the breakdown field in liquid Ar as a function of stressed area. There does not appear to be a significant difference between breakdown behavior in liquid Ar and liquid Xe (Tvrznikova *et al.*, 2019).

It is remarkable that the breakdown in liquid Ar was proposed to have a two-phase nature (Auger *et al.*, 2016): namely, while at its first stage the charge multiplication occurs directly in the liquid, at the second stage the streamer propagation actually takes place in the elongated gas bubble. This hypothesis is supported by Fig. 4.33, showing photon emission spectra of EL, in the visible range, at the first (field emission) and the second (streamer) stage for a breakdown in liquid Ar. These spectra feature the presence of both a liquid and a gas phase in the breakdown. Indeed, the curve with a

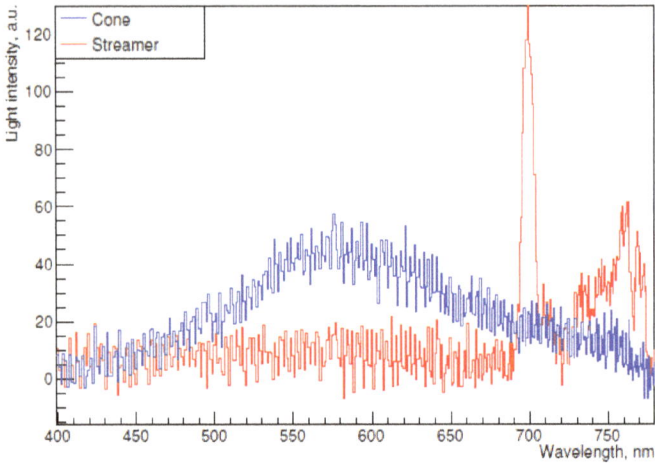

Figure 4.33: Photon emission spectra of EL at the first (field emission cone) stage and the second (streamer) stage for a breakdown in liquid Ar: the curves are that of blue with a broad continuum and that of red with distinct peaks, respectively. The first stage curve, with a broad continuum, is similar to the scintillation spectrum of liquid Ar as observed by Heindl *et al.* (2010), while the second stage curve, featuring distinct peaks around 700 and 750 nm, is attributed to Ar*(3p^{5}4p^1)Ar*(3p^{5}4s^1) transitions in gaseous Ar. Redrawn from Auger *et al.* (2016).

broad continuum is similar to the scintillation spectrum of liquid Ar observed by Heindl *et al.* (2010), while the curve featuring distinct peaks around 700 and 750 nm is attributed to atomic photon emission in the NIR due to Ar*(3p^{5}4p^1) $\rightarrow$ Ar*(3p^{5}4s^1) transitions in gaseous Ar (see Fig. 4.4 and Eq. (4.8)).

Chapter 5

Two-Phase Detectors for Dark Matter Search

5.1. Introduction

This chapter is devoted to the very unusual experimental field which had the great feedback on the development of a two-phase detection technique. We are talking about the era of dark matter or WIMP detectors. WIMPs are Weakly Interacting Massive Particles, a hypothetical type of particles introduced to explain the hidden mass of the Universe. Before this, such type of detectors was studied episodically for purely academic purposes by several groups of experimentalists, exclusively in Russia, where this type of detectors was invented in the 1970s (Chapter 1). Nowadays, this instrument has reached superior sensitivity among all types of detectors used for this purpose due to a combination of its unique properties. To understand the reasons for this situation we need to give a brief introduction of the basic principles of WIMP detection technique.

Experimentalists are looking for WIMP elastic scattering off atomic nuclei. This process must have a very low value of cross-section because of the only possibility of the weak interaction of these particles with a baryonic matter (in cases of electromagnetic and strong interaction they would have been observed already). The signals produced by WIMPs in a detector are also characterized by a rather small energy deposition (from zero to several-keV region). The spectrum profile can be determined from the kinematics of elastic collisions of WIMPs of the certain masses with detector nuclei taking into account Maxwellian velocity distribution of the WIPMs in our galaxy and the Earth motion in the interstellar space. The resulting spectrum of energy deposition in the detector medium has a shape very close to an exponent with a decay constant of ≤ 10 keV depending on the atomic mass of detector medium (usually called *target*) and depending on the WIMP mass which may be from several GeV to several hundred of GeV. Unfortunately, background signals of various origins in detectors have similar exponential shapes of pulse height distributions. This fact significantly complicates the process of disentangling the weak WIMP signal from backgrounds. Standard representation of dark matter search results is usually performed as an exclusion plot in a parameter space σ_p and M_W, interaction cross-section and WIMP mass, correspondingly. Most of the dark matter search results can be found on Internet publications. The plots are produced separately for so-called spin-dependent and spin-independent interactions. In the first case, the cross-section is proportional to $J(J + 1)$, where J is a nucleus spin; in the second case, it is proportional to A^2, where A is a mass number of nuclei. In order to compare the results obtained with detectors having various elemental compositions, a so-called reduced-to-proton cross-section σ_p is used:

$$\sigma_p = \frac{m_{\text{red}}^2(p, W)}{C m_{\text{red}}^2(N, W)} \sigma_N,$$

where $C = A^2$ for the case of a spin-independent interaction.

The dark matter search experiments are so-called low-background ones. They are carried out deep underground (at a depth ≥ 1 km) in special underground laboratories in order to reduce the cosmic muon flux and the muon induced neutron flux by many orders of magnitude.

A serious breakthrough in reduction of radioactive background in dark matter detectors took place after introduction of so-called methods of active suppression of electromagnetic background (electron recoils from γ-rays and beta-particles originated in ambient or intrinsic radioactive sources) by means of simultaneous measurements of the energy deposition produced by a particle in two or more different ways or by analyzing the measured scintillation pulse shape.

These methods allow one to select only the events when the interactions were with atomic nuclei of a detection medium with production of a nuclear recoil and to reject with very high efficiency the events with production of electrons (from γ- and beta-interactions).

The possibility to distinguish between different kinds of ionizing particles was one of the basic features of liquid noble gas detectors that stimulated their use in the DM search experiments. Note, that the first proposals of the noble liquid detectors (liquid xenon, LXe) were based not on a two-phase emission technique but on a traditional pulse-shape analysis method (Davies *et al.*, 1994) and on the simultaneous measurement of scintillation and ionization signals in the single-phase liquid (Benetti *et al.*, 1993). In the second case, for measurement of ionization signal, the use of electroluminescence in the liquid phase in the vicinity of the extra thin wires was proposed. In the first case, the LXe as a scintillator as well as other noble liquids (LAr, LKr) allows one to perform a pulse-shape analysis because it has different intensities of the singlet (short lived) and triplet (long lived) excited states of dimers for the particles with different ionization power. This method was successfully used in the first dark matter single phase liquid xenon detectors such as ZEPLIN-I (Alner *et al.*, 2005; Kudryavtsev, 2005) in Boulby mine and the DAMA/LXe in Gran-Sasso (Belli *et al.*, 1991).

The second method turned out to be very complicated and hasn't been realized in dark matter experiments. Instead, a two-phase emission method progressed that allowed one to realize several remarkable properties of a liquid noble gas detector at once. It should be noted that in application to the WIMP search problem, it was proposed for the first time in the Russian publication (Barabash and Bolozdynya, 1989) (with the participation of one of the authors of this book) to combine the high density of the detection substance (a target) in the liquid phase and the high sensitivity to the very small value of ionization (down to single ionization electrons) in the gas phase. However, instead of a liquid noble gas medium the authors considered 2,2,4-trimethylpentane (isooctane) or 2,2,4,4-tetramethylpentane which are liquids at room temperature. This fact, nevertheless, doesn't make any difference with respect to a two-phase detection principle itself. In the subsequent paper (Bolozdynya *et al.*, 1995), a conception of a "wall-less" two-phase xenon detector was introduced with application to several tasks of low-background physics such as neutrinoless double-beta decay, reactor antineutrino detection via v–e elastic scattering, and WIMP search for. In such a detector, the ionization is transformed to the light signal by electroluminescence (called often as proportional scintillation), and detection of both scintillation and ionization is performed with the use of the same photodetector array, placed above the electroluminescent region. The event coordinates are obtained from the time interval between scintillation and electroluminescence (vertical), and from distribution of the signals among the photodetector array (in horizontal plane).

In 1998, at the Dark Matter 98 conference, two different basic conceptions of a two-phase emission WIMP detector were proposed (Wang, 1998; Akimov *et al.*, 1998) that have grown later to the first designed dark matter detectors ZEPLIN-II and ZEPLIN-III (the first conceptual design of the ZEPLIN-II detector was published also in (Cline *et al.*, 2000). This is because the physicists who developed detectors for this very new and unexplored field knew practically nothing about the detection properties of LXe medium with respect to the small energy depositions produced by nuclear recoils. The authors of the ZEPLIN-II design considered that the track from nuclear recoils is so dense that it is practically impossible to extract the ionization electrons by applying the reasonable electric field. Thus, the selection of WIMP events was based on its inherent signature, as thought, of the presence of scintillation signal only. The events that have both scintillation and electroluminescent signals were considered as those produced by electrons, i.e., background events. The important feature of a detector design in this case must be the absolute absence of the dead regions for charge extraction. The authors of the ZEPLIN-III design, on the contrary, relied on the possibility of detection of ionization (at least, a tiny amount of it) from nuclear recoils. For this purpose, the detector had a frying pan-like geometry: the sensitive region was very thin, in order to apply as maximal electric field strength as possible. It was thought that at the strong electric field, one can extract from a nuclear recoil track at least a single electron with a non-zero probability. Later, the measurements of ionization yield for nuclear recoils in LXe performed with neutrons by several experimental groups had shown that this value is relatively high. For the first time, it was demonstrated by the Russian group (ITEP, Moscow), a participant of the ZEPLIN-III collaboration. However, this result is not widely known because it was published in the *IDM2002* conference records only (Akimov *et al.*, 2003). Even in the ZEPLIN-II experiment itself, the clear ionization signal from nuclear recoils was observed during calibration of the detector by the Am–Be neutron source.

Before going to the detailed description of the dark matter two-phase emission detectors, note that only the use of xenon targets was considered at the beginning. This is mainly because other quite heavy noble gas media such as Ar and Kr contain radioactive isotopes: ^{39}Ar in argon induced by cosmic rays and ^{85}Kr in krypton artificially produced by the atomic industry. From other points of view, the noble gases are excellent media for the low-background experiments: they are quite dense, they don't have their own long-lived radioactive isotopes and cannot contain U and Th progenies, they allow to build scalable to the large masses detectors with 3D positioning and self-shielding, and they allow particle identification and active discrimination of γ- (or electron) background.

5.2. Xenon Detectors

Historically xenon-based two-phase emission detectors have been considered first as a new much promising technology for dark matter search (Bolozdynya *et al.*, 1995). Since that time this technology has been used to achieve the best limits on the existence of massive WIMPs as dark matter candidates at least in the last ten years as a result of a series of fundamental experiments based on this technology. This section describes the detectors used in these experiments.

5.2.1. *ZEPLIN program*

The ZEPLIN program started from the ZEPLIN-I detector which was a single-phase one, however this single-phase detector from the very beginning was considered as a demonstration of the possibility of operation of liquid xenon-based detectors. The abbreviation ZEPLIN came from ZonEd Proportional scintillation in LIquid Noble gases that assumed exploring not only a single-phase detector. We do not stop here on description of the ZEPLIN-I experiment because it is not in the scope of this book.

5.2.1.1. *ZEPLIN-II detector*

ZEPLIN-II detector became the first two-phase emission detector for dark matter search to operate in the world (Alner *et al.*, 2007). In the ZEPLIN II detector, seven quartz-windowed VUV sensitive photomultipliers D742QKFLB are situated above the surface of liquid xenon and look from above to the electroluminescent gap and the entire sensitive volume of LXe (Fig. 5.1). This gap is formed between the LXe surface and the upper of the optically transparent grids marked as "extraction field grids" in Fig. 5.1. The electric field in the liquid between the extraction field grids is 4.2 kV/cm that provides a 90% extraction efficiency of electrons from the liquid surface to the gas phase. The electric field strength in the target volume was chosen rather small, $\sim$1 V/cm applied by a "cathode grid". Such value is enough to pull out the significant part of ionization electrons from the electron tracks (produced by γ-s or beta-particles), and as has turned out to observe later, from the nuclear recoil tracks (see Chapter 3 for details). The 14-cm deep liquid xenon target volume is defined by a thick PTFE ring which also serves as a reflector for the VUV 175 nm scintillation light and provides a support structure for the "drift field rings" (Fig 5.1) which ensure a uniform electric field inside the target volume. The inner PTFE walls of the vessel were made narrowed to the bottom in order to minimize charge trapping on walls, thereby allowing the entire charge to be collected to the surface of the liquid. The LXe target mass was 31 kg ($\sim$10 l), with a fiducial mass of 7.2 kg after

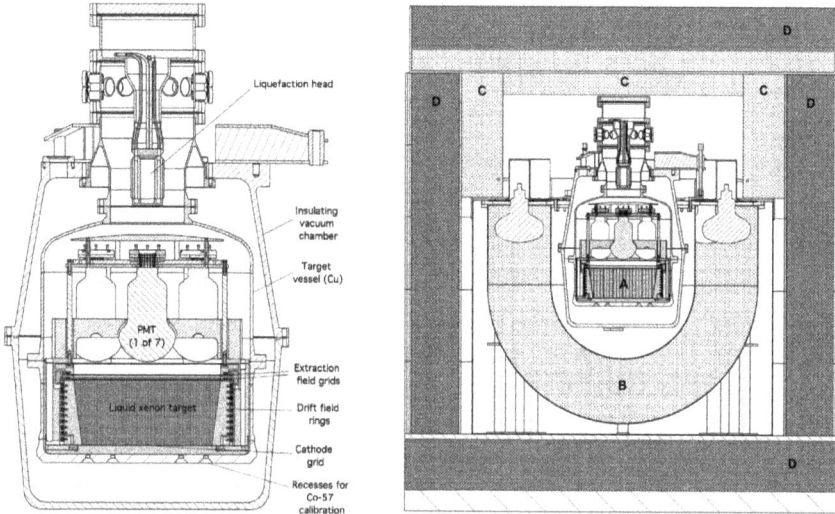

Figure 5.1: Schematic of the ZEPLIN-II detector. The liquid xenon volume is shown, viewed from above by 7 quartz-window photomultipliers. The electrode arrangement defines a drift region between the cathode grid and the lower extraction grid where the field is parallel and uniform (this is obtained with the help of lateral field-shaping rings embedded in the PTFE walls). The extraction region (where electroluminescence is generated) is defined by the two grids located either side of the liquid surface. Xenon liquefaction occurs on the liquefaction head, with liquid dripping onto a copper shield which deflects it away from the photomultiplier array and the active volume. Redrawn from Alner *et al.* (2007).

all spatial cuts were applied. The xenon medium was constantly purified by circulation with passing through a SAES getter PS11-MC500 purifier with a flow rate of 3 slpm. This ensured sufficient purity of xenon (>100 μs) to allow charge collection with relatively small losses throughout the whole target volume.

The ZEPLIN-II detector has been installed into the liquid scintillator veto/neutron shield and lead γ-ray shield and exposed in the Boulby salt and potash mine (Cleveland, UK) at a vertical depth of 1070 m (2805 m of water equivalent). At this depth, the cosmic ray muon flux is reduced by a factor of about 10^6 to a level of $(4.09 \pm 0.15) \times 10^{-8}$ muons/cm^2/s. The total exposure taken in analysis (after applying fiducial and stability cuts) was 225 kg $\times$ days.

The ZEPLIN-II team proposed a scenario for the analysis of experimental data, which was subsequently followed by all other groups in WIMP search for. The scenario is as follows. The scintillation and electroluminescent signals (S1 and S2, correspondingly) obtained after event reconstruction procedure are used to produce a scatter plot of log(S2/S1), or S2/S1 on logarithmic scale, vs. S1 (Fig. 5.2). The first term reflects an identification parameter of an event, while the second one, the event energy obtained from scintillation and expressed in photoelectrons or keVee. The different S2/S1 ratio for different species results in different 3D distribution of the corresponding populations on this scatter plot.

At the first step, the 3D profiles of populations corresponding to electron and nuclear recoils obtained in calibration runs with γ-ray and neutron sources are obtained. These populations are centered at different lines on log(S2/S1) and S1 space. In each energy bin (vertical slice), the electron and nuclear recoil populations (bands) have practically a Gaussian profile and are centered at different log(S2/S1) values. However, despite the different central values, these distributions overlap rather significantly with each other. In order to achieve the high value of separation of these species, roughly half of the nuclear population is sacrificed by setting the upper threshold on the log(S2/S1) value. The leakage of the electron events below this threshold is rather small, of an order of 10^{-3} of the total number of electron recoil events. A so-called WIMP-search (or acceptance) window is defined on the log(S2/S1) and S1 space. For the ZEPLIN-II analysis, it was defined by a nuclear recoil population median on top (50% acceptance) and by 5 and 20 keVee on the left and right, correspondingly. On bottom, the window boundary was set at S2/S1 equal 40. This window is made "blind" for the experimental data obtained during the science run, i.e., nobody can plot or "see" somehow the events in this region.

At the second step, the leakage of the events from the electron band to the acceptance window is calculated for the science run, and the parameters of analysis are tuned in order to minimize this number. This is performed on a slice by slice basis by extrapolation

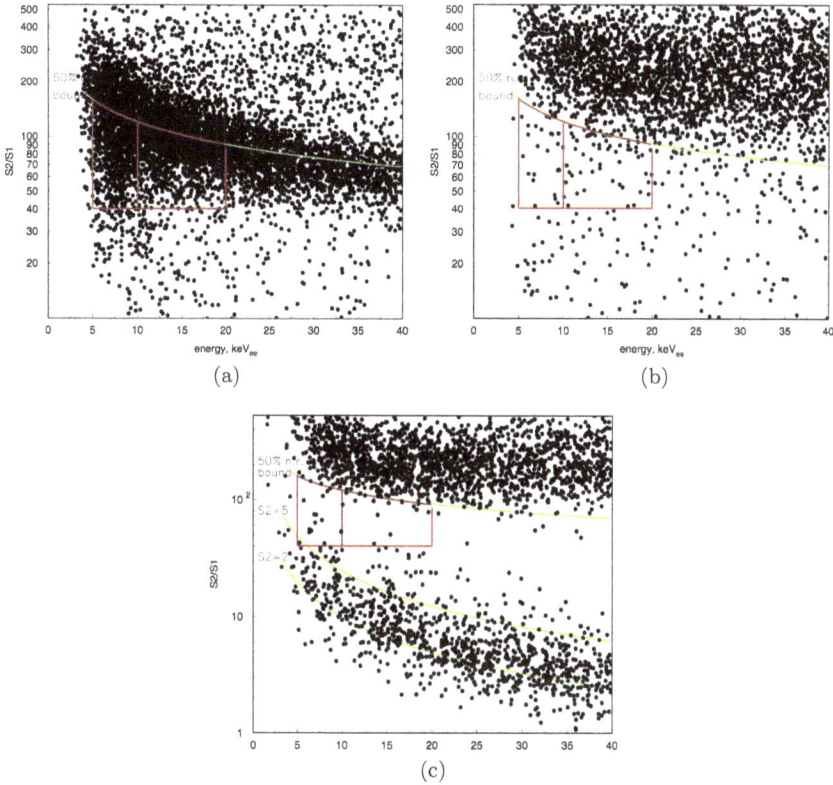

Figure 5.2: Calibration using neutrons from an AmBe source (a) and Compton scattered ^{60}Co γ-rays (b) and scatter plot (c) for the science data. In the plot (b), presence of events below the main electron recoil band is explained by a high trigger rate during calibration by a ^{60}Co source. The acceptance window is shown by red color, subdivided to two windows, 5–10 keVee and 10–20 keVee. In plot (c), there are 29 events in the acceptance window, and some additional band below this window (see text). Redrawn from Alner *et al.* (2007).

of Gaussian tails to the blind regions. After that, the parameters are frozen.

The third step is the so-called box-opening procedure, at which the total number of real events in the acceptance window is obtained. If this number is in consistency with the predicted one the upper limit on the WIMP-nucleus cross-section is set. A Feldman–Cousins

(Feldman and Cousins, 1998) method is usually used to set limits in such cases.

The expected number of events in the ZEPLIN-II acceptance window was 28.6 ± 4.3 events, the observed number was 29. With the use of the Feldman–Cousins method, the presence of the real WIMP events within this observed number was estimated to be 10.4, at 90% confidence level. This obtained number was used to set an upper limit on the reduced-to-proton cross-section of the spin-independent WIMP–nucleon interaction. The limit is $6.6 \cdot 10^{-7}$ pb for the WIMP mass 65 GeV.

However, in spite of the seeming detector perfectness one very unpleasant drawback revealed during its operation. This drawback did not allow achieving the planned sensitivity of the experiment and the experiment was stopped. That trapezoidal shape of the PTFE volume which was designed to minimize trapping of the electrons on walls, on the opposite, played a fatal role for the positive ions. During operation of the detector not only the Xe ions from particle tracks were accumulated on the PTFE surface but also the positive ions that were the products of beta-decays in the radon induced decay chain. The radon was supposed to come from the used SAES getter. These ions produced nuclear recoils flying to the active volume in subsequent alpha-decays when the alpha-particle was directed towards the PTFE wall. Because of the small energy deposited by these nuclear recoils, the spatial resolution of the detector was poor, and such events cannot be rejected by fiducial cut.

5.2.1.2. *ZEPLIN-III detector*

Detailed description of the ZEPLIN III detector can be found in Akimov *et al.* (2007) (see Fig. 5.3 taken from the review (Chepel and Araújo, 2013). The detector contained 50 kg of xenon, 12 kg and ∼6 kg of which were in the active region (between the electrodes) and fiducial volume, correspondingly. As mentioned above the detector geometry was a frying pan-like in order to apply the high electric field. It was expected also that this geometry will help to achieve the enhanced light collection of scintillation. For this reason,

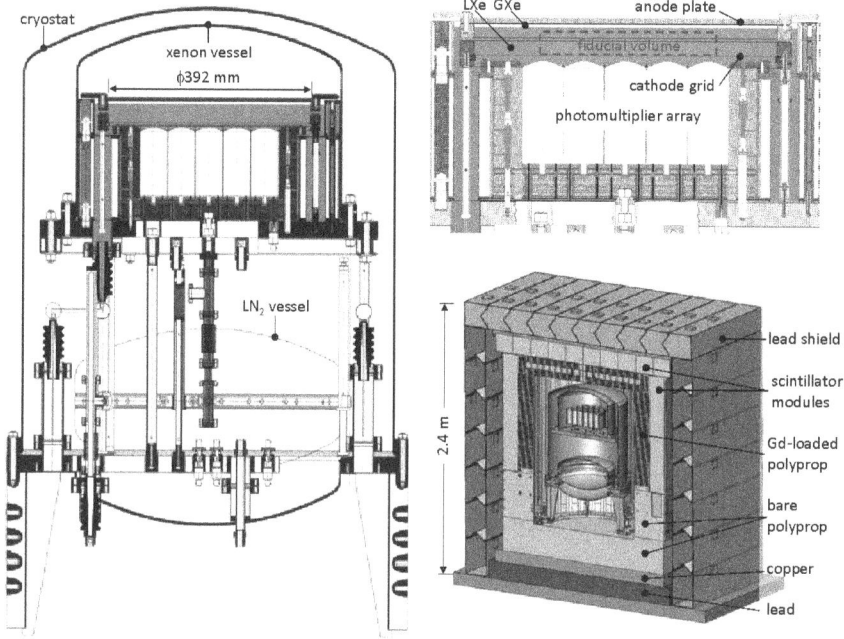

Figure 5.3:   The ZEPLIN-III dark matter detector. Redrawn from Chepel and Araújo (2013).

31 pcs. of 2-inch PMTs were submerged in the liquid xenon. In fact, the light collection yield significantly improved with respect to that one of ZEPLIN-II: from 0.55 to 1.8 phe/keV at the applied electric field and from 1.1 to 5.0 phe/keV at zero field (for ^{57}Co source). The PMTs viewed the active volume from the bottom through two transparent wire grids (with a wire diameter of 0.1 mm and a pitch of 1 mm). Those grids were: an electric field screen (the closest to the PMTs one, which was kept at the same potential as the potential of the photocathodes), and a cathode ("−") of the active volume. The "+" HV was applied by a flat polished copper plate anode located 40 mm from the cathode and 4 mm above the LXe surface. The electrode system produced an electric field of 3.9 kV/cm in the liquid and 7.8 kV/cm in the gas. The side metal surface is divided by half in the middle of height. The bottom one is connected to the cathode electrode, and the top one, to the anode. Thus, the

fields in the near side regions were configured so that any charge produced near the side surface of both the cathode and anode rings will be collected to the side surface of the anode ring under the surface of the liquid xenon. Thus, this charge does not produce an EL signal. Such configuration of electric fields eliminated background events produced by radon progenies plated on the active volume sides observed in the ZEPLIN-II detector. Another innovative feature of this configuration of electrodes was that the liquid xenon surface contacted the metal anode on the side. This allowed the electrons, which were not emitted to the gas phase, to be swept out from the surface. Since only xenon-friendly and low-outgassing materials were used in the ZEPLIN-III detector (PTFE was practically avoided) the xenon purification system was designed without constant gas circulation through the hot getter. The gas was purified several times before and during the detector filling. The resulting purity was quite moderate: the electron lifetime was ∼14 μs at the beginning. However, due to the cleanness of the detector this parameter actually improved to 45 μs by the end of the run (Majewski *et al.*, 2012).

The detector was placed in plastic and lead shields. The dark matter search was carried out during two science runs. In the first run, the plastic shield was made purely passive with the use of polypropylene bars; in the second run, the upper part of the plastic shield (see Fig. 5.3) was made active: the inner close to detector part was manufactured from gadolinium loaded polypropylene and the outer part, from plastic scintillator viewed by PMTs. These two parts operated as a neutron veto to tag the single-vertex events produced in the active volume by neutrons (the real WIMP events must have the only single interaction in the LXe fiducial volume). The neutrons were captured by Gd with high efficiency with production of the energetic gammas which were detected by plastic scintillators in the outer part.

However, the main difference between these two runs were the background conditions. In the first 83-day long science run in 2008 (847 kg · days of total exposure) the photomultipliers 9829QA specially produced by ETEL on the basis of commercial model D730Q were used. These PMTs had bialkali photocathodes with special metal strips ("fingers") deposited on quartz windows under photocathodes to increase their electrical conductivity at low temperature conditions. The average quantum efficiency (at cold) for 175-nm xenon light was 30% (Araújo *et al.*, 2004). In the second 319-day long dark matter run in 2010–2011 (1344 kg · days), those tubes were replaced by another specially developed by ETEL model with 40 times lower radioactivity per unit. As a result, the overall event rate of electron recoil background was reduced to 0.75 kg^{-1}day^{-1}keV^{-1} at low energies (Araújo *et al.*, 2012). Unfortunately, the poorer performances of these PMTs (lower quantum efficiency and the large gain dispersion) did not allow to significantly improve the results of the first science run.

In both datasets, a handful of events were observed in the acceptance windows, that was consistent with background expectations in both cases. The combined result on cross-section of a WIMP-nucleon spin-independent interaction was $3.9 \cdot 10^{-8}$ pb at 90% confidence level for 50 GeV WIMP mass (Lebedenko *et al.*, 2009; Akimov *et al.*, 2012a), that was by a factor of ∼17 lower than the limit set by the ZEPLIN-II experiment.

5.2.1.3. *Event position reconstruction*

The Portugal group, a member of the ZEPLIN collaboration, has developed a procedure of event position reconstruction in a horizontal plane based on the Maximum likelihood method (Neves *et al.*, 2007; Lindote *et al.*, 2007; Solovov *et al.*, 2012). This method was applied first for liquid xenon dark matter detectors; however, it can be used for any types of detectors which utilize two-dimensional arrays of photosensors. The traditionally used technique for event position reconstruction in such types of detectors is the method originally introduced by H.O. Anger to obtain γ-ray interaction position in thin scintillation crystals for medical imaging applications (Anger, 1958). In this method, the position of scintillation is determined from the distribution of signal amplitudes between the photosensors in an array using the center-of-gravity (or centroid) algorithm. However, this method suffers with systematic uncertainties, especially at the side parts of the photosensor array. To compensate for these uncertainties special corrections can be introduced (a so-called "corrected centroid" algorithm). But this does not improve the situation completely.

In the proposed method, the likelihood function is built with the use of the measured photon distribution across the photosensor array and a set of so-called light response functions (LRF) for each individual photosensor. If the LRF functions are known, the position of the event can be found in a straightforward way by maximizing the likelihood function. However, the method can work even in the case when LRFs are unknown beforehand. Figure 5.4 (taken from Solovov *et al.*, 2012) demonstrates the iterative process for the ^{57}Co calibration events when, at the 1-st stage, LRFs for the S2 signals

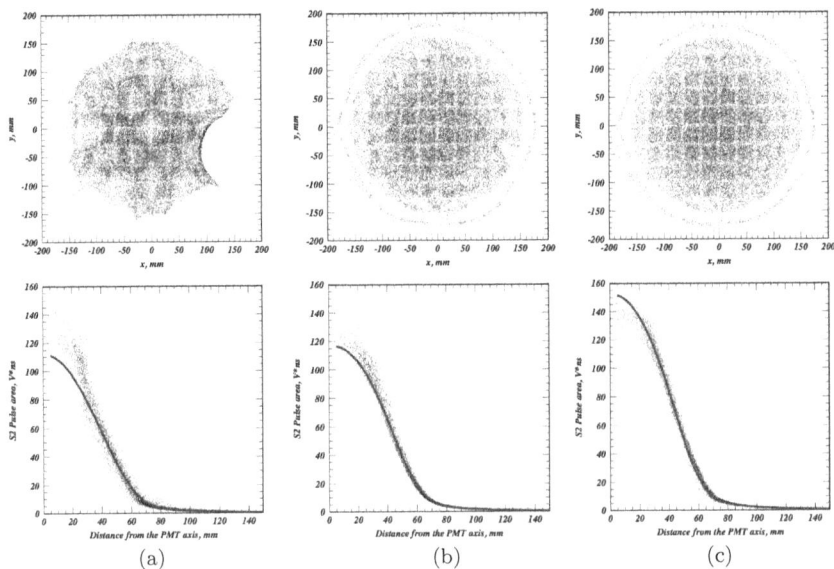

Figure 5.4: Example of the iterative reconstruction procedure in the ZEPLIN-III detector for the ^{57}Co calibration data: (a) initial position estimates obtained by centroid, (b) first iteration, (c) final (fifth) iteration. On top: the evolution of the distribution over the detector plane of the event reconstructed positions. On bottom: the evolution of the response of one of the PMTs vs. the obtained distance from its center (dots) and of the corresponding LRFs derived from these distributions (curve). Redrawn from Solovov *et al.* (2012).

are taken from the centroid reconstruction procedure (Fig. 5.4(a)). A rectangular structure clearly seen at all shown iterative stages is a shadow from the copper grid (30 mm pitch, 5 × 5 mm bars) placed on the top side of the anode electrode (the ^{57}Co γ-ray source was placed on top of the detector cryostat, at the vertical axis). One can see that the distribution of the reconstructed event positions is strongly distorted. Obviously, the strongest distortion takes place at the location of the dead PMT (in the second science run). The next iterations significantly improve the uniformity of the event position distributions: at the final (fifth stage) the shadow pattern is regular, and the distortion is even not visible. The obtained final light response functions fit perfectly the dependences of PMT responses vs. the distances from their centers (shown in Fig. 4(c) on the example of one of the PMTs in array). The method allows one not only to reconstruct the event positions but also to obtain a relative gain in each PMT channel and to reconstruct the true event energy.

On the basis of this method, the group has developed a software package ANTS2 (Morozov *et al.*, 2016) which can be used now for any Anger camera type detectors.

5.2.2. *XENON program*

Another line of dark matter experiments with two-phase xenon emission detectors is carried out by the XENON collaboration. This line is based on the earlier experience of the Columbia University group headed by E. Aprile. Before switching to the dark matter search activity this group successfully developed LXe TPC (time projection chamber) for γ-ray imaging and study the cosmic γ-rays by the balloon launched γ-telescope having a liquid xenon camera (LXeGRIT (Aprile *et al.*, 2000). The initial conception of this line was presented in publication (Aprile *et al.*, 2002), see also (Aprile *et al.*, 2005b). The idea was to start with a 100-kg-scale liquid xenon TPC, and to build a tonne-scale detector by a modular design. This was caused mainly by the reason that the author's previous experience was only with 30-kg LXeGRIT TPC. The detector is very similar to the ZEPLIN-II one by design except that there was an additional photo-sensitive layer made of CsI, a so-called CsI photocathode, to reach the high light collection efficiency for the scintillation light. Such a device had been tested previously by this group in different liquid noble gases, and the quite high quantum efficiency of it has been demonstrated (Aprile *et al.*, 1994). In this case, the signal recorded by PMTs should consist of three separate signals: the first one, scintillation; the second one, electroluminescent; and the third one, again electroluminescent but produced by the electrons emitted from the CsI photocathode. The last one must have the fixed time interval from the scintillation pulse equal to the total drift time through the whole volume. Although the operation of CsI photocathode in LXe was successfully demonstrated in a 3-kg detector prototype at Columbia University (Yamashita, 2005) this idea will never be used in the real experimental setups of this group. Instead of a CsI photocathode, an additional PMT array which viewed the active volume from bottom was used in further detectors.

5.2.2.1. *XENON10 detector*

After the 3-kg above-ground detector prototype, the XENON10 was the first detector operated in an underground laboratory. The detector was deployed at the Gran Sasso underground laboratory in Italy and operated from October 6, 2006 to February 14, 2007. Although this detector was considered as a prototype of a larger 100-kg detector, it has produced a scientific result, an experimental limit on WIMP-nucleon interaction (Angle *et al.*, 2008). The result came in the late 2007 very soon after release of the ZEPLIN-II result (Alner *et al.*, 2007).

The detailed description of the XENON10 experiment was published later (Aprile *et al.*, 2011). The detector schematic drawing is shown in Fig. 5.5. The main difference of this construction from those of the ZEPLIN detectors is the use of two arrays of PMTs instead of one. Note, that all xenon based dark matter detectors built later will follow this basic conception. The bottom photomultipliers situated in the liquid are most sensitive to the scintillation signal, while the top array is placed in the gas phase, to the electroluminescent one.

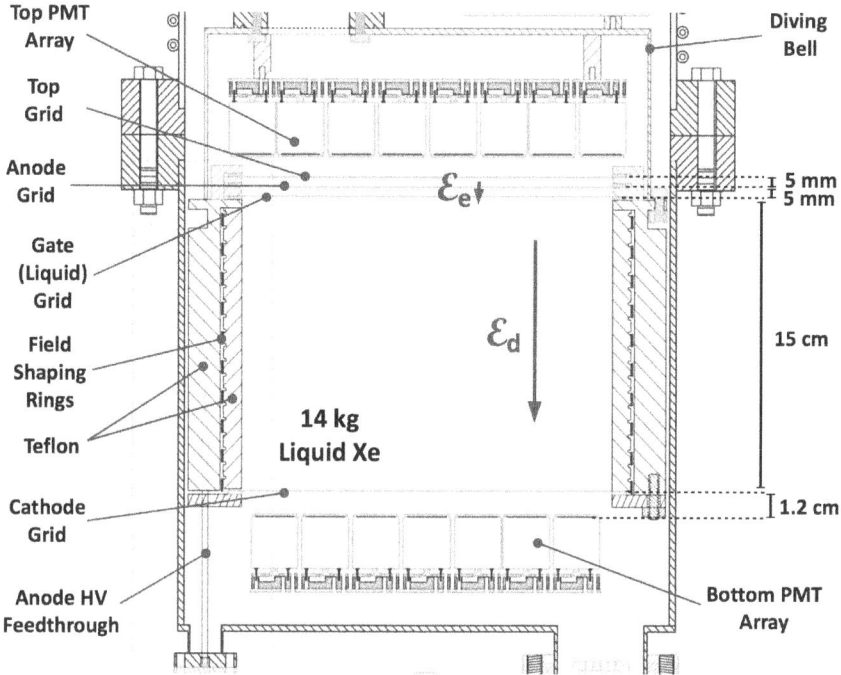

Figure 5.5: Schematic drawing of the XENON10 dark matter detector. Redrawn from Aprile *et al.* (2011).

The active volume of XENON10 was defined by a PTFE cylinder with an inner diameter of 20 cm and a height of 15 cm. The total active mass of LXe inside this cylinder was about 14 kg. A PTFE cylinder was used as a VUV light reflector and as an electrically insulating support for the electrodes and field shaping rings. Four optically transparent electrodes were used: two in the liquid (cathode and gate) and two in the gas (anode and top electrodes). The electrodes were meshes, they were made from 0.203 mm thick electropolished 304 stainless steel with a bar width of 0.182 mm and 2.0 mm × 2.0 mm square holes. A 5-mm gap separated the gate mesh from the anode and the anode from the top mesh. The LXe level was between the anode and gate mesh (2.5 mm above the gate) and determined the emission field. Thus, the electroluminescence took place in a narrow 2.5 mm gas gap. The top mesh was used to protect the PMT photocathodes from the strong electric field produced by the anode. Field-shaping rings, made of 0.5 mm thick copper and spaced by 0.76 cm, were mounted outside the inner cylindrical part of the PTFE cylinder to ensure a uniform electric field across the drift volume. The liquid level was kept constant by the use of a "diving bell". The overpressure was provided by the return of the gas circulating through the purification system. The inclination of the detector was measured by four custom-made capacitors in order to monitor leveling the detector by special leveling feet.

The XENON10 detector was equipped with two arrays of $1'' \times 1''$ square low-radioactive photomultipliers Hamamatsu R8520-06-AL specially designed to work in the LXe low-temperature conditions. There were 48 and 41 PMTs in the top and the bottom array correspondingly. The radioactivity of these PMTs was specially measured with an ultra-sensitive HPGe detector at the Gran Sasso laboratory. The measured activities in mBq/PMT of ^{238}U, ^{232}Th, ^{40}K, and ^{60}Co were (0.25 ± 0.04), $(0.21 \pm 0/05)$, (9.3 ± 1.1), and (0.59 ± 0.05), correspondingly. The signals from each PMT were amplified by Phillips PS776 $\times$ 10 fast amplifiers and passed through a 30 MHz filter. Then they were digitized by Struck 3301 fast ADC VME modules. After identification of the S1 and S2 signals the coordinates of events were obtained by two developed algorithms: a minimum χ^2 algorithm and a neural network based algorithm. The achieved event reconstruction accuracy was about 1 mm. This accuracy allowed to select a fiducial volume (containing only 5.6 kg of LXe) corresponding to 2 cm LXe self-shielding in radial direction and 3 cm each on top and on bottom.

The detector was surrounded with a passive shield of high-density polyethylene (HDPE, 20 cm, the closest inner part) and then of 20-cm lead. The lead was supplied from $5 \times 10 \times 20$ cm bricks. The inner 5-cm low-radioactive layer had an activity in ^{210}Pb of (17 ± 5) Bq/kg, while the outer 15-cm one, 560 ± 90 Bq/kg.

During the dark matter run, the following XENON10 detector performances were obtained. The XENON10 experimental group has obtained the excellent purity of the LXe by constant circulation through the hot getter: the electron lifetime was achieved

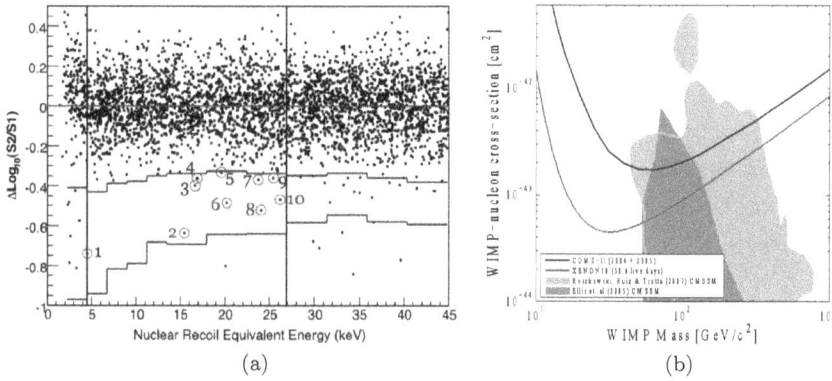

Figure 5.6: Result of blind analysis of XENON10 experimental data (a) and the exclusion plot for spin-independent WIMP-nucleon cross-section (b). Redrawn from Angle *et al.* (2008).

2.2 ± 0.3 ms. Such LXe purity allowed to have practically the uniform ionization charge response of the detector in a vertical direction. The scintillation light collection yield was about 3.1 phe/keV at 0.73 kV/cm electric field (for ^{57}Co source). This is practically equivalent to the value obtained for ZEPLIN-III taking into account the much weaker electric field in XENON10.

All these efforts and achieved good detector performances resulted in a rather small number (in amount of 10) of events in the WIMP search window shown in Fig. 5.6(a) for the 58.6 live days of experimental data. All these events were considered as a background, and a corresponding limit on spin-independent WIMP-nucleon cross-section was set (see Fig. 5.6(b)). The value of cross-section in the minimum of the curve at a WIMP mass of 30 GeV is $4.5 \cdot 10^{-8}$ pb. The obtained limit is only slightly inferior to the limit obtained later in the ZEPLIN-III experiment.

5.2.2.2. *Measurement of ionization yield of LXe for nuclear recoils*

The XENON collaboration provided the first careful study of ionization yield of LXe because of the importance of this knowledge (Aprile *et al.*, 2006b). Indeed, as mentioned above, the big difference in ZEPLIN-II and ZEPLIN-III designs was caused by the lack of knowledge on this at that time. The simultaneous measurement of scintillation and ionization yield (the first one is also referred to as relative scintillation efficiency L_{eff}) was carried out with the use of different test chambers, at Columbia University and at Case Western Reserve University, USA. Both chambers were two-phase emission detectors with a PMT readout. The chambers were calibrated by ^{57}Co, ^{133}Ba, and ^{137}Cs γ-sources, and by neutron sources: AmBe (in Columbia university) and ^{252}Cf (in Case Western university), with lead shielding to attenuate gammas from the sources. In addition, an internal ^{210}Po source was deposited on the center of the cathode plate of the Case chamber, providing 5.3 MeV alpha particles and 100 keV ^{206}Pb nuclear recoils. A broad study was done for all these types of particles at different electric field strength values. The results of both groups were consistent taking into account systematic and statistical error of measurements. Both groups obtained a rather unexpectable result that the ionization yield grows up with the decrease of energy deposited by nuclear recoil. Typical behavior for other detecting materials such as Si crystal, for example, is opposite and is consistent with the Lindhard prediction (Lindhard *et al.*, 1963a, 1963b). Later, the Case Western university group published the results of more detailed study with the lower energy threshold (Shutt *et al.*, 2007). In this publication, it was pointed out that not only the ionization yield from nuclear recoils increases with a decrease of energy, but this also takes place for electron recoils at the low deposited energies (≤ 10 keVee) in opposite to the commonly known trend at higher energies (Fig. 5.7). This discovery was very "helpful" because this meant that the electron and nuclear recoil bands do not intercross at low energies as thought before. In fact, separation between them even increases with energy decrease (see Chapter 2 for details). The authors also introduced a very convenient representation of experimental results from a two-phase emission detector. Because the S1 and S2 signals are anticorrelated (Conti *et al.*, 2003) they proposed a combined energy $E = (n_\gamma + n_e) \cdot W$ was "the best measure of the event energy", where n_γ and n_e are the number of photons and electrons obtained from S1 and S2 signals,

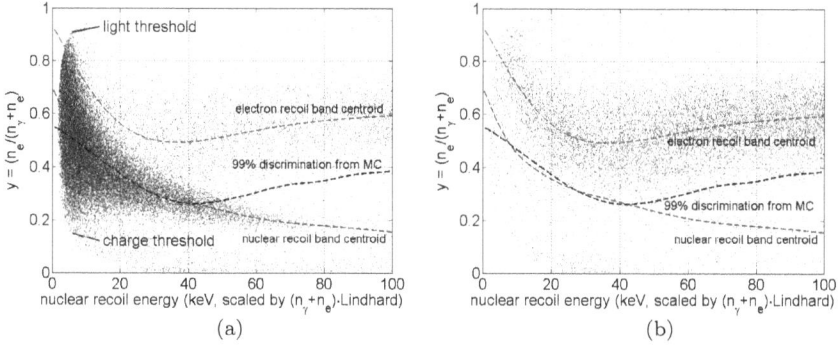

Figure 5.7: Electron and nuclear recoil bands obtained with ^{252}Gf neutron (a) and ^{22}Na gamma (b) sources at an electric field of 4.0 kV/cm. Redrawn from Shutt *et al.* (2007).

and W (the "W-value") is the average energy required to create any excitation (photon or electron). This value was found to be 13.46 ± 0.29 eV. With this new measure, scaling between nuclear recoil energy (keVr) and electron recoil energy (keVee) is given only by the "Lindhard factor" which is equal to ~ 4. Also, they proposed a discrimination parameter $y = n_e/(n_e + n_\gamma)$ which is the fraction of the total signal seen in the charge channel. This parameter (usually in logarithmic scale) will be used later by all experimental groups in their representations of experimental data.

5.2.2.3. *Chromatographic technology for purification of xenon from krypton*

Presence of the beta-decaying anthropogenic ^{85}Kr isotope in xenon is one of the most serious problems in the xenon based dark matter WIMP search experiments. This isotope has an anthropogenic origin and is released into the atmosphere from nuclear power plants. Xenon gas is separated from the air during its production together with other accompanying noble gases including krypton. Therefore, the presence of ^{85}Kr in modern xenon is inevitable. Although the abundance of this isotope in natural krypton is only $\sim 2 \cdot 10^{-11}$ and the concentration of natural krypton in xenon is typically rather small ranging from 10^{-9} to 10^{-6} mol/mol depending on manufacturer, ^{85}Kr is, nevertheless, very problematic for the WIMP search experiments because of the high decay rate ($T_{1/2} = 10.76$ years).

There are known methods of krypton removal from xenon such as distillation, adsorption-based chromatography and centrifugation. However, there were neither existing by that time ready-made systems that can be used in the experiment for preparing the xenon medium. The method used by the XENON group is based on the faster diffusion of krypton through a charcoal chromatographic column than the diffusion of xenon. The principle of operation of the system developed by Case Western Reserve University was as follows (Bolozdynya *et al.*, 2007). The purification was performed by automatically repeated cycles, each consisting of three phases: Xe feed, Kr purge and Xe recovery. At the beginning, the system was filled with helium as a carrier gas circulating through the column. Xenon, initially contaminated with krypton, was injected by a short bunch into the charcoal 10-kg column (C-column). Krypton and xenon appeared at the output of the column at different times: xenon appeared approximately in 10 min after the Kr pulse ended. The circulation path after the column was set through a Kr trap cooled by liquid nitrogen, and the recirculation was switched to the Xe condenser in 15–30 s before the appearance of xenon at the output. After the end of the xenon phase continued 156 min, the system was switched again to the recirculation through the Kr trap, and the process repeated. Once the Xe condenser is full, the multi-cycle run is paused, and xenon is fully removed. Xenon in a total amount of 25 kg was processed in 120 cycles. The initial krypton level of 5 ppb was measured directly with the mass spectrometer, while the final one was estimated to be < 3 ppt.

5.2.2.4. *Scintillation and ionization yield measured with neutrons*

There were no precise data on relative scintillation efficiency L_{eff} of LXe for nuclear recoils at the time of publishing of the paper (Angle *et al.*, 2008) on the first limit of WIMP-nucleon interaction obtained by XENON10. The L_{eff} was taken to have a constant value equal to 0.19 in the whole energy range of the WIMP search window, i.e., from 4.5 to ~25 keVr, that followed as an average value from the measurements at energies >10 keVr by several experimental groups performed by that time using tagged neutron scattering (Aprile *et al.*, 2005a; Akimov *et al.*, 2002; Bernabei *et al.*, 2001; Arneodo *et al.*, 2000).

The XENON group proposed an alternative method to obtain L_{eff} (Sorensen *et al.*, 2009) with substantially less uncertainties in comparison to those obtained in tagged neutron scattering measurements. The full shielded XENON10 detector located at the Gran Sasso underground laboratory was irradiated by an AmBe neutron source with the known activity. The source was delivered to the detector through a 7-mm hole in the shield and positioned next to the detector cryostat, behind an additional 5 cm of lead shielding. The data analysis was performed for the same fiducial volume as in the WIMP search run. A detailed Monte Carlo simulation of the nuclear recoil spectrum in the LXe target was done using the precise model of the whole setup (detector plus shield). The experimental data were fitted then with the Monte Carlo prediction in which the L_{eff} value was used as a free parameter. The same procedure was done for the charge yield of LXe for the nuclear recoils. Another procedure used by this group was based on comparison of measured and predicted multiplicity for neutron events (see details in Sorensen *et al.*, 2009). With the use of these methods, the range of measured scintillation efficiency and ionization yield of LXe was extended down to about ~4 keVr. It was obtained that the best fit corresponds to the L_{eff} that decreases towards the lower energies. For the ionization yield the trend is opposite: it continues to increase with decreasing energy.

5.2.2.5. *XENON100 detector*

The complete detector description is given in Aprile *et al.* (2012b), see Fig. 5.8 from this paper. The active target, 62 kg of liquid Xe, is enclosed in a vertical hollow Teflon cylinder with ~30-cm inner diameter and ~30-cm height with field shaping rings mounted inside the walls. This cylinder is surrounded by an LXe veto of 99 kg.

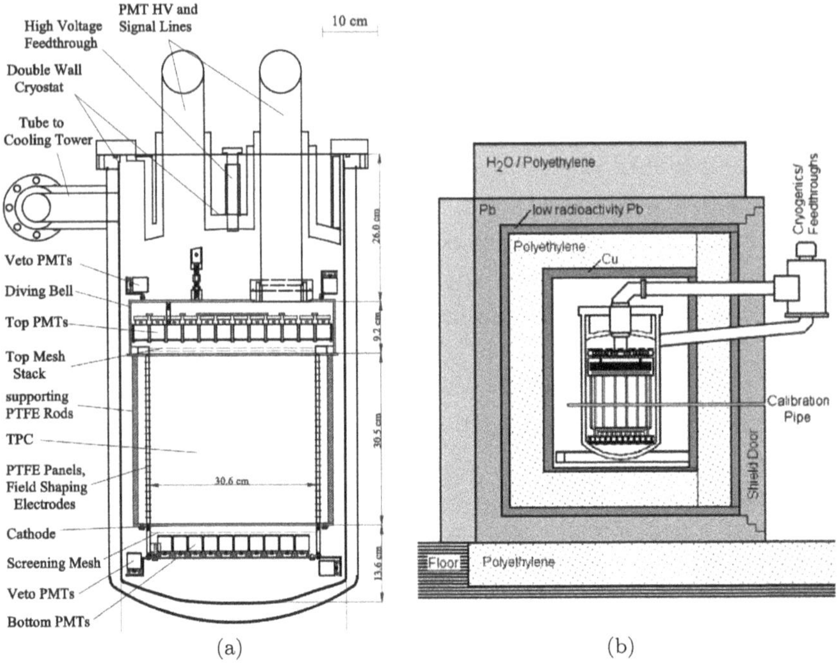

Figure 5.8:   Schematic drawing of XENON100 detector (a), and the detector installation inside the passive shield (b). Redrawn from Aprile *et al.* (2012b).

Both volumes are instrumented with R8520-06-Al 1" square photo-multiplier tubes. The active volume is viewed by 178 photomultipliers in two arrays, 98 in the gas phase and 80 in the liquid. The last one that viewed the active volume from the bottom consisted of PMTs having the highest quantum efficiency (>30% for most of them) and packed as tightly as possible. This was done in order to ensure the highest light collection efficiency of this PMT array to the S1 signal. The top array for readout of the S2 signal consisted of 89 PMTs having lower quantum efficiency (~25% on average) and packed less tightly as in the bottom one. The overall detector light collection efficiency for the scintillation light was (2.20 ± 0.09) phe/keV at the operating electric field of 0.53 kV/cm and 4.3 phe/keV at the zero field at 122 keV. These values are practically the same as those for the ZEPLIN-III and XENON10 detectors. The veto volume was viewed by 64 PMTs of the same type having quantum efficiency < 25% on average.

For purification of the xenon medium for the XENON100 detector from Kr, a specially designed custom-made rectification column incorporated into the gas supply system was used. This column had a higher production rate than the previous one used for the XENON10 experiment (0.6 kg/h in comparison to 0.1 kg/h). The system has the similar design as that developed by the XMASS collaboration described in Abe *et al.* (2009). The concentration of Kr in the detector was estimated using a second decay mode of ^{85}Kr: ^{85}Kr (beta, 173 keV) $->$ ^{85m}Rb (gamma, 514 keV) $->$ ^{85}Rb, with a 0.454% branching ratio. The lifetime of the intermediate state (^{85m}Rb) is 1.46 μs. This provides a clear delayed coincidence signature of such events. With the use of this method, the Kr concentration in the commissioning run was estimated to be $143 + 135/ - 90$ ppt. Although it was increased then by a factor of 5 because of a small leak in the gas system, this did not have a serious impact on the scientific result of the first observation run. Further xenon distillation allowed the XENON group to achieve even the lower value of Kr concentration.

The detector was placed in a passive shield (Fig. 5.8) which was an improved version of the XENON10 one. The detector was surrounded (from inside to outside) by 5 cm of OFHC (oxygen-free high-thermal conductivity) copper, followed by 20 cm of polyethylene, and 20 cm of lead (with an innermost 5-cm low-radioactive layer). The entire shield rested on a 25-cm thick slab of polyethylene. An additional outer 20-cm water or polyethylene layer was added on top and on three sides of the shield in order to reduce the neutron background.

A final WIMP search result of the XENON100 experiment (Aprile *et al.*, 2016) was based on combination of three WIMP search runs (477 live days from January 2010 to January 2014; $1.75 \cdot 10^4$ kg $\cdot$ day). Analysis was done using a conventional procedure with selection of a WIMP search window described above. The experiment achieved a 90% confidence level limit on WIMP-proton spin-independent interaction with a minimum of $1.1 \cdot 10^{-9}$ pb at

50 GeV WIMP mass. The experiment improved the limits set by 10-kg scale detectors of the previous generation (ZEPLIN-III and XENON10) by more than one order of magnitude. One should mention that the lowest value of electromagnetic background (before electronic recoil band rejection) of $\sim 5 \cdot 10^{-3}$ events $\cdot$ keVee^{-1} $\cdot$ kg^{-1} $\cdot$ day^{-1} was achieved in this experiment.

5.2.2.6. *XENON1T and XENONnT detectors*

The XENON1T experiment at the Gran Sasso underground laboratory is the largest currently operating WIMP detector in the world (to the moment of writing this book). The illustration of the XENON1T detector is given in Fig. 5.9 (see a complete description of the detector in Aprile *et al.* (2017)). The detector is functionally similar to XENON100. The main exception is that instead of a layered shield from copper, lead and polyethylene the detector is placed in a water tank having a diameter of 9.6 m and a height of 10.2 m. The detector cryostat is suspended inside this tank

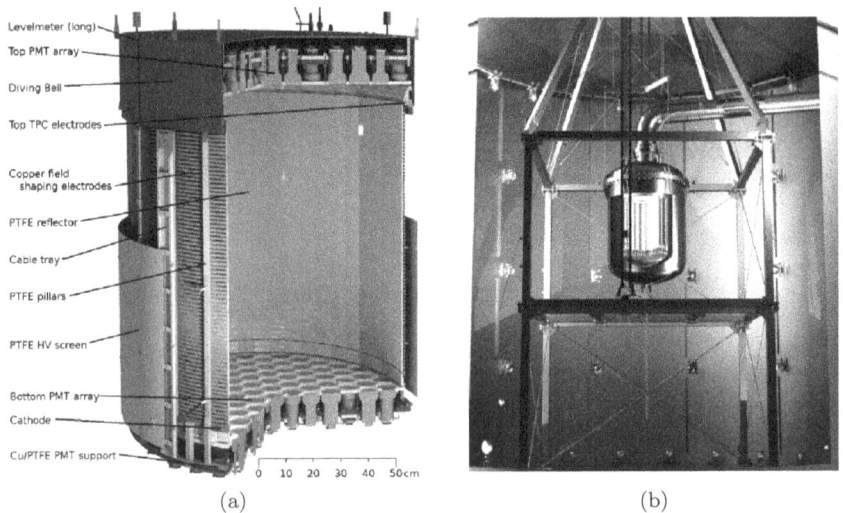

(a)   (b)

Figure 5.9: Internal structure of XENON1T detector (a), and a view of the detector suspended inside the passive water shield (b). Redrawn from Aprile *et al.* (2017).

at a special metal frame manufactured from the low-radioactive preselected materials.

The basic technical characteristics of the XENON1T detector are as follows. The XENON1T cryostat is manufactured from a low-radioactive stainless-steel. Its inner vessel has a height of 1.96 m and a diameter of 1.1 m. It is enclosed by an outer vessel of 2.49 m height and 1.62 m diameter that is large enough to accommodate the detector of the upgrade stage XENONnT. The inner vessel contains 3.2 t of liquid xenon in total with 2.0 t in the active target TPC which has 97 cm in high and 96 cm in diameter and ∼1.3 t in fiducial volume. It is planned that for XENONnT the total amount of xenon will enhance to ∼8 t and to ∼4 t in fiducial volume (Moriyama, 2019).

As in previous designs, there are five electrodes (from top to bottom): top screen, anode, gate, cathode and bottom screen. The first three are made from electropolished etched meshes which are spot-welded to circular stainless-steel supporting frames. The last two are parallel wire electrodes made by pre-stretching the wires on an external structure and fixing them by wedging between the upper and lower parts of the frames. There are 74 field shaping ring electrodes with a cross-section of ∼10 × 5 mm^2 made from low-radioactivity (OFHC) copper between the gate and the cathode electrodes. The inner part of the field shaping ring cage is coated by a PTFE reflector.

A total of 248 Hamamatsu R11410-21 3″ PMTs are used to read out the signals from the TPC. There are 127 PMTs radially installed in the top array and 121 PMTs in the bottom one packed as tightly as possible array to maximize the light collection efficiency of S1 signals. The PMT signals are digitized 100 MHz frequency and zero-length-encoding. As in previous designs, the top array is installed inside a diving bell in order to maintain and precisely control the LXe level. There are four custom-made parallel-plate capacitive level meters installed inside the diving bell to measure possible tilt of the TPC. They have a precision of level measurement of ∼30 μm.

The following basic characteristics of the detector were achieved. The electron lifetime has reached a plateau value of ∼650 μs during

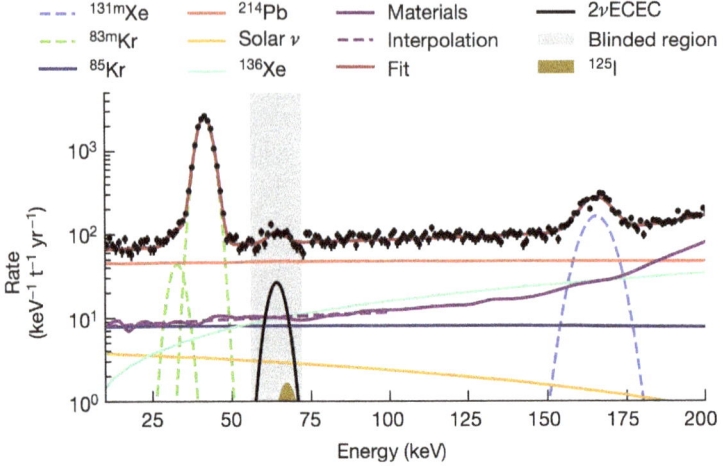

Figure 5.10: The energy spectrum obtained from XENON1T detector (177.7 days exposure and 1.5t inner fiducial mass) and simulation with different background and calibration sources (see legend). Redrawn from Aprile *et al.* (2019b).

the second science run (Aprile *et al.*, 2017) that is quite moderate taking into account that the total drift time is 673 μs. The light collection efficiency for scintillation is (8.02 ± 0.06) phe/keV at an electric field of 0.125 kV/cm. However, this value is measured with the use of ^{83m}Kr metastable isotope (41.5 keV line in Fig. 5.10), and it cannot be compared directly with light collection efficiencies of previous detectors measured with a ^{56}Co (122 keV) γ-ray source. The Kr concentration in Xe has been reduced by the cryogenic distillation column (see above) to a level of 0.66 ± 0.11 ppt. As a result, the ^{85}Kr decay contributes ~1/10 of the total background rate of electron recoils in the WIMP search energy region.

The XENON group reported (Aprile *et al.*, 2017) that they have achieved the ever lowest background rate of electron recoils of $(82^{+5}_{-3} \text{(sys)} \pm 3 \text{(stat)})$ events $\cdot$ keVee$^{-1} \cdot$ t$^{-1} \cdot$ yr^{-1} (~0.2 $\cdot$ 10^{-4} events $\cdot$ keVee$^{-1} \cdot$ kg$^{-1} \cdot$ day^{-1}) in a WIMP search energy region that is by a factor of ~25 lower than in the previous XENON100 experiment. The 90% confidence level limit set on spin-independent WIMP-proton cross-section after analysis of ~1 t$\cdot$ yr exposure WIMP search data is $4.1 \cdot 10^{-47}$ cm^2 (minimum value at 30 GeV WIMP mass). At the moment of writing this book, the XENON1T detector has been decommissioned, and preparations for the next XENONnT is underway.

5.2.2.7. *Observation of double electron capture with XENON1T detector*

The large masses of detectors used for experimental search for WIMPs opens the possibilities to use these detectors for studying other fundamental physics processes. One of them is a double electron capture (two-neutrino mode — 2νECEC) in the ^{124}Xe isotope which is represented in a natural mixture of Xe isotopes. The XENON group reported on the first observation of this process (Aprile *et al.*, 2019b). Note that this was the first physical process of fundamental importance observed by the detectors built for another goal, search for dark matter. Although this process belongs to a "standard physics", its observation is very difficult because of the very large half-life ($\sim$10^{22} yr) of the ^{124}Xe isotope with respect to this decay mode and small amount of this isotope in ^{nat}Xe ($\sim$1%). These make this process extremely rare and very difficult for observation. This discovery became possible due to the record background count rate of the detector (see above) and the ever-largest xenon detector mass.

In the 2νECEC process, two protons in a nucleus are simultaneously converted into neutrons by absorption of two electrons from one of the atomic shells (mostly the K-shell) and the emission of two electron neutrinos. After this, filling the vacancies results in a cascade of X-rays and Auger electrons. Two neutrinos, not detectable within the detector, carry away the energy practically equal to the nuclear binding energy Q ($\sim$1 MeV). A practically point-like energy deposition is resulted from only X-rays and Auger electrons (64.3 keV). A combined S1 and S2 signal was used in the analysis to achieve a better energy resolution for the 2νECEC peak. The XENON group analyzed the dark matter search data recorded between February 2, 2017 and February 8, 2018. A blind analysis was used: the interval from 56 to 72 keV was blinded and unblinded only after the data quality criteria, fiducial volume and background model had been fixed. Energy calibration in this region was performed by ^{83m}Kr and ^{131m}Xe isotopes. The latter one had occurred after calibration of the detector by neutrons.

The energy spectrum in the energy range from 15 to 200 keV obtained during 177.7-day exposure and within 1.5-t inner fiducial mass is shown in Fig. 5.10 together with Monte Carlo simulation taken into account for different background and calibration sources. The blinded region is indicated by grey color. After unblinding this region, a peak with (126 $\pm$ 29) events corresponding to 2νECEC of ^{124}Xe was identified. The most dominant background in this region is the β decay of ^{214}Pb which is a daughter of ^{222}Rn. This isotope is emanated from inner surfaces of the detector to the liquid or gaseous xenon. Other less dominant sources are the β decay of ^{85}Kr and the two-neutrino double β decay of ^{136}Xe. Note, that even the solar neutrino interaction with xenon is considered as a background. The significance of the obtained signal is 4.4 σ and the corresponding half-life is (1.8 $\pm$ 0.5 (stat.) $\pm$ 0.1 (syst.)) $\cdot$ 10^{22} years.

5.2.3. *LUX experiment*

The Large Underground Xenon (LUX) experiment (Akerib *et al.*, 2013b, 2014, 2017d) explored the detector of practically similar size as that of the XENON100 but designed for the goal to have at least one order of magnitude higher sensitivity. To achieve this, the detector was placed in a water tank to shield it from the outer radioactivity and to provide a muon veto. The muons were tagged by detecting the Cherenkov light with twenty 10″ PMTs. The experiment was carried out in the Davis Cavern at the Sanford Underground Research Facility (SURF), Homestake gold mine, South Dakota. This is the same place where the famous Davis solar neutrino experiment was carried out. The LUX detector was assembled above ground in a specially built laboratory. In July 2012, the detector was moved into the newly recommissioned Davis Underground Laboratory located at 4850 feet below ground.

The general rendering view of LUX setup and the detector details are shown in Fig. 5.11 (Akerib *et al.*, 2013b). The detector

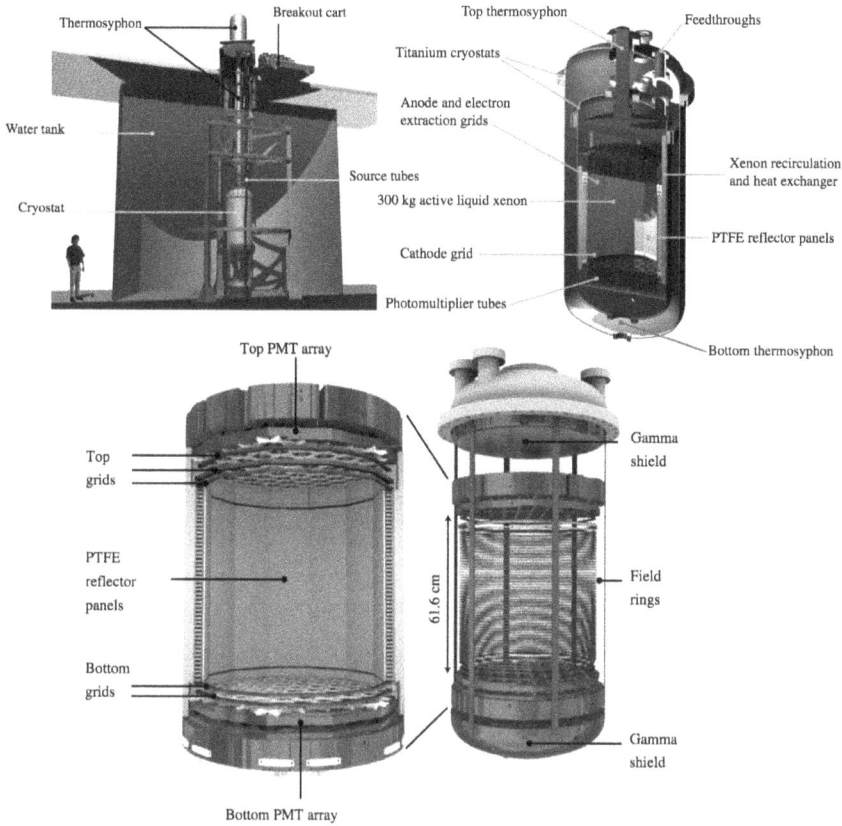

Figure 5.11: Rendering of the LUX systems: a general view (top left), the LUX detector inside titanium cryostat (top right), the LUX inner structure (bottom right), and TPC structure (bottom left). Redrawn from Akerib *et al.* (2013b).

was suspended in the middle of the water tank having 7.6 m in diameter and 6.1 m in height. There were special pipes on sides to deliver radioactive sources for detector calibration and for gas and electrical communication and for cooling on top of the detector. The two-phase emission detector was placed in a cryostat shown in Fig. 5.11(on right). The cryogenic system of the cryostat was based on a thermosyphon technology (gravity assisted heat pipe, see description below). The detector TPC is shown in Fig. 5.11 (on bottom, left). It had a drift distance of 49 cm and a diameter of 50 cm. The total amount of liquid xenon in this active volume was about 250 kg. The side walls were made from Teflon panels. The detector volume was viewed by two arrays of 61 Hamamatsu R8778 2″ low-background PMTs, specially designed for operation in the LXe environment at 170 K. The photocathode of this PMT is sensitive to 175 nm LXe scintillation with a typical quantum efficiency (QE) of 33%.

The electric fields were applied in the detector by circular grid electrodes and field-shaping rings. Below, the grids are described starting from bottom to top. The lowest one was located 2 cm above the bottom PMT array to screen the photocathodes from the high voltage potential of the cathode. It had ∼0.2-mm diameter stainless steel wires spaced with a pitch of 1 cm (98% optical transparency). The cathode grid was located 4 cm above the screening electrode and is identical in construction, except that it had a pitch of 5 mm and a 96% open area. There was an extraction grid 49 cm to the top from the cathode which had 50-micron stainless steel wires spaced with a pitch of 5 mm, resulting in a 99% open area. The next one was an anode grid with 30-μm wires spaced by 0.5 mm and sited 15 mm above the extraction grid. The liquid xenon level was maintained between these two grids, 5 mm above the extraction grid. The next to top grid was located 4 cm above the anode and 2 cm below the top PMT array. Its function and geometry were identical to that one at the bottom PMT array, except for the wire diameter (50 mm) resulting in a 99% open area. The large open area of the grids resulted in 8.8 phe/keVee for scintillation for 122 keV γ-rays at zero field (much higher than the corresponding value in XENON100).

The cathode HV system was designed for applying as high as 100 kV (drift field ∼2 kV/cm), and the field in the electrolumines-cence gap between the anode and the LXe surface was expected to be 10 kV/cm. However, in the first data taking run of WIMP-search exposure in 2013, the electric field in the drift field was only as high as 0.181 kV/cm and the electroluminescence field was 6 kV/cm (this value corresponded to electron extraction efficiency of 0.65 ± 0.01) (Akerib *et al.*, 2013b). In preparation for the next WIMP-search data taking run (in 2014–2016 years), the grid electrodes: anode, gate, and cathode underwent a so-called "conditioning" in cold Xe gas.

During this procedure, the HV to each electrode was set just above the threshold value of sustain discharge, and the discharge was maintained during several days. This resulted in the increased HV in the electroluminescent region and the rise up of the extraction efficiency to 0.73 ± 0.04. However, this treatment had resulted in a dramatic drawback. It was observed that the electron drift trajectories were significantly altered from the near-vertical paths seen previously. This was explained by a buildup of the negative electrical charge on the Teflon TPC walls. The charge resulted from the exposure to the VUV which was emitted from a discharge during the conditioning procedure. The VUV photons produced in this process created the electron–hole pairs in Teflon. The holes, as having higher mobility, were removed more rapidly from the surfaces, thus, resulting in a buildup of the net negative charge. This curving of trajectories was measured during calibrations and taken into account in 3D reconstruction by applied corrections.

The complete dataset with a total exposure of $4.35 \cdot 10^4$ kg · day (by a factor of ~2.5 larger than that of XENON100) included the data from the first and the second WIMP-search runs (the second one continued from September 2014 till May 2016). The experiment achieved a background of $(3.6 \pm 0.4) \cdot 10^{-3}$ events keVee^{-1} · kg^{-1}· day^{-1} in a fiducial volume of ~100 kg (Akerib *et al.*, 2015). This is by ~30% lower than the background achieved in the XENON100 experiment. The LUX experiment set a 90% confidence level limit for the spin-independent interaction of 1.1×10^{-10} pb at 50 GeV · c^{-2} (Akerib *et al.*, 2017d) which is by one order of magnitude lower than that of XENON100 published a year before. However, the authors did not discuss the reasons for such dramatic improvement of the cross-section limit.

5.2.3.1. *Thermosyphon cooling technology*

The LUX collaboration proposes and utilizes a new cooling and thermostabilization method called "thermosyphon technology" (Bolozdynya *et al.*, 2009). This method allows one to "deliver the cold" to the cryostat much more efficiently than by a traditional thermal conductivity.

A thermosyphon is a vertically oriented closed pipe which can be divided functionally into the following parts: on top, a condenser immersed in a bath of free-boiling liquid nitrogen; on bottom, an evaporator thermally attached to the detector and a passive adiabatic part between these two parts. The pipe is filled with a variable amount of gaseous nitrogen. Effective heat removal from the detector is achieved by phase transition of nitrogen from the liquid to the gas inside the thermosyphon. The gas condenses in the top part of the pipe and trickles down to the evaporator. There it boils up removing heat from the detector and rises upwards to the condenser where the process repeats. Thermal conductivity of a thermosyphon can be very high, of an order of ∼several tens of $kW/(k \cdot m)$.

The LUX collaboration used this cooling technology because of its convenience when the cold detector is located inside the large water tank.

5.2.3.2. *Calibration of liquid noble gas detector by a low-energy* ^{83m}Kr *source*

The idea of xenon detector calibration by an ^{83m}Kr source was proposed and tested for the first time by the experimental group from Yale University, a member of the LUX collaboration (Kastens *et al.*, 2009; Kastens *et al.*, 2010). The short-life ^{83m}Kr metastable isotope is produced as an intermediate state in the decay of ^{83}Rb (86.2 days half-life). Since this isotope has the very short half-life of 1.83 h it is ideal for calibration of low-background detectors. The ^{83m}Kr isotope decays in two steps, emitting a 32.1-keV and then, with 154 ns decay time, a 9.4-keV conversion electron. Since krypton is a noble gas having much lower temperature of triple point than that of xenon it is perfectly soluble in the liquid xenon and allows a possibility to perform a uniform calibration of the whole detector active volume.

The ^{83}Rb isotope was infused in the amount of 700 nCi by syringe in a zeolite located in the bottom arm of a VCR cross incorporated in a gas system. Then the zeolite was baked at 80° C for several days and, since the Rb is quite violate, with careful monitoring of the amount of ^{83}Rb in it by measuring its activity in γ-ray lines (521, 530, and 553 keV) with NaI detector. The ^{83m}Kr was injected into the detector (a test chamber) by circulating xenon with a diaphragm pump at 2 L/min through the VCR cross with zeolite for 5 min.

Later, the ^{83m}Kr source was used by many experimental groups for calibration of liquid noble gas detectors, including LUX collaboration (Akerib *et al.*, 2017b).

5.2.3.3. *In situ measurement of ionization yield of LXe for Xe nuclear recoils at sub-keV energies*

All previous measurements of ionization yield of LXe for nuclear recoils were carried out in separate test chambers. The LUX collaboration performed the first calibration of the WIMP-search detector *in situ* (Akerib *et al.*, 2016b).

Calibration was done with the use of a neutron generator (D-D, producing 2.45-MeV neutrons) outside the 8-m water tank of the LUX setup. The neutrons from this generator were delivered to the detector via a narrow (4.9 cm inner diameter) air-filled pipe which displaced water thus producing neutron collimation. This 3.77 m plastic conduit was suspended by stainless steel wire rope on top of the water tank and was removed during the WIMP-search run. After the 3D reconstruction procedure, the double scatter events were selected, and the neutron scatter angles at the first interaction points were defined (in the range from 7° to 79° that corresponds to the energies from 0.3 to 30 keVnr). The nuclear recoil energies were defined then by kinematics. This method allowed the LUX group to reach the lowest ever nuclear recoil energy of 0.7 keVnr. The results demonstrated nearly flat behavior of the ionization yield vs. the nuclear recoil energy in the vicinity of $\sim$1 keV, however, the uncertainties of this measurement were quite large. However, the recent measurements performed by the experimental group from the Lawrence Livermore National Laboratory with participation of the members of the LUX experiment (Lenardo *et al.*, 2019) and carried out at TUNL (Triangle Universities Nuclear Laboratory) with a specially designed test chamber have demonstrated the decrease of the ionization yield in the energy range from 0.3 to 0.6 keV.

5.2.3.4. *Tritium calibration of the LUX dark matter*
         experiment

Before LUX, electron equivalent energy scales of all two-phase
detectors were calibrated in the regions of energy higher than the
WIMP-search energy range. The source with the closest energy
was ^{83m}Kr. Since the cascades of electrons in ^{83m}Kr decay are not
spatially separated, calibration of S2 with this source can be done
only for 41.5 keVee total energy. This source is soluble in LXe in
a detector and provides spatially uniformly distributed events. All
other sources usually used for calibration: ^{57}Co, ^{133}Ba, ^{137}Cs are
external ones. This means that the distribution of events from them
are distributed non uniformly across a detector sensitive volume due
to self-shielding.

The LUX collaboration first time performed electron recoil
calibration in the keV-range by a tritium source (Akerib *et al.*,
2016a). Tritium (T) is the lightest beta decaying isotope with a
Q value of 18.6 keVee which spectrum is well known. It has a broad
peak at 2.5 keVee and a mean energy of 5.6 keVee. Tritiated methane
(CH_3T) was used as the host molecule to deliver tritium activity into
LUX. Compared to molecular tritium (T_2), this chemical compound
is more preferable because it does not adsorb onto surfaces like T_2.
Since diffusion of tritium into the detector plastic components was a
serious concern, the LUX collaboration carried out special benchtop
tests which had shown that injection and removal of CH_3T can be
done without any risk to the LUX experiment.

Two CH_3T sources with total activities of 3 and 200 Bq were
prepared. Each source was contained in a 2.25-L stainless steel bottle
and mixed with two atmospheres of purified xenon. The gas from the
bottle was injected into the standard LUX circulation line through
a special SAES methane purifier (model MC1-905F). After passing
through the detector the CH_3T gas was removed in a certain time by
a standard noble gas SAES purifier (model PS4-MT15-R1). Before
injection of CH_3T into the system the efficient methane removal
by the SAES getter had been checked by injection of $\sim$1 ppm
natural methane. Then, in August 2013 an initial injection of 20 mBq

of CH$_3$T was performed followed 5 days later by an injection of 800 mBq. In both cases, the count rate increased during one hour and then decreased with a time constant equals approximately 6 hours. At the end of the WIMP-search run in December of 2013, a total of 10 Bq of tritium was injected and removed with the same procedure. The obtained events were uniformly distributed across the sensitive volume of the LUX detector. These events were used for characterization of the electron recoil response of the LUX detector.

As mentioned by the developers of this technique, it can be successfully used in tone scale experiments such as LZ (see below) because in detectors the obtaining of electron recoil response by an external source will be absolutely impossible.

5.2.4. *LUX-ZEPLIN (LZ) experiment*

In this experiment, which is currently at a preparation stage, the efforts of the ZEPLIN and LUX collaborations are joined together in order to build a 10-ton detector, the largest two-phase emission dark matter detector in the near future. The goal of the LZ-experiment is to achieve projected cross-section sensitivity of spin-independent WIMP–nucleon interaction of $1.5 \cdot 10^{-12}$ pb (at 90% confidence level) for a WIMP mass of 40 GeV/c^2. The detector will operate in the same water tank where the LUX detector was. The detailed technical description of the detector is given in (Akerib *et al.*, 2020a). A cutaway rendered drawing of the experimental setup is shown in Fig. 5.12. The basic detector subsystems are the two-phase xenon detector (1), the outer veto liquid scintillator detector (2), the diffuse reflective surface with PMT readout (3), and the water tank (4). The principal difference of the current detector design from the previous ones is that the high voltage is supplied (5) directly to the bottom part of the TPC. This potentially increases the maximal possible electric drift field in the detector. The nominal 300 V/cm drift field is planned in the active region of the detector. This requires application of an operating voltage of -50 kV to the cathode electrode. The designed maximum operating voltage is -100 kV.

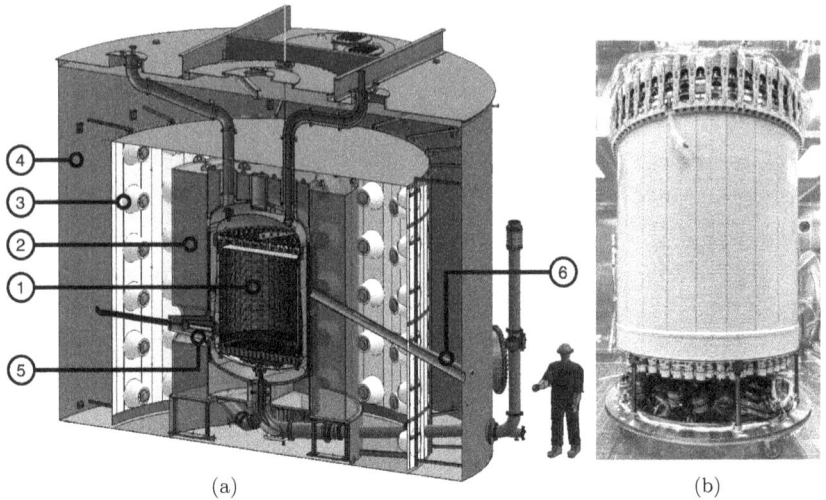

(a)   (b)

Figure 5.12:   The LZ experimental setup (a): the two-phase xenon emission detector (1), outer detector based on gadolinium loaded scintillator (2), the diffuse reflective cylindrical surface equipped with PMTs (3), the water tank (4), the high voltage supply (5) and the conduit for neutron calibration (6). A photo picture of the inside structure (b). Redrawn from Akerib *et al.* (2020a).

The active region (TPC) has 145.6 cm distance between cathode and gate electrodes, and 145.6 cm in diameter, and is viewed by two arrays of ultra-low PMTs Hamamatsu R11410-22 specially designed for operation in LXe. There is a "top" array located in the gas phase containing 253 PMTs and the bottom one with 241 units close-packed in a hexagonal pattern. The active region contains 7 tons of LXe.

The TPC part is surrounded by a LXe skin veto. The side region of the skin has 4 cm of LXe at the top part and 8 cm at the cathode level for increased electrical isolation of the cathode TPC region from the inner cryostat walls. The skin region is viewed from above by 93 1″ Hamamatsu R8520-406 PMTs. In addition, this region is viewed from the bottom TPC part by 20 2″ Hamamatsu R8778 PMTs. The dome region of the skin at the bottom part of the detector is instrumented with 18 2″ R8778 PMTs. These PMTs are mounted horizontally below the bottom PMT array, with 12 and 6 of them looking radially outward and inward, correspondingly.

Special care was paid to fabrication of the grid electrodes of the detector because of the ever-largest grids of ~1.5 m in diameter. The collaboration has chosen fabrication of custom-made grids from 304 stainless steel ultra-finish woven wires with a tension of each wire of ~250 g. The woven mesh has several advantages over wires stretched in a single direction. It produces more uniform load on a ring, and this allows one to minimize the ring mass. The next advantage is that the regions of nonuniform electric fields are smaller; and finally, penetration of electric fields behind the mesh is smaller. The gate and cathode meshes are fabricated from 75 and 100 μm wires, correspondingly, with a 5-mm pitch. The anode mesh is fabricated from 100 μm wires with 2.5 mm pitch. This small pitch is chosen in order to minimize the S2 resolution of the detector. The distance between the anode and gate meshes is 13 mm with LXe dividing this distance to 8 mm of gas and 5 mm of liquid. With such large sizes of electrodes, electrostatic deflection of the mesh planes becomes important. The deflection of the mesh planes vs. electric field strength was measured for the fabricated meshes, and the obtained values are within the expectations obtained from electrostatic modelling. It is expected that the 13-mm between the anode and the grid electrodes will decrease by ~1.6 mm in the central part at the 11.5 kV nominal operating voltage. The combined effect of field increase and gas gap reduction will result in ~5% decrease of the electroluminescent yield in the central part of the detector. Obviously, this effect can be corrected using the known event coordinate.

Xenon prepared for the LZ experiment was purified from ^{85}Kr by the chromatographic method used previously in the Xenon10 and LUX experiments and described above. It was estimated that in order to achieve the projected cross-section sensitivity the concentration of Kr/Xe must be reduced to <0.3 ppt (parts-per-trillion (g/g)). The entire amount of Xe for the LZ experiment has been purified to this level before the detector filling. This was done by the Kr removing system containing 800 kg of activated charcoal.

The inner cryostat vessel is suspended inside the outer cryostat vessel, cooled by a set of LN thermosyphons, and thermally isolated by an insulating vacuum. Both the inner and outer vessels are fabricated from low radioactivity titanium.

The cryostat is surrounded by an outer liquid scintillator detector (2) in Fig. 5.12. The concept of using a neutron active veto was brought to LZ by the ZEPLIN-III members of the LZ collaboration. However, in contrast to ZEPLIN-III the LZ neutron veto system serves for tagging the neutrons originating inside the LZ detector itself in (α, n) reactions on the light elements in construction materials (predominantly in PTFE). In the case of LZ, the outer neutrons are practically completely absorbed by the water passive shield. The veto detector is based on a gadolinium loaded liquid scintillator (linear alkyl benzene (LAB)). Neutrons are captured by Gd or by hydrogen (with a probability of ~10%) and produce gammas which are detected by a liquid scintillator. The liquid scintillator is enclosed in a segmented transparent acrylic vessel which is viewed by the PMTs situated on the diffuse reflective cylindrical surface (3). The water layer between this surface and the acrylic vessel serves to shield the liquid scintillator detector from the radioactivity of the PMTs.

5.2.5. *PandaX program*

The Particle and Astrophysical Xenon Experiments (PandaX) dark matter search program was the next one of four in the world with the use of a two-phase emission detection technique. This program is carried out at the China Jinping underground laboratory (CJPL). The CJPL laboratory is located in the middle of an 18-km tunnel under 2400 m of rock overburden (6500 m.w.e) in the Sichuan province of south–west China. This is the deepest underground laboratory in the world. The PandaX collaboration which was formed from several China institutions and two US universities successfully uses the technology developed by other groups.

5.2.5.1. *PandaX-I and PandaX-II detectors*

In contrast to the ZEPLIN, XENON, and LUX programs, the PandaX program started from the stage of building a 250-kg

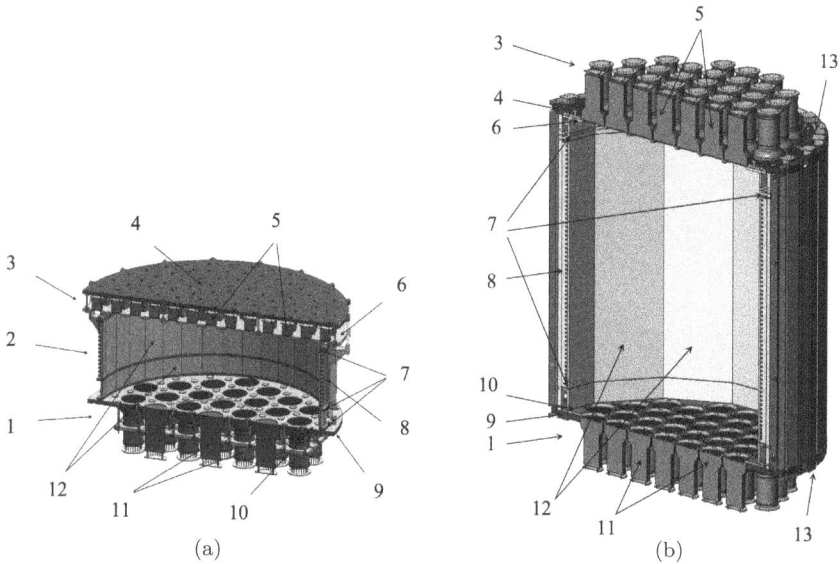

Figure 5.13: The PandaX-I (a) and PandaX-II (b) detectors; 1 — bottom PMT array, 2 — field cage, 3 — top PMT array, 4 — top copper plate, 5 — PMTs Hamamatsu R8520, 6 — top Teflon reflector, 7 — electrodes: anode, grid, and cathode, 8 — copper shaping rings, 9 — bottom copper plate, 10— bottom Teflon reflector, 11 — PMTs Hamamatsu R11410, 12 — Teflon side reflecting panels, 13 — skin layer PMTs Hamamatsu-R8520-406. Redrawn from Cao *et al.* (2014).

detector (PandaX-I). In fact, this was a demonstration stage and simultaneously the data taking run, quite short, however, 17.4 live days in 2014 with 37 kg of LXe in fiducial volume. The result of this run was published in (Xiao *et al.*, 2014). The 90% confidence level limit for the spin-independent interaction was set at 3.7×10^{-8} pb at WIMP mass 49 GeV· c^{-2}.

The design of the PandaX-I detector is shown in Fig. 5.13(a). The scheme of the detector was similar to that of other dark matter two-phase emission detectors with a traditional set of electrodes: the anode, the gate grid, the cathode, and the screen electrodes in front of the photocathodes all with 60-cm internal diameter of rings. The gate and the cathode grids and the screen electrodes were fabricated from stainless steel (304 or 316L) 200-μm diameter wires tensioned with ~290 g and pressed between stainless steel rings by screws.

With a pitch of 5 mm these electrodes had an optical transparency of 96%. The anode grid electrode was made of photo-etched mesh with crossing bars of 200 μm width and 5 mm spacing having an optical transparency of 92%. The electrode cage formed by the cathode and gate grids and the field shaping rings had ~60 cm in diameter and 15 cm in height. The electric field strength in this active volume (~120 kg of LXe) was ~1 kV/cm and it was limited by the maximum HV between the gate and the cathode of ~15 kV before breakdown. The distance between the anode and the gate grids was 8 mm with the level of LXe set in the middle. With 5 kV voltage between the anode, which was grounded, and the gate, the extraction field was 8.3 kV/cm that corresponds to an electron extraction efficiency of 90%. The active volume was viewed by two arrays of PMTs with 143 of Hamamatsu R8520-406 type in the top array and 37 of Hamamatsu R11410-MOD PMTs type in the bottom one. The detector was surrounded with a passive shield made of a 5 layered structure of (from the outside to inside) 40 cm polyethylene, 20 cm lead, 20 cm polyethylene, and 5 cm copper. Inside that, there was a cylindrical 5-cm thick copper vessel which served both for shielding and as the wall of the cryostat.

After demonstration of the successful operation of PandaX-I, the collaboration started building the second stage detector PandaX-II (Tan *et al.*, 2016a). The design drawing of the PandaX-II detector is shown in Fig. 5.13(b). The PandaX-II detector was scaled up in height PandaX-I, and it reused most of the elements and the infrastructures of PandaX-I. The detector was put in the new inner vessel manufactured from stainless steel with much lower radioactivity. This resulted in reduction of the ^{60}Co activity by more than one order of magnitude. The mass of LXe was increased to 580 kg. The inner diameter of the TPC was slightly increased, from 60 to 64.6 cm, while the height of the active volume was increased by a factor of four, to 60 cm. Similarly to PandaX-I, the screen and the cathode electrodes, and now the anode, were fabricated from 200-μm wires with 5-mm spacing. The diameter of gate wires was reduced from 200 to 100 μm with the same 5-mm spacing kept. The distance between the anode and the gate was increased from 8 to 11 mm with the LXe level set in the middle that resulted in a lower extraction field than in PandaX-I at the same $\sim$5 kV voltage applied between them (with the anode grounded). The operational electric field strength in the active volume achieved in the commissioning run was 393.5 V/cm (at -29 kV on the cathode), which is significantly less than that for PandaX-I. The active volume was viewed by two arrays of Hamamatsu-R11410 3'' photomultipliers with 55 PMTs in each one.

In comparison to PandaX-I, the PandaX-II detector was equipped with two circles of Hamamatsu-R8520-406 1-inch PMTs with 24 pieces in each one. These PMTs were located at the same levels as the main PMTs (13 in Fig. 5.13(b)), and they viewed the region filled with LXe outside of the field shaping rings, a skin layer.

A brief physics commissioning run was taken in late 2015 (Run 8, 5845 kg-day exposure) with a dominated background from ^{85}Kr. The PandaX group used this background for electron recoil calibration. The observed ^{85}Kr background was estimated to be 0.082 mBq/kg that corresponds to 437 ± 13 ppt in LXe. Presumably, krypton in xenon was introduced by an air leak in the previous filling of the detector. A 90% confidence level limit was set on WIMP-proton spin-independent interaction with a minimum of $2.97 \cdot 10^{-9}$ pb at 44.7 GeV/c^2 WIMP mass.

After krypton removal by distillation method, the PandaX-II started the data taking run (Run 9, March to June 2016; 79.6 live days; Tan *et al.*, 2016b). During this run, the detector was calibrated with a low-intensity Am–Be neutron source and with tritium source by injecting tritiated methane using the method developed by the LUX group. The electron lifetime, the most important characteristic of a two-phase detector, reached 940 ± 50 μs towards the end of the run (compared to the maximum drift time of 350 μs) with the average value of $\sim$700 μs. In the next Run 10 (April to July 2017; 77.1 live days; Cui *et al.*, 2017), the electron lifetime was improved to an average of 850 μs. The electron extraction efficiency from liquid to gas phase was $54.5 \pm 2.7\%$ and $57.7 \pm 1.9\%$ for Run 9 and Run 10, correspondingly. This is by $\sim$40% lower than that in the PandaX-I run. The total background rate had reduced from

$(1.96 \pm 0.25) \cdot 10^{-3}$ in Run 9 to $(0.79 \pm 0.16) \cdot 10^{-3}$ events. keVee^{-1} · kg^{-1} · day^{-1} in Run 10. The necessity to interrupt the Run 9 was caused by the fact that the tritium rate plateaued at $\sim 2\mu$Bq/kg in spite of effective removal by getter indicating tritium attachment on detector surfaces with its further slow emanation. Between the Run 9 and Run 10 the detector was emptied and reconditioned, and the xenon was re-distilled once again for further krypton reduction.

After analyzing the combined statistics of two datasets Run 9 and Run 10 the PandaX collaboration reported (Cui *et al.*, 2017) the final cross-section limit on spin-independent WIMP-nucleon interaction of $8.6 \cdot 10^{-11}$ pb at 40 GeV/c^2 at 90% confidence level.

5.2.5.2. *PandaX-4T detector*

While PandaX-III will be a gas phase detector with the aim of searching for a neutrinoless double beta decay of the ^{136}Xe isotope with enriched xenon, the two-phase emission detector of the next generation for WIMP search will be PandaX-4T (Zhang *et al.*, 2019).

The PandaX-4T (Fig. 5.14) is a scaled-up version of the PandaX-II detector, and its design is very similar to that of the XENONnT and LZ detectors. The cryostat is planned to be manufactured from a low background stainless steel (SS). The total mass of liquid xenon inside the cryostat is ~6 tons, with 4 tons inside the sensitive volume which has a transverse diameter of 1.2 m. The drift distance between the gate and cathode electrodes is 1.2 m. The distance between the anode and the gate is 10 mm with the liquid xenon surface in the middle. All electrodes are made of SS rings with 0.2 mm SS mesh fixed on them. The design drift field is 400 V/cm in the liquid xenon, and the field in the gas phase is 6 kV/cm.

The sensitive volume is viewed with 3$''$ PMTs Hamamatsu R11410-23. There are 169 top PMTs placed in a concentric circular pattern, and 199 bottom PMTs in a compact hexagonal structure. Similarly, as in the PandaX-II detector, there are two circles with 1$''$ Hamamatsu R8520-406 skin PMTs in amount of 72 in each viewing the space between the field shaping rings and the cryostat wall.

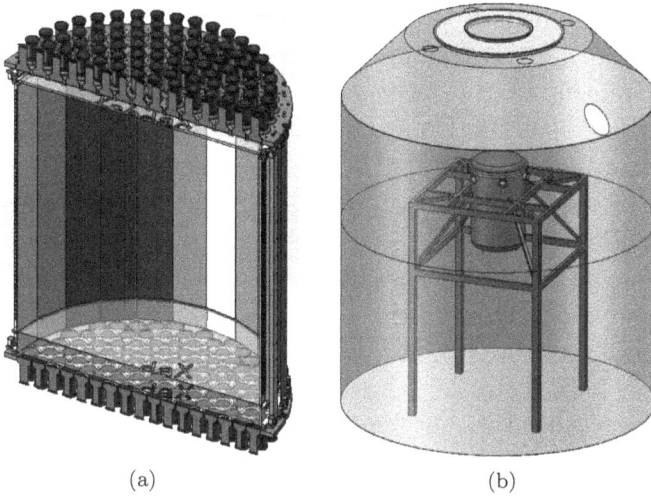

(a)   (b)

Figure 5.14: The PandaX-4T detector; (a) — the inside structure of the detector, (b) — the water tank with suspended PandaX-4T detector inside. Redrawn from Zhang *et al.* (2019).

The PandaX-4T detector will be suspended in the center of a 13-m tall, 10-m diameter water tank with an ultrapure water to suppress the external gammas and neutrons. The experiment will be located in the B2 experimental hall in the second phase China Jinping underground Laboratory (CJPL-II). The projected sensitivity with 5.6 ton-year exposure to spin-independent WIMP-nucleon interaction is $6 \cdot 10^{-12}$ pb for WIMP mass of 40 GeV/c^2.

5.2.6. *DARWIN project*

Although the possibility of construction of detectors bigger than XENONnT, LZ, and PandaX-4T will be clear only after the successful operation of them, physicists have already started to think about the next generation of the dark matter detectors. The next generation DARWIN (DARk matter WImp search with liquid xenoN) project belongs to it (Aalbers *et al.*, 2016). Although the main task of the experiment is search for WIMPs, there are other important physical tasks such as: search for solar axions, galactic axion-like particles, and the neutrinoless double-beta decay of ^{136}Xe, as well as measurement of the low-energy solar neutrino flux with <1% precision, and detection of galactic supernovae via coherent neutrino–nucleus interactions. The DARWIN experiment aims at a ∼10-fold increase in sensitivity (down to ∼10^{-13} pb) to spin-independent WIMP–nucleon interaction compared to the projects of the previous generation. This will be possible due to the increase of the total mass to 50 tons (40 tons in the TPC). To contain such an amount of xenon the size of the TPC will be of 260 cm in diameter and height, more than by a factor of 2 larger than the size of the PandaX-4T TPC. A conceptual design is very similar to that of the other ton-scale detectors; however, the specific technical decisions are not specified yet. Operation of the xenon two-phase TPC of such sizes is a challenging task for the present moment.

The current basic option for photodetectors is Hamamatsu R11410 (future versions of this model). Nevertheless, several alternative light readout options are under consideration now because PMTs, even low-radioactive ones, still produce the major part of the radioactive background. Those are large area SiPMs, SiGHTs, and GPMs. There is a significant progress in development of the silicon photomultipliers: surface mount boards with sizes of 5 × 5 cm are becoming commercially available and are under consideration now by the DarkSide collaboration (see below). A Silicon Geiger Hybrid Tube (SiGHT) is another promising photodetector which utilizes a hemispherical photocathode biased at ∼−3 kV and SiPM in the middle of the structure to count the photoelectrons. The entire structure is made from ultra-clean synthetic fused silica to guarantee the very low levels of radioactivity. A Cryogenic Gaseous Photomultiplier (GPM) combines a high-QE CsI-photocathode and cascaded GEMs or THGEMs (thick GEMs). The CsI photocathode could be deposited directly on the surface of THGEM.

In addition, new challenges may arise in scaling up the "traditional-size" two-phase TPC to the sizes of more than 2 m. Even in the LZ with the grid electrode size of ∼1.5 m, there is the significant bending of the grids leading to the decrease of the distance between the anode and the gate in the middle part of the detector. Because of this the DARWIN collaboration considers the single-phase operation option with amplification in a Liquid Hole-Multiplier (LHM; see Chapter 4). In fact, this will be a two-phase detector, but at a microscopic level. The large drift time of electrons requires the extra purity of the LXe medium: the electron lifetime before capture by electronegative impurities must be continuously maintained at a level of more than 5–10 ms.

To meet all these potential challenges further R&D programs are required.

5.3. Argon Detectors

The WIMP search with two-phase argon emission detectors is represented with substantially less number of experiments. There are two main reasons for this. The first one is that the emission spectrum of scintillation and electroluminescence for LAr peaks at a wavelength of ∼125 nm, significantly shorter than that for LXe. There were no commercially produced low-background photodetectors for such wavelengths with a large photocathode area. In all liquid argon detectors, the VUV light was detected with the use of a wavelength shifter. Till now, the commonly used chemical compound for this purpose, is Tetraphenyl butadiene (TPB, see Chapter 8). However, traditional vacuum deposition of TPB to the PMT windows and detector walls (usually made from Teflon) does not guarantee the stable layer of it. The second reason is the presence in natural argon of the cosmogenic radioactive ^{39}Ar beta-decaying isotope resulting in the activity of the natural argon of ∼1 Bq/kg (Benetti *et al.*, 2007). Another reason that the experimental groups were focused on xenon was that the masses of WIMPs were predicted preferably in the region of hundred GeV/c^2. The use of heavy elements in this case is preferable for the efficient energy transfer from dark matter particles to atomic nucleus.

On the other hand, the liquid argon medium has an undeniable advantage before the liquid xenon for nuclear recoil detection. That is pulse shape discrimination (PSD) capability. The difference between decay time of triple and singlet components in LAr is much bigger than in LXe (see Chapter 3). As in LXe, the ratio between these components depends on the type of particle. This allows one to reject effectively the gamma and beta backgrounds. One should mention that the natural argon is represented practically only by the ^{40}Ar isotope which has the zero spin. Thus, the argon target is sensitive only to the spin-independent interaction.

5.3.1. *WArP and ArDM programs*

WIMP Argon Program (WArP) was the first experiment that used liquid argon (LAr) for direct WIMP search at Gran Sasso. The first stage of research was carried out with a 2.3-liter chamber, 2.6 kg in active volume (Benetti *et al.*, 2008). The conceptual design of the detector was very similar to that of the ZEPLIN-II detector: seven 2″ TPB coated PMTs viewed the active volume defined by the PTFE walls that narrowed towards the bottom. A diffusive TPB coated reflector layer was placed on the surface of the PTFE walls. The chamber was entirely surrounded by a passive lead and polyethylene shield. Both ionization to scintillation ratio analysis and pulse shape analysis were used in the detector.

From the experimental data collected during the engineering run (~100 kg-days of statistics) a 90% confidence limit of ~10^{-6} pb for the spin-independent WIMP-nucleon cross-section was set. This was actually the first result obtained with a liquid argon detector. With this study the WArP group has proved that detection technology with a two-phase argon can be used for the WIMP search.

The next stage of the WArP program would be the detector with 140-kg TPC and projected sensitivity of $5 \cdot 10^{-9}$ pb for a WIMP mass of 100 GeV (Acciarri *et al.*, 2011). We will not provide a description of the detector here because the project did not enter the stage of WIMP search. Those who are interested in detail we refer to the description of the WArP detector in Acciarri *et al.* (2011). However, one important feature of this detector design should be mentioned. The LAr TPC is placed inside an active veto shield of LAr (with a minimum thickness of 60 cm) that completely surrounds the inner detector. This is similar to a skin layer in the LXe detectors, but much bigger one. The total amount of LAr in this shield is about 5600 L. Readout of this volume is performed by PMTs positioned on the walls of this inner cryostat vessel. A significant contribution of the WArP collaboration to the progress of liquid argon dark matter detectors was the discovery together with scientists from the US of the possibility to obtain argon from the underground sources containing significantly less ^{39}Ar (Acosta-Kane *et al.*, 2008).

A similar situation took place with another WIMP search project, Argon dark matter (ArDM). Initially, the conceptual design was different from that of other detectors only in the part of charge readout (see, for example, Marchionni *et al.*, 2011). The thick gas electron multiplier (THGEM or LEM) situated above the liquid argon surface was considered as a basic option for charge collection and amplification. Later, because of technical difficulties in maintaining stable operation of this system, the collaboration returned back to a traditional readout scheme by PMTs. The ArDM was commissioned in a single phase operation mode at the Canfranc underground laboratory in 2015 (Calvo *et al.*, 2017) and has reached two-phase operation in 2017 (http://darkmatter.ethz.ch). However, by that time the LAr dark matter community switched to the DarkSide experiment. Now the ArDM group joined the DarkSide-20k project (see below).

5.3.2. *DarkSide program*

The work on this project started from building of the DarkSide-10 prototype (Alexander *et al.*, 2013a). The prototype was tested at the Gran Sasso underground laboratory. The 10-liter chamber was viewed by two arrays of visible light Hamamatsu R11065 3″ PMTs with 7 tubes in each one. The TPB wavelength shifter was used to detect the VUV light with these phototubes. TPB was deposited by vacuum evaporation onto the reflector lining the acrylic cylinder and the inner surfaces of the fused silica windows located in front of the PMT entrance windows. Thus, practically all surfaces of the TPC faced to the LAr were coated with TPB. A ~1 mm layer of LAr optically coupled these windows with the PMTs windows (for both top and bottom arrays). A conductive optically transparent indium tin oxide (ITO) was deposited on the inner surfaces of the top and the bottom windows. All these measures resulted in the significantly high light collection efficiency of ~9 phe/keVee at zero field. However, none was reported about two-phase operation of this prototype in this publication.

After successful prototyping, the DarkSide-50 has been designed, built and was successfully operated at the Gran Sasso underground laboratory (Agnes *et al.*, 2015). For the moment of writing this book the detector is still in place. The LAr TPC (Fig. 5.15) has an active volume (~35 in diameter and ~35 cm in height) formed by the cylindrical PTFE wall, the cathode, and the extraction grid. This gives an active mass of ~46 kg of liquid argon. The active volume is viewed from the top and from the bottom by two arrays of Hamamatsu R11065 3-inch low-background PMTs (with 19 PMTs in each one). The cylindrical wall is a 2.54-cm-thick PTFE reflector. Similarly as it was done in DarkSide-10, there are two fused silica

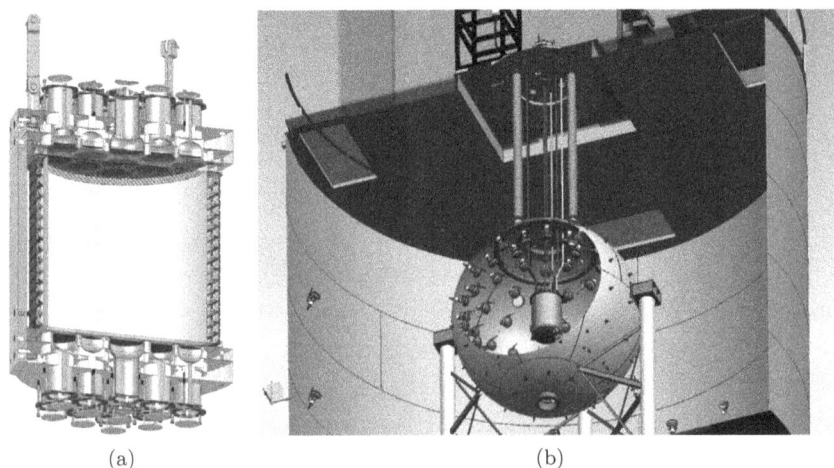

(a)   (b)

Figure 5.15:   Schematic drawing of DarkSide-50 detector: sectional view of the TPC (a); the detector installed inside the sphere with liquid scintillator and suspended inside the water shielding tank (b). Redrawn from Agnes *et al.* (2015).

windows in front of the PMT photocathodes coated with optically transparent conductive 15-nm-thick ITO layers from both sides. These layers play the role of (from top to bottom): a screening electrode of the top PMT array, an anode, a cathode, and a screening electrode of the bottom PMT array. The use of conductive optically transparent electrodes is the basic difference of the DArkSide-50 TPC from the TPCs of other two-phase detectors considered above. Only the extraction grid submersed 5 mm below the LAr surface is a "traditional" hexagonal mesh etched from a 50-μm-thick stainless steel foil and has an optical transparency of 95%. The top window, with the anode on the bottom surface, has a cylindrical rim extended downward to 10 mm to form the "diving bell" which maintains the necessary level of liquid argon and defines the gas gap size. The PTFE reflector and the top and bottom fused silica windows are coated with a TPB wavelength shifter from the inside of the active volume. There are field shaping copper rings outside the PTFE reflector to produce the uniform electric field in the active volume. The anode is kept at the ground potential while the cathode is biased negative: the grid and the cathode potentials are -5.6 and -12.7 kV, correspondingly. The closest to PMTs ITO layers on the fused silica windows are kept at the potential equal to that of the PMT photocathodes.

Calibration of the LAr response to electron recoils was obtained from the fit of the entire spectrum from ^{83m}Kr and ^{39}Ar. At the operation electric field strength of $\sim$200 kV/cm, the light collection efficiency was obtained to be 7.0 ± 0.3 phe/keVee, and 7.9 ± 0.4 phe/keVee at the zero field. Calibration of the LAr response to nuclear recoils in S1 and S2 channels was performed in a separate experiment called SCENE (Alexander *et al.*, 2013b; Cao *et al.*, 2015) for the recoil energies from 10.3 to 57.3 keV.

Differently from other two-phase WIMP search detectors, in DarkSide-50, the electroluminescent signal is used only for the event of precise coordinate measurements in the horizontal plane and for rejection of the multiple events caused by neutrons. Discrimination of the electron background events is performed by the PSD analysis. A fraction of scintillation signal area within the 1-st 90 ns from the beginning of the scintillation pulse (called f_{90} parameter) is used for event identification. This parameter is significantly different for electron and nuclear recoils.

The cryostat with TPC is suspended in the middle of the Liquid Scintillator Veto (LSV) detector which is a 4.0-m-diameter stainless steel sphere (Fig. 5.15) equipped by PMTs and filled with 30 t of borated liquid scintillator which is very sensitive to neutrons due to the neutron-capture ^{10}B$(n, \alpha)^7$Li reaction. The LSV detector is placed in the middle of the 11-m-diameter, 10-m-high cylindrical tank filled with high purity water (water Cherenkov detector, WCD). This tank was originally part of the Borexino Counting Test Facility.

The first WIMP search run (November 2013 to May 2014) was performed with atmospheric argon filling with an exposure of ~1400 kg-days. The obtained 90% CL upper limit from this run on the WIMP-nucleon spin-independent cross-section is $6.1 \cdot 10^{-8}$ pb for a WIMP mass of 100 GeV/c^2. As expected, the main contribution to the electron recoil background came from the beta-decay of ^{39}Ar.

The next, technical run, was performed for the first time with the underground argon. The exposure of (2616 ± 43) kg-days of data was accumulated during 70.9 live days (Agnes *et al.*, 2016). The background from ^{39}Ar had been reduced by three orders of magnitude. The upper limit obtained in the analysis of the results of this run was 2.0×10^{-8} pb, i.e., the previous limit was lowered down three-fold. The analysis was non-blind in order to keep the same set of analysis parameters as for the run with atmospheric argon.

Finally, the 532.4-live-days run with the underground argon was performed from August 2015 to October 2017 with 16,660 $\pm$ 270 kg-day exposure (Agnes *et al.*, 2018b). The 90% confidence upper on spin-independent cross-section was set at $1.14 \cdot 10^{-8}$ pb for a WIMP mass of 100 GeV/c^2 after analysis of this dataset.

In 2018, the collaboration DarkSide has started the wide R&D program aimed at building a 20-ton liquid argon two-phase detector (Aalseth *et al.*, 2018). The decision was taken to skip the stage of a ton-scale detector and to begin developing a multitone one in order to be competitive with the LXe experiments. However, a ton-scale detector DarkSide-proto is being built for prototyping of various detector systems. The DarkSide-20k program is now a joint effort of the leading dark matter projects which use liquid argon detectors (single-phase or two-phase) such as ArDM, DarkSide-50, DEAP-3600, and MiniCLEAN). The ArDM setup, for example, will be used for testing of the radioactivity of the ^{39}Ar depleted argon (Aalseth *et al.*, 2020b).

Although the final design of the detector is not elaborated yet, it will have the following basic characteristics. The total mass of liquid argon inside the cryostat is $\sim$35 ton, with 23 ton inside the TPC (active). The TPC has a shape of an octagon prism with a width of 290 cm and a height of 239 cm. Basically, this TPC is a scaled-up version of the DarkSide-50 one. The difference is that the optical windows with transparent ITO electrodes will be made of acrylic instead of fused silica, because large fused silica discs of such diameter are not available. Another very important difference is readout: SiPMs will be used instead of PMTs. They are combined to 5×5 cm^2 tiles with a total number of tiles of $\sim$5000. The HV potential of the cathode must be as low as $\sim$$-50$ kV to ensure the same as in the DarkSide-50 electric field (200 V/cm) in the active volume. As DarkSide-50, the DarkSide-20k is surrounded with a spherical neutron scintillation veto and placed in a 15-m-diameter water tank.

It is expected to reach the cross-section limit (90% C.L.) to the spin-independent interaction of $\sim$$10^{-12}$ pb for WIMP mass of $\sim$100 GeV/c^2 and 200 ton-year exposure, i.e., the DarkSide-50 limit will be improved by four orders of magnitude. In the longer term, the ultimate aim of the DarkSide Collaboration is to develop a liquid argon detector with 300-ton active (200-ton fiducial) mass called Argo.

In parallel to the R&D and design work with DarkSide-20k, there is an ongoing activity with mass production of the detection medium (^{39}Ar depleted argon) for this and the future experiment Argo. There are two projects for this purpose, Urania and Aria (see Aalseth *et al.*, 2020b and references therein). The Urania plant will extract and purify the underground argon from the CO_2 wells at the Kinder Morgan Doe Canyon Facility located in Cortez, Colorado state, USA, at a production rate of 100 kg/day with a depletion factor of $\sim 1.4 \cdot 10^4$. The aim of the Aria project is to provide further purification of the underground argon extracted by Urania. The Aria plant consists of two 350-m tall distillation columns with different diameters, Seruci-I and Seruci-II, capable of separating the argon isotopes by means of cryogenic distillation. The columns are installed in a 350-m deep shaft of the former Seruci coal mine in Sardinia island, Italy. The Seruci-II column will be able to produce 150 kg/day of depleted argon.

Chapter 6

Neutrino Detection with Two-Phase Emission Detectors

6.1. Introduction

Over 50 years of development the technology of two-phase emission detectors has gone the way from miniature detectors for methodological studies with a mass of working medium of several grams to detectors with a mass of working medium of hundreds tons for a study of the fundamental properties of neutrinos.

6.2. Two-phase Emission Detectors for CEνNS

Two-phase emission detectors are considered as very promising instruments for detection of neutrinos. They may play a major role in detection and study of neutrino interactions characterized by very small energy depositions in detection media. Even the smallest signal produced only by one ionization electron can be reliably detected. Due to this, the attractive possibility of the use of this technique is the detection of a coherent elastic neutrino-nucleus scattering (CEνNS). This process has been predicted in the framework of the Standard Model of particles and interactions more than 40 years ago (Freedman, 1974). It was pointed out that a neutrino interacts coherently via exchange of Z-boson with all nucleons in a nucleus. This takes place when the momentum transfer to a nucleus is significantly smaller than the nucleus radius (in dimensionless units).

The cross-section of this process is approximately given by the formula (Drukier and Stodolsky, 1984):

$$\sigma \approx 0.4 \cdot 10^{-44} N^2 \left(E_v\right)^2 \text{ cm}^2,$$

where N is the number of neutrons in a nucleus and E_ν is the neutrino energy (in MeV). The formula is valid for the neutrino energies of up to $\sim$50 MeV. This process has a very large cross-section in comparison to other known neutrino interactions due to the N^2 dependence. CEνNS plays a very important role in the dynamics of supernova bursts because $\sim$99% of their energy is released in a form of neutrino radiation. Thus, the precise experimental measurement of the cross-section of this process is very important for astrophysics. Study of the CEνNS process is considered not only as a confirmation of the Standard Model, but also as a probe for new physics beyond it. The CEνNS process is also considered as a potential instrument for nuclear reactor monitoring purposes.

There was no experimental evidence for this process until recently because of technical difficulties in its detection: the energy transferred from a MeV-energy neutrino to a recoil nucleus is of an order of hundreds of eV. To observe such low-energy interactions, a detector with a mass of more than several kilograms having a sub-keV energy threshold and working in a low-background environment is required which is rather challenging. Only in 2017, the process was observed by COHERENT international collaboration (Akimov *et al.*, 2017a) at an intense pion decay-at-rest source (with neutrino energies up to 50 MeV) of the SNS accelerator facility, Oak Ridge, USA.

The tiny ionization signals produced by neutrinos in the CEνNS interaction are in the keV and sub-keV energy range (for the reactor antineutrinos). Many experimental groups developed detectors for the WIMP search, usually proposed experiments with the similar technique for the detection of CEνNS. Although the detection technique is practically the same, there is a big difference in much smaller signals and in background conditions because all available artificial intense neutrino sources are located in places above ground or with a moderate shield from the cosmic rays. In this chapter, we mostly consider the application of two-phase emission technique for

CEνNS detection, and as well, the use of this technique for detection of other neutrino interactions.

6.2.1. *Liquid argon emission detector*

Despite the fact that in the early 2000s the experimental dark matter search groups were focused on the development of liquid xenon based detectors, the first proposal of an experiment to observe CEνNS was based on an argon two-phase emission detector (Hagmann and Bernstein, 2004). This is also surprising because the cross-section of neutrino interaction with the argon target is significantly smaller in comparison with the xenon target. Apparently, the reason for this was the confidence that it will be much easier to detect the nuclear recoils in liquid argon since they will have $\sim$3 times higher kinetic energy than the xenon recoils due to the lower mass of the argon atomic nucleus. A group from the Lawrence Livermore National Laboratory, USA carried out quite an extensive research work in this direction using a small test chamber (Sangiorgio *et al.*, 2012). The operation of a two-phase argon detector in the keV-energy region and at energies less than 1 keVee was demonstrated with this chamber filled by the liquid argon with a dissolved radioactive isotope ^{37}Ar (K and L electron capture, 2.82 and 0.27 keV, respectively). Another very important step was the first direct measurement of a specific ionization yield for argon nuclear recoils produced by neutrons in liquid argon and having a kinetic energy in the keV-energy range (6.7 keV) (Sangiorgio *et al.*, 2013). It was demonstrated experimentally that the produced ionization is quite significant, and its value varies from 3.6 to 6.3 electrons/keV for an electric field in the range from 0.24 to 2.13 kV/cm. This result demonstrated the possibility of CEνNS detection with the use of a liquid argon two-phase emission detector.

6.2.2. *Liquid xenon emission detectors*

Proposals of experiments on detection of the CEνNS process using two-phase xenon emission detectors appeared (Akimov *et al.*, 2009; Santos *et al.*, 2011) after the observation of the very small signals in such detectors which were interpreted as signals produced by single ionization electrons (Edwards *et al.*, 2008; Burenkov *et al.*, 2009). At that time, there was still no experimental data on a specific ionization yield (the number of ionization electrons per 1 keV of the deposited energy) for nuclear recoils in liquid xenon in the keV and sub-keV energy ranges. The ionization yield had been measured down to $\sim$4 keV of xenon recoil energy by that time. Nevertheless, its increasing trend with the energy decrease, which was observed in this energy region (see Section 5, and Shutt *et al.*, 2007; Sorensen *et al.*, 2009), was very encouraging (see explanation of this effect in Section 2.4.3). The LUX dark matter detector in situ calibration (see above) has shown a continuous growth of the yield down to 0.7 keV energy of nuclear recoils in xenon. However, the recent measurement performed by the Livermore group in the energy range 0.3–0.6 keV (Lenardo *et al.*, 2019) has shown that the yield tends to decrease below 1 keV (decreasing from $\sim$7 e/keV at $\sim$1 $\div$ 2 keV to $\sim$3.5 e/keV at 0.3 keV). At present, the physics of the interaction of charged particles with liquid noble gases is fairly well understood, parameterized, and included (with the recent result as well) in the NEST package (Noble Element Simulation Technique; Szydagis *et al.*, 2011) for modeling of the scintillation and ionization response of the liquid xenon and argon for nuclear recoils and electrons (gamma rays) at various energies.

6.2.2.1. *ZEPLIN-III study of the possibility to detect CEνNS*

Among all dark matter search groups only the ZEPLIN-III collaboration had carried out an extensive study (Santos *et al.*, 2011) of the possibility to detect CEνNS by the ZEPLIN-III detector (after completion of the underground WIMP search run) or with a ZEPLIN-III-like detector. It was pointed out for the first time that although a two-phase emission detector is sensitive to the single ionization electrons, these single electrons, which have different origins and spontaneous appearance, a so-called single electron noise, may seriously obstruct the detection of CEνNS. A detailed study of this background was performed during two dedicated underground runs with the ZEPLIN-III detector. These runs differed one from each other by the background conditions: the total rate of background events was significantly reduced in the second run due to the replacement of the PMTs with the new less radioactive ones. It was observed that the background of spontaneous single electrons reduced proportionally to the reduction of the overall radioactive background count rate and that the distribution of the single electron events is practically uniform over the XY plane. The distribution of the areas (charge integrals) of these single electron signals demonstrated a very nice peak at ∼30 photoelectrons. The ZEPLIN-III collaboration presented calculations of the count rate of the CEνNS events in an experiment at a 3-GW nuclear reactor and at the ISIS spallation neutron source, and the events caused by single electron background signals and their accidental coincidence (overlapping). However, the count rate of the single electron noise signals was assumed in this estimation to be fairly low, only ∼10 Hz, equal to the rate during the first WIMP data taking run, before the replacement of the PMTs to the new low-radioactive model. As we will see below, the count rate of the spontaneous single electron events in an experiment carried out above ground is by several orders of magnitude higher due to the huge charge produced per unit of time by cosmic muons passing through a detector.

6.2.2.2. *RED-100 detector*

The Russian Emission Detector (RED) project started from the study (Akimov *et al.*, 2012b, 2016) of the single electron noise in an above ground conditions with a two-phase xenon test chamber called RED-1 which was originally designed as a reduced-size prototype for the ZEPLIN-III dark matter detector (see above). The active volume was a cylinder with 22-mm height and a diameter of 105 mm. It contained 0.6 kg of liquid xenon. The thickness of the gas gap between the anode and the liquid xenon surface was 5 mm. As in ZEPLIN-III, the anode was a flat polished electrode with a rim submerged to the liquid xenon in order to collect the electrons that failed to emit into the gas phase and, as a result, trapped under the interphase surface playing a role of the potential barrier. The liquid working volume was viewed from the bottom by seven $1''$ MgF_2-windowed PMTs.

The investigation of the possibility to observe the CEνNS effect started from the study of the noise caused by the spontaneous appearance of electrons in the electroluminescence region (Akimov et al., 2012b). It was experimentally demonstrated that after passing a muon through the detector volume, the cloud of electrons which failed emission to the gas phase starts to move under the liquid xenon surface towards the detector edge. The direction of this moving coincides with the direction of the detector inclination, i.e., the cloud moves along the tangential component of the electric field caused by this inclination. The point of the spontaneous electron emission from the surface was observed to move with this cloud (see details in Akimov et al., 2016). It was observed also that the count rate of spontaneous single electron events becomes higher when the planes of the anode and the liquid xenon surface are parallel (Akimov et al., 2012b). A trivial explanation of this phenomenon is the longer stay of the electrons under the surface before arriving at the edge in this case. On the basis of this study, it was concluded that the main contribution to this noise comes from the electrons failed emission to the gas phase.

The RED-100 detector design and construction works started in 2012 (Akimov et al., 2013a). Monte Carlo simulations have shown that the detection of CEνNS is feasible at a nuclear power plant with a LUX- or XENON100-like detector placed in $\sim$20-m vicinity from a nuclear reactor core. It was shown that muon induced neutrons which is the main source of background in the detector is significantly suppressed due to the reduction of the muon flux by concrete blocks of a power station building and a spent fuel storage pool (specifically for Kalinin nuclear power plant — KNPP, with a total overburden thickness of $\sim$50 m.w.e.) above the detector. Additional neutron flux reduction is performed by a passive shield surrounding the detector. However, the value of single electron noise was underestimated. It was assumed to be only 100 Hz in a pessimistic case and that it can be neglected for the events having several ionization electrons. Later, the first laboratory tests with the RED-100 detector have shown that this background is nearly four orders of magnitude higher at a ground level laboratory. This resulted in significant changes in the detector electrode system construction (see below).

At the beginning, the RED-100 detector had been designed very similarly to the LUX and XENON100 WIMP search detectors. A detailed detector description can be found in (Akimov *et al.*, 2017b). A cutaway 3D view of the RED-100 detector is shown in Fig. 6.1(a). A schematic drawing of the detector TPC (latest version; see Akimov *et al.*, 2020) is shown in Fig. 6.1(b). A PTFE-made light collection dodecagon prism with a diameter of ~380-mm and a height of 415 mm is placed inside an electrode drift system. This prism is an electrically insulating support for the electrodes and field shaping rings. There are PTFE insulation inserts between the field shaping rings and the cryostat wall which also serves for xenon displacing. The TPC is viewed by two arrays of 19 Hamamatsu R11410-20 PMTs. The total mass of liquid xenon in an active part of TPC is ~160 kg (with a fiducial mass of ~100 kg). All electrode grids are hexagonal meshes etched from a 200-μm-thick stainless-steel foil with a 4-mm cell size. The following operation potentials are applied to the electrodes: +4.5 kV to the anode (A), −4.5 kV to the extraction grid (G2), and −12 kV to the cathode (C). The PMT screening top (T) and bottom (B) electrodes are grounded. The distance between the anode and the extraction grid is 19 mm. As one can see from the scheme in Fig. 6.1(b), the anode electrode has a rim submersed to the liquid xenon as it was done in the ZEPLIN-III and RED-1 detectors. For this reason, the layer of the liquid xenon above the extraction grid is sufficiently thick (10 cm), and the electroluminescence gap has the remaining 9 mm. During the laboratory test in 2019 the following main detector characteristics were obtained. The record purity among all two-phase detectors of such scale was achieved: the lifetime of free electrons before capture by electronegative impurities reached several milliseconds (at the operation field equal to 217 V/cm) by the end of one-month period. This was the result of a thorough preparation of the detection medium with the procedure described in Akimov *et al.* (2019). The extraction efficiency was obtained to be 0.54 ± 0.08 at 3 kV/cm

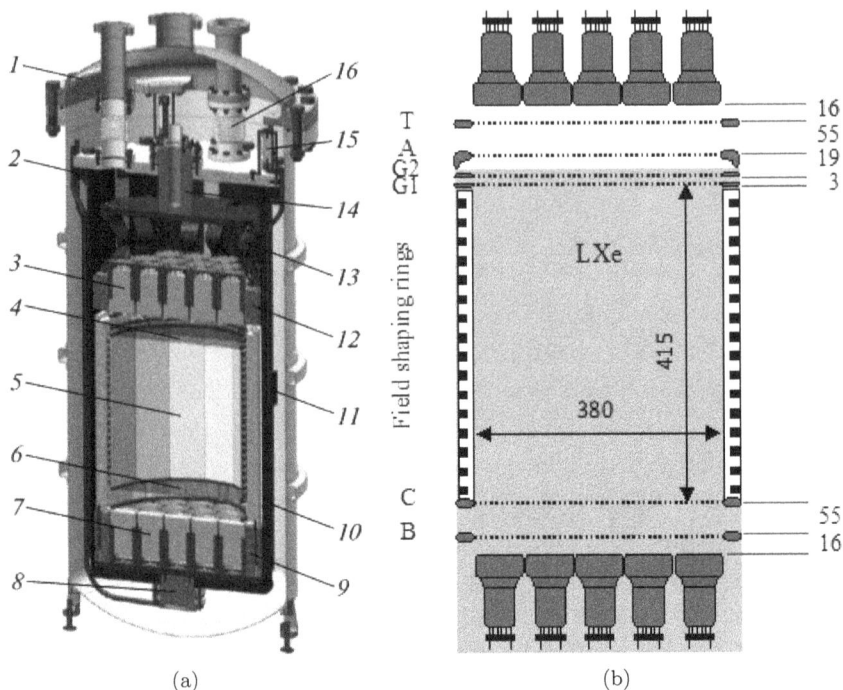

Figure 6.1: Schematic drawing of the RED-100 detector (a) and structure of the drift cage (b): 1 — external vessel of the cryostat, 2 — internal vessel of the cryostat, 3 — top array of 19 Hamamatsu R11410-20 photomultipliers, 4 — anode and gate electrodes, 5 — drift cage with Teflon reflecting walls, 6 — cathode, 7 — bottom array of 19 Hamamatsu R11410-20 photomultipliers, 8 — cold head of the bottom thermosyphon, 9 — copper housing of the bottom PMT array, 10 — Copper screen of the internal vessel of the cryostat, 11 — cold head of the side thermosyphon, 12 — copper housing of the top PMT array, 13 — flexible heat bridge, 14 — top cold head for xenon condensation, 15 — Vespel made stand supporting cold vessel inside the external vessel of the cryostat, 16 — connection for cable channel; T and B — top and bottom grounded grids, A — anode grid, G1 — electron shutter grid, G2 — extraction grid (gate), C — cathode grid; sizes of the drift volume and distances between grids are shown in mm. Redrawn from Akimov *et al.* (2017b, 2020).

extraction field (in liquid). This moderate value of the extraction field and efficiency resulted from quite a large distance between the anode and the extraction grid. Nevertheless, because of the large electroluminescent gap of 9 mm the distribution of single electron signals demonstrated a very nice peak at ~30 photoelectrons (at the detector central part).

However, the RED-100 detector has two principal differences from the dark matter detectors. The first one is that the PMT system operation is adjusted to the aboveground conditions defined mainly by the high cosmic ray muon background. The point is that the energy deposition from cosmic muons passing the detector is very high (~250 Mev in average). This causes the very intense S2 light signal, and special measures are required in order to prevent PMT photocathodes degradation with time. For this purpose, a 300-V positive pulse is applied to the cathode of each PMT in order to switch off the electric field between the PMT photocathodes and the first dynodes (Akimov *et al.*, 2014). The pulse generator is triggered from the muon scintillation signal. The second one is the use of a so-called electron shutter (Akimov *et al.*, 2018; see details in Akimov *et al.*, 2020). This is an additional grid (G1) placed 3 mm below the extraction grid (G2) and kept at the same potential as G1 (−4500 kV). A 300-V positive pulse (the same as the pulse going to the PMT system) is applied to G1 through a capacitor. The shutter becomes closed for the period of time equal to the pulse duration (1 ms or more). This prevents passing through it the electrons from the muon track and substantially reduces by this way the total charge trapped under the surface. The shutter was introduced to the electrode system after the first laboratory test with RED-100 had demonstrated an enormous count rate of spontaneous single electron events of an order of several hundred Hz. The use of the electron shutter has reduced the rate of such events, but not radically, by a factor of ~3. It turned out that not only the electrons trapped under the liquid xenon surface contribute to the single electron noise, but there is also another component of the unknown origin. This component has a characteristic decay time much longer than of the first one, of an order of milliseconds. This long component was also observed after the events with the large deposited energy caused by alpha-particles in the test with a two-phase chamber described in (Sorensen and Kadmin, 2018). The presence of this long component after passing of cosmic muons was also observed in the tests with the RED-1 chamber, however, no explanation was given (Akimov *et al.*, 2016). As discussed in Sorensen and Kadmin (2018), this component might be explained by trapping ionization electrons by the electronegative impurities with a subsequent release of them. It was pointed out on the possible relation of the intensity of the second component to the liquid xenon purity, as the rate decreased with time while the LXe purity increased. The same trend was observed in the latest laboratory test with RED-100 (Akimov *et al.*, 2020), however the variation was not in inverse proportional dependence. Similar qualitative anticorrelation of the single electron event rate with purity was observed in the DarkSide50 experiment (Pagani, 2017). And finally, this anticorrelation is confirmed in the very recently appeared publication on the electron background noise study in the LUX detector (Akerib *et al.*, 2020b). Thus, further R&D studies are required to understand and mitigate this component of single electron noise.

Nevertheless, as estimated in (Akimov *et al.* 2020), the experiment on detection of the CEνNS process is possible even with this significant level of single electron noise. The point is that at KNPP, where this experiment is planned, the muon flux is reduced by a factor of ∼5 by the ∼50 m.w.e. of overburden. It is expected that the single electron noise will be reduced proportionally. This level is still high, but selection of single-point multielectron events must suppress the rate of accidental coincidences of spontaneous single events significantly. This can be done using different light distribution patterns over the top PMT array for these two different types of events. Estimations made in (Akimov *et al.* 2020) have shown that setting a threshold of ∼4 electrons and applying a point-like events selection criterion allow one to suppress the level of accidental coincidence events down to the level of the CEνNS event rate.

The laboratory tests with the RED-100 two-phase detector have shown that simple application to the CEνNS detection of the technology developed for the very similar task of detection nuclear recoils in WIMP search experiments is not trivial. Much more difficult background conditions caused mainly by the presence of cosmic muons require special measures with changes in a detector construction and further studies of the properties of a two-phase emission detector. Hopefully, these studies will result in knowledge that allows one to build in the future a perfect two-phase emission detector for successful studies of the CEνNS process.

6.3. DUNE Project

In this section, we will talk about the project of the giant liquid argon detector that uses two-phase emission technology, the Deep Underground Neutrino Experiment (DUNE) project for the study of high energy neutrino oscillations. In this experiment, four 12-kiloton liquid argon TPC modules will be deployed in the Sanford Underground Research Facility (SURF) in South Dakota, USA. The goal of DUNE is to search for the leptonic CP violation and to determine the ordering of the neutrino masses using a powerful neutrino beam produced at the Fermilab accelerator. Additional plans include setting new limits on the proton lifetime in a variety of possible decay channels and collection of the high statistics of neutrino events from atmospheric and astrophysical sources.

Currently, there are considering two options for the neutrino detector: single-phase detector and two-phase detector (Abi *et al.*, 2018, 2020). The DUNE project is considering four detector modules placed in two caverns at SURF by two modules in each cavern. At least two of them will be single phase ones. For such large-scale detectors, a single-phase technology was pioneered in the ICARUS project and now is better elaborated than a two-phase one. However, the latter one offers the advantage of additional gas amplification in a gas phase. This amplification is planned to be done by a THGEM technology (or LEM — Large Electron Multiplier). The final decision on the use of a two-phase technology for the detector modules will be taken later, relying on the results of the prototyping studies. If accepted, the detector module will be built as a single active volume with a 60-m length, a 12-m width, and a 12-m height, with an anode at the top, a cathode near the bottom, and a PMT array

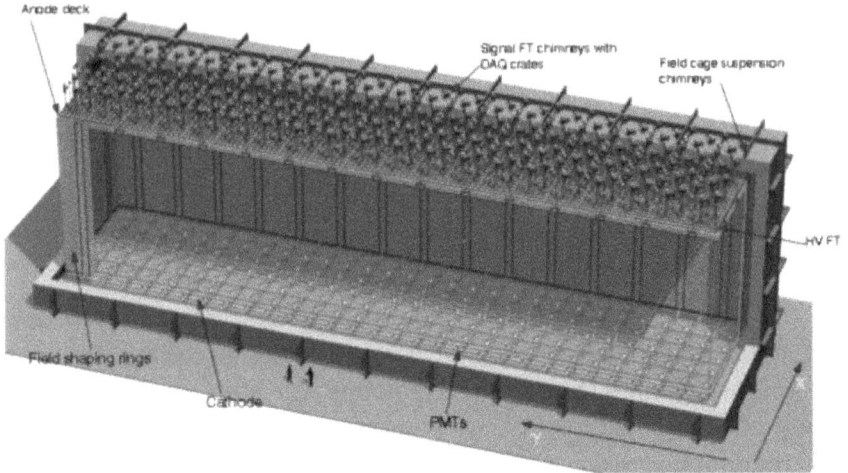

Figure 6.2:   Cutaway view of the DUNE two-phase argon neutrino detector module. The detector scale can be estimated from the figures of two people standing nearby the detector module. Redrawn from Abi *et al.* (2020).

located underneath the cathode. The active volume (see Fig. 6.2) is surrounded by field shaping electrodes. A fiducial volume of the detector module is 10.643 kton. The TPC will be placed in a cryostat with sizes of 62.0 m (length) × 14.0 m (width) × 14.1 m (height). It is interesting that for such large-scale cryostats, thermo insulation is performed with the use of low thermal conductivity material between the cryostat walls (without vacuum).

A two-phase detector prototype ProtoDUNE-DP (Dual-Phase) consisting of a 6 × 6 × 6 m³ active volume LAr TPC (1/24th or 1/24-th part of the planned two-phase detector module) is now under construction at CERN. A pilot version, a 4-ton demonstrator, having sizes of 3 m (length) × 1 m (width) × 1 m (height) is currently at a phase of beam tests at CERN (see Fig. 6.3; Cuesta, 2019; Aimard *et al.*, 2018). The first tests reported the excellent purity of the LAr medium (electron lifetime of ~4 ms). The most critical part of the detector is signal readout. It is performed by charge

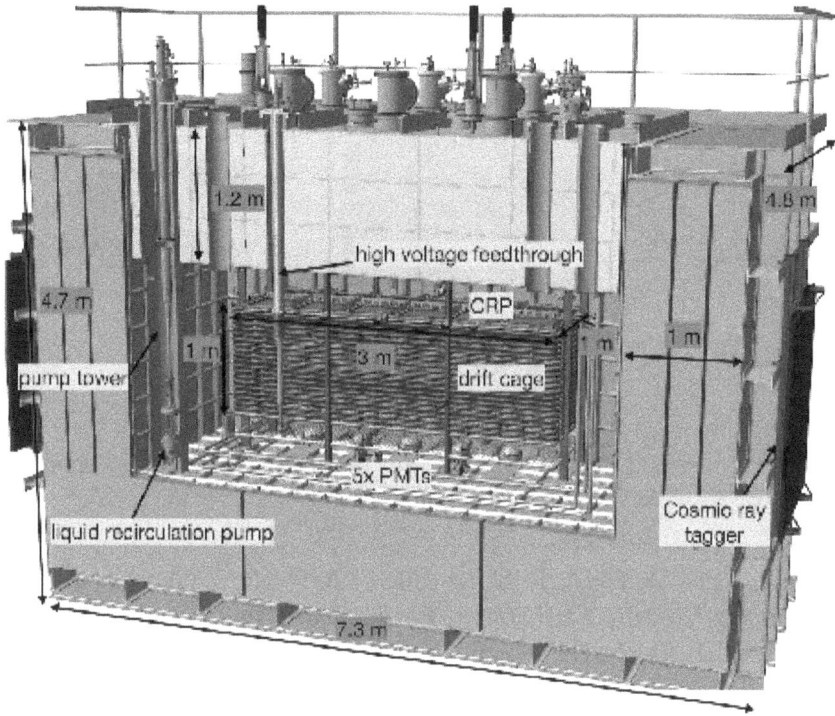

Figure 6.3: Cutaway view of the $3 \times 1 \times 1$ m^3 two-phase emission liquid argon TPC in a "passive" cryostat. Redrawn from Cuesta (2019).

readout plane (CRP) independent units each of 0.5×0.5 m^2. A schematic layout of the CRP unit elements is shown in Fig. 6.4. Each unit is individually leveled vertically and horizontally. The extraction grid is made of 100-μm diameter 3-m long stainless-steel wires. The wires are soldered to printed circuit board lamellas in groups of 32 on a pair of independent tensing pads fixed on a mechanical holder. The wire pitch is 3.125 mm that matches the pitch of the anode readout strips in order to provide a uniform extraction field. The anode is grounded in order to simplify signal readout, and other electrodes are biased negatively as shown in Fig. 6.4. The anodes and THGEMs

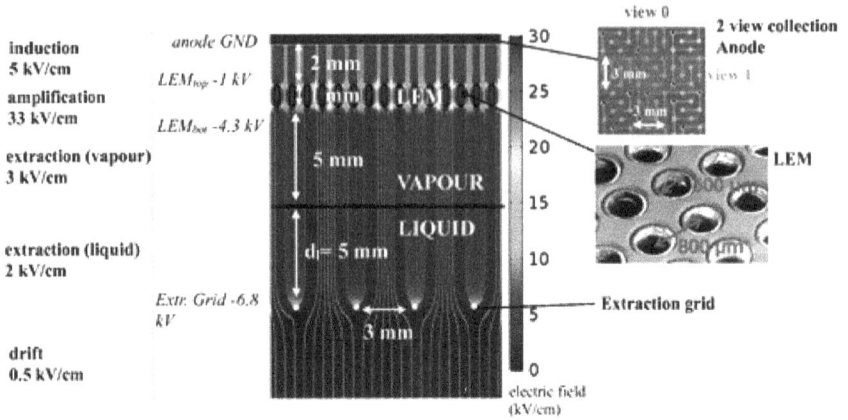

Figure 6.4: Schematic drawing of CRP at the top of the LAr TPC in ProtoDUNE-DP detector. Redrawn from Cuesta (2019).

are combined in single units, so-called sandwiches. The scintillation light is detected by TPB coated PMTs placed beneath the cathode.

If built, the DUNE two-phase module will be the biggest two-phase emission detector ever constructed.

Chapter 7

Imaging Two-Phase Emission Detectors

7.1. Introduction

The last decade has been marked by a rapid development of the emission detector technology for detecting rare processes involving a very few particles. Meanwhile, in the first two decades of the development of emission detectors, the great potential of using emission technology to visualize topologically complicated events as well as 2D distributions of radiation fields for applied tasks such as nuclear medicine was clearly demonstrated. This much promising area of the technology application is still waiting for further development. In this chapter, we briefly recall the main achievements in the direction that has been covered in more details in the previous monographs (Barabash and Bolozdynya, 1993; Aprile *et al.*, 2006a; Bolozdynya, 2010).

7.2. Spark Emission Chamber

A new method of controllable recording of traces of ionizing particles in condensed matter, using the effect of electrostatic emission of electrons from condensed noble gases into equilibrium gas phases, was first time clearly formulated in the article (Dolgoshein *et al.*, 1970). This work describes the operation of a two- and three-electrode plane parallel ionization chamber partially filled with liquid argon and an anode located in the gas phase (Fig. 7.1). The anode and the intermediate electrode in the three-electrode version were made of wires with a diameter of 50 μm with a pitch of 0.6 mm.

A flat alpha source and a corona discharge from a tungsten tip mounted on a cathode in liquid argon were used as ionization sources. Electrons in the gas phase were recorded either by means of a spark discharge at the anode wires or by registering electroluminescence using photoelectron multiplier during the drift through the gas of electrons extracted by an electric field from the liquid. In the case of using a photomultiplier, a scintillation flash was also observed, and the time of electron drift through liquid argon was measured by the delay between scintillation flash and the onset of electroluminescence. The gas phase consisted of a mixture of 50%Ar + 50%Ne. Electrons arising from the ionization of liquid argon by alpha-particles were extracted into the gas phase by an electric field of 3–7 kV/cm strength. The probability of electron emission from liquid argon in these fields was observed close to 100%.

To register electrons using a spark discharge, high-voltage pulses with amplitude of 40 kV and duration of 100 ns were applied to the anode. If at the moment of voltage pulse supply near the wire there were electrons drawn by a constant electric field from the liquid, then a spark discharge developed near this wire. An optically transparent window was installed above the anode and spark discharges near the wires were recorded using a photo camera. By superimposing images of individual sparks, a 2D image of the distribution of the alpha-active isotope at the detector cathode was recorded. Thus, the distribution density of the radioactive isotope emitting alpha particles became visible (Fig. 7.1).

In the development of this technique, attempts were made to obtain a gas gain at the anode wires. However, the stable gain has been found with a value not more than 500. To create conditions for more stable gas amplification, the wired anode was immersed in liquid argon and the anode wires were heated by 0.1–1A electric current. A gas shell around heated wires in the liquid limited the development of avalanches by the size of bubbles of the boiling liquid. In this mode, it was possible to obtain gas amplification with a multiplication factor up to 10^4. However, at the same time it was found that the dead time has increased significantly (10 ms vs. 0.1 ms in the gas), which was probably associated with

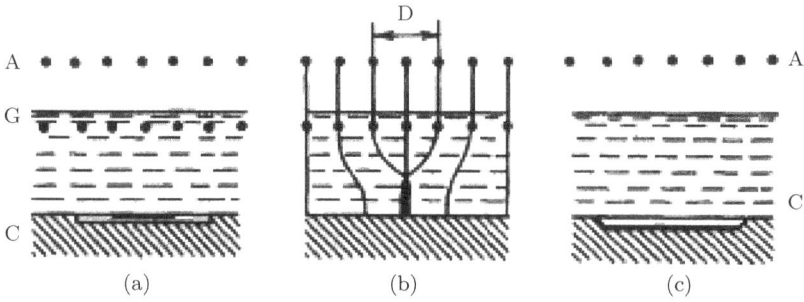

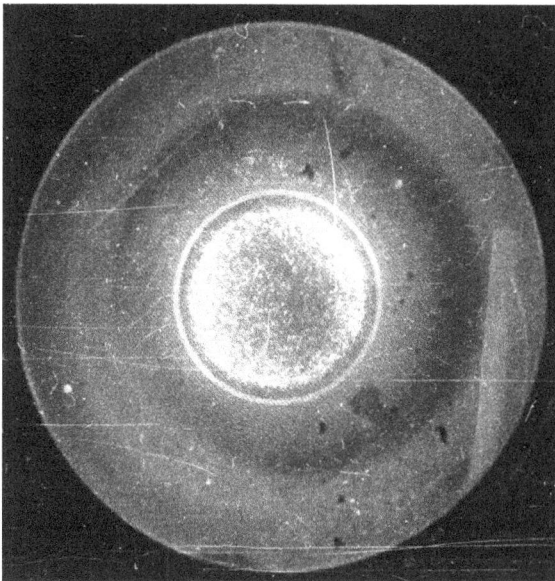

Figure 7.1: The first spark emission chamber schematic drawing (top) and a spark alpha-source image recorded (bottom): three (a, b) and two (c) electrodes configurations with alpha-source (a, c) and tungsten tip (b) installed on the cathode; A — anode; C — cathode; B — grid; AC distance is 1 cm; thickness of the liquid layer is 4 mm. The top picture is redrawn from Dolgoshein *et al.* (1970). The bottom picture is courtesy of B. U. Rodionov.

localization of positive ions inside bubbles. Using pulsed high voltage, the gas amplification was raised up to 10^6 (Dolgoshein *et al.*, 1973). Nevertheless, this technique has not found further practical applications, because it requires huge energy consumption for heating the anode wires and boiling liquid argon around them.

7.3. Emission Streamer Chamber

Initially, emission detectors were created as devices for visualizing and picturing tracks of elementary particles of high energies. The first emission streamer chamber was built in the late 1970s by a group headed by Boris U. Rodionov at the Moscow Engineering Physics Institute (Bolozdynya *et al.*, 1977a). A solid krypton of 5 mm thick at a temperature of 78 K was used as a working medium in this detector with a field of view 12.5 cm in diameter and a gas gap of 1 cm wide that was filled with neon at a pressure of 1 bar. Electrons generated in solid krypton by relativistic particles have been extracted in the gas phase by a constant electric field of 1.5 kV/cm strength. Particles that passed through the solid krypton were selected from a beam of relativistic particles using a telescope of scintillation counters. The counters generated a signal that triggered the Arkadyev–Marx high-voltage pulse generator to supply high-voltage pulses with an amplitude of 100 kV and a duration of 60 ns to the anode. The gridded anode was suspended in the center of the optical window above the krypton layer covering the bottom of the camera. Tracks formed by streamers have been pictured by photo camera through the wired anode (Fig. 7.2).

The detector was tested at the secondary beam of the ITEP proton accelerator for visualization of pion tracks with a pulse of 3 GeV/c passing through solid krypton. The pictured tracks were looking as chains of streamers of ~0.5 mm in diameter and 2 mm in length with a density of streamers along tracks to be about ~1 mm^{-1}. Thus it has been demonstrated an improved spatial resolution of this device compared to gas-filled streamer cameras. The study of the properties of the emission streamer chamber confirmed the observation made for the first time when studying the properties of a cryogenic streamer gas chamber (Gorodkov *et al.*, 1974; Sidorov, 1975): despite the increased density of the cold working gas compared to gas at room temperature, the threshold field strength for creating streamers at the same pressure remained unchanged at 2 kV/cm. It has been hypothesized that this effect is associated with an increase in the content of dimer molecules of noble gases with decreasing

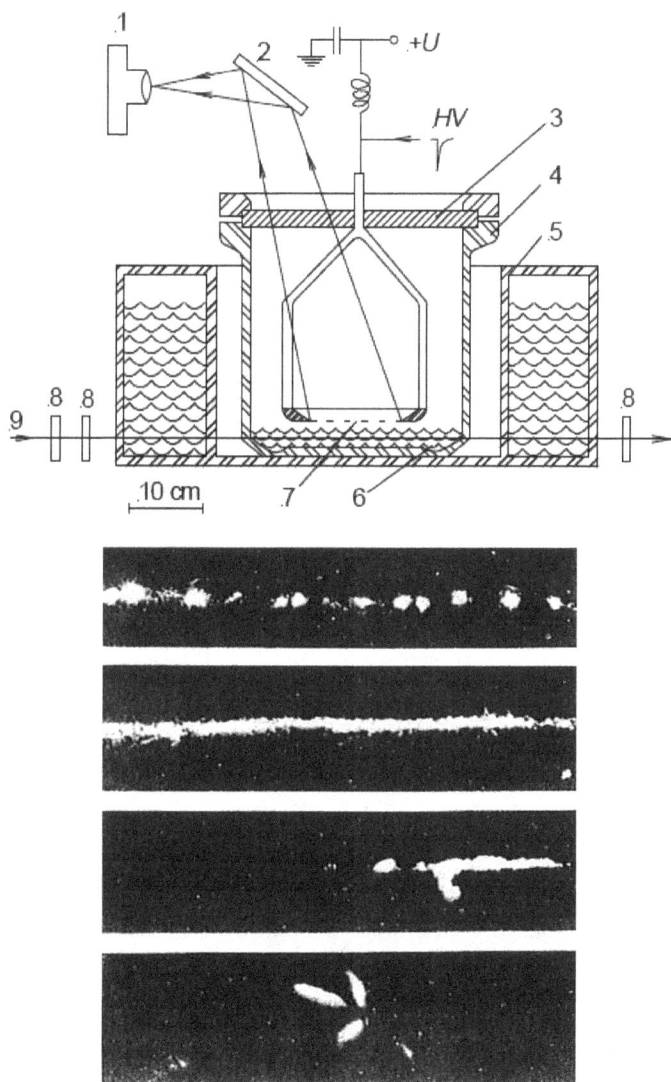

Figure 7.2: Emission streamer chamber (top) and detected images of relativistic particles in gas phase and solid krypton and delta-electron and products of interaction of relativistic particle with the bottom of the chamber (pictures from top to bottom): 1 — photo camera; 2 — mirror; 3 — optical window; 4 — stainless steel vessel; 5 — liquid nitrogen cryostat; 6 — solid krypton; 7 — gridded HV electrode with 12 cm wired section; 8 — telescope of scintillation detectors; 9 — track of relativistic particle. The top picture is redrawn from Bolozdynya *et al.* (1980); the bottom picture is borrowed from A. Bolozdynya' archive.

temperature. Since the ionization potential of dimers is reduced by 1–2 eV compared with atoms, the average ionization potential of the medium decreases in proportion to the increase in gas density, as a result of which the intensity of the visualizing field remains practically unchanged.

An analysis of the data revealed abnormal tracks with a very low streamer density along tracks: one streamer per 1–2 cm of track length. More detailed studies showed that this was a two-phase medium memory effect associated with the capture of part of the electrons under the liquid-gas interface. Anomalous tracks had been detected exactly in the same places where the tracks of relativistic particles with a normal ionization density have been registered recently. However, the idea of detecting particles with an anomalously low ionizing ability (down to single electrons at a few centimeters of the track length) using emission detector was first formulated at that time (Bolozdynya et al., 1980).

In the 1980s, the Nadezhda large emission chamber was built at ITEP, the working medium of which was to be liquid krypton with a diameter of 50 cm and a thickness of 20 cm, and the tracks should be visualized using a cryogenic streamer camera of 1.5 m diameter (Fig. 6.6, Bolozdynya, 2010). It was supposed to use this device to study the multiple productions of neutral pions during the annihilation of antiprotons in heavy nuclei. The ability to visualize tracks with a high ionization density by the streamer camera was experimentally tested using an ultraviolet laser. The emission section filled with liquid krypton was used to accurately measure the radioactivity of krypton and to study the feasibility of obtaining efficient electron emission from large mass liquid krypton (Anisimov et al., 1989a, 1989b). However, the chamber was not tested in the complete assembly due to problems with financing scientific projects in the USSR in the 1990s.

7.4. Electroluminescence Emission Chamber

An *electroluminescence emission camera* with array of photomultipliers (Fig. 7.3) has been developed for two-dimensions gamma ray

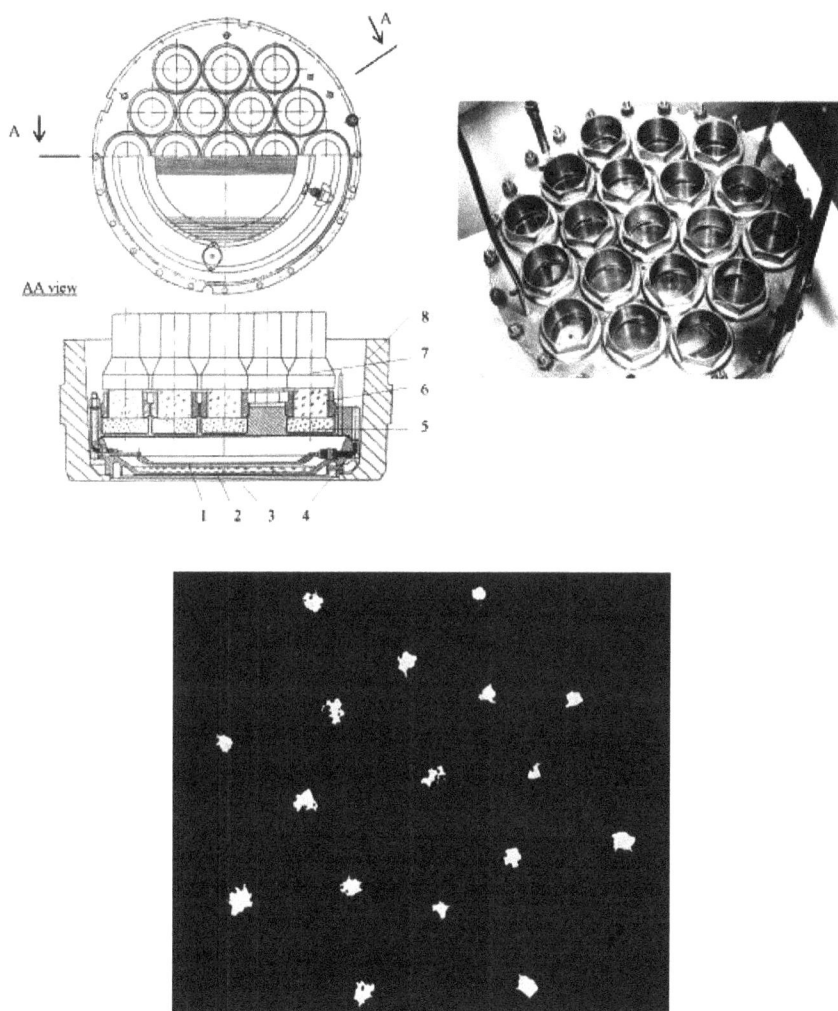

Figure 7.3: Electroluminescence emission gamma camera (top) and digital image of a lead mask over 22 cm field of view in ^{57}Co gamma rays: 1 — gridded anode; 2 — cathode as bottom of the camera; 3 — vacuum insulation on gamma-ray window; 4 — thermal insulation; 5 — glass window coated with wave-length shifter; 6 — acrylic light guide; 7 — 3″-diameter glass window PMT FEU-110; 8 — lead shielding. Redrawn from Egorov *et al.* (1983) (top) and borrowed from A. Bolozdynya' archive (bottom).

imaging in nuclear medicine at the beginning of 1980s (Bolozdynya *et al.*, 1981, 1985; Egorov *et al.*, 1983). The stainless steel vessel of the detector enclosed a 30-cm diameter anode with 24.5-cm diameter gridded central part. The flat grid consisted of Nichrome wires of 50 micron diameter parallel-stretched with 1 mm pitch. Nineteen 7-cm diameter glass windows coated with 0.5 mg/cm^2 p-terphenyl wave-length shifters have been installed in the lid of the vessel in hexagon order. Every window has been viewed with an individual glass photomultiplier of FEU-110 installed with an acrylic light guide. Xenon or krypton was purified passing through the hot (900 K) calcium absorber and chromo-silicate oxygen adsorbent based on silica gel activated with chromium salt. The gas has been stored in a stainless steel tank (80 cm diameter and 150 cm tall) inside the surface of which was sputtered with titanium getter generated by the head of titanium sublimation pump installed on the top flange of the tank. The detector was cooled down with nitrogen vapor circulating in the jacket surrounding the vessel. The gas was condensing in the layer of up to 1 cm thick at the thin bottom of the vessel serving as a cathode. The detector was tested with krypton and xenon working media.

Coordinates of electroluminescence flashes have been determined as the position of center-of-gravity of signals acquired from the PMT array operating in *Anger camera* mode (for details see, for example, Kalashnikov, 1985). Storage oscilloscope Tektronix 603 has been used to collect all data and to display images of an alpha-source placed on the cathode and a lead multi-hole collimator illuminated with gamma radiation of ^{241}Am or ^{57}Co gamma sources placed outside the detector; the images have been pictured with a photo camera. Internal position resolution was measured to be 2.5 mm FWHM measured for alpha-source installed on the cathode in solid krypton and 3.5 mm FWHM with a detector filled with liquid xenon and irradiated with 59.6 keV (^{241}Am) gamma source. A pulse height distribution of summarized over the PMT array analog signals has been analyzed with AI-256-6 pulse analyzer; the energy resolution of

15%FWHM has been measured with ^{57}Co gamma source at 2 kV/cm electric field strength in 4-mm thick solid xenon and 16%FWHM at 4 kV/cm electric field in 1.5-mm thick sample of liquid xenon (Bolozdynya *et al.*, 1982).

With this detector, electroluminescence of the liquid xenon has been observed at a uniform electric field at >10 kV/cm field strength. Note, that operations with krypton were essentially affected by high count rate $(\sim 10^5 \mathrm{s}^{-1})$ associated with radioactive decays of ^{85}Kr radionuclide presenting in natural mixture of krypton isotopes at the level of $\sim$10–12 relative concentration (Anisimov *et al.*, 1989b). Operations with solid working media were possible only for limited time ($\sim$10 min) because of polarization of the working media due to low mobility of positive ions and holes. De-polarization was achieved by the application of the reverse electric field of the same strength for about the same time as operation in the normal mode.

Work on the development of the emission gamma camera led to the idea of using the registration of two signals from one event: a scintillation signal (also called signal S1), which occurs at the moment the detected particle interacts with the condensed working medium of the detector, and the subsequent electroluminescent signal (or signal S2), arising during the drift through the gas phase of ionization electrons extracted by the electric field from the bulk condensed phase (Fig. 1.1). Using these two signals in recording quasi-point events allows us to determine the position of the interaction point in 3D space in order to create a "wall-free" detector for recording rare events such as interactions of neutrino and dark matter particles with baryonic matter (Bolozdynya *et al.*, 1995). This possibility of reconstructing a 3D picture of particle interactions in a dense noble gas using scintillation and electroluminescence signals was first demonstrated with compressed xenon (Bolozdynya *et al.*, 1997a). About that time, it was proposed to use a two-phase electroluminescent emission chamber with two signals from one event to search for rare decays such as neutrinoless double beta decay (Bolozdynya *et al.*, 1997b).

This idea has been very fruitful. Currently, the best results in the search for massive particles of hypothetical dark matter have been achieved using such a nominal technology (Chapter 5). The application of such detectors in neutrino physics also looks very promising (Chapter 6).

Chapter 8

Recent Developments in Two-Phase Emission Detector Techniques

8.1. Introduction

The technology of liquid noble-gas detectors, rapidly developing in the last three decades, was described in detail in a number of books (Fastovsky *et al.*, 1972; Barabash and Bolozdynya, 1993; Aprile *et al.*, 2006a). Among the possible working media for two-phase emission detectors, noble gases argon and xenon occupy a special place. This is due to the fact that in condensed argon and xenon the ionization yield of electrons from the tracks of ionizing particles is quite high. The excess electrons formed in this case have high mobility and can be effectively heated by a moderate electric field in order to be emitted into the gas phase with a probability close to 100%. In addition, condensed noble gases have relatively high scintillation and electroluminescent yields, which makes it possible to organize the three-dimensional spatial sensitivity of two-phase emission detectors to weakly ionizing particles.

Accordingly, in this chapter we will focus mostly on the recent developments in two-phase Ar and Xe detector techniques.

8.2. Cryogenic PMTs

All noble gases scintillate in the vacuum ultraviolet (VUV), in particular around 128 and 175 nm for Ar and Xe, respectively (see Fig. 4.4). Accordingly, the Xe light in two-phase Xe detectors can be directly detected by cryogenic photomultiplier tubes (PMTs) with a quartz window. In contrast, a MgF_2 window for PMT is required to directly detect scintillations in Ar. While quartz-window PMTs can be manufactured with relatively large diameters and in ultra-low background implementation, this is not the case for PMTs with MgF_2 windows. Therefore, for detection of Ar scintillation the common solution is to use a wavelength shifter (WLS) in front of the quartz-window PMTs. Tetraphenyl butadiene (TPB), which emits at around 420 nm (see Fig. 8.9), is a WLS choice for liquid Ar detectors (Aalseth *et al.*, 2018). The review of cryogenic PMTs by 2013 can be found elsewhere (Chepel and Araujo, 2013). In this section, the most recent developments of cryogenic PMTs relevant to the current dark matter search experiments based on two-phase Ar and Xe are presented.

A characteristic problem of the operation of PMTs at cryogenic temperatures is the increase in photocathode resistances at low temperatures. This effect is particularly important for bialkali photocathodes, which, having the highest quantum efficiency among visible-range photocathodes, have also the highest photocathode sheet resistance. This effect can lead to a decrease of PMT sensitivity due to saturation of the photocathode current under high photon flux: see Fig. 8.1 demonstrating the sensitivity reduction at high pulse rate for PMT with bialkali photocathode. Nevertheless, technical solutions were found that allowed PMTs with bialkali photocathodes to be used at cryogenic temperatures in rare-event experiments. These may include radial metal strips or semitransparent metal (e.g., Pt, see Fig. 8.1) film deposited under the photocathode.

Figure 8.2 and Table 8.1 present, respectively, the images and properties of the most advanced cryogenic PMTs with bialkali photocathodes, developed by Hamamatsu for two-phase detectors operated at LAr ($-186°C$) and LXe ($-110°C$) temperatures. Note that these PMTs are of moderate lateral size (not of large area), since they are supposed to form the PMT matrix, thus providing good position resolution. These PMTs are $1''$ square R8520, $2''$ round R8778 (with underlying metal strips), $2''$ round compact R6041 and $3''$ round R11410 and R11065. These were characterized in a number of works: R6041 in (Bondar *et al.*, 2015b; Bondar *et al.*, 2017), R11410 in (Akerib *et al.*, 2013a; Lyashenko *et al.*, 2014;

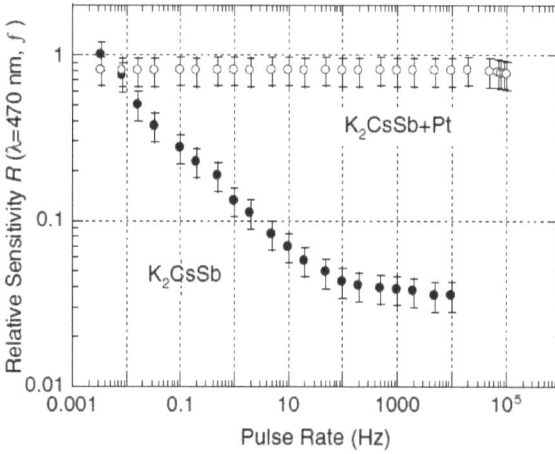

Figure 8.1: Relative PMT sensitivity as a function of the light pulse rate for bialkali K$_2$CsSb photocathode and for that with semitransparent Pt film underneath (K$_2$CsSb+Pt) at 77 K and 470 nm light source. Redrawn from Ankowski *et al.* (2006).

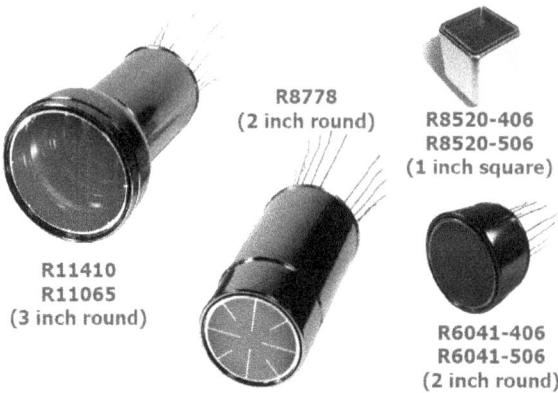

Figure 8.2: Moderate-size cryogenic PMTs with bialkali photocathodes of Hamamatsu production (Hotta, 2014): 3″ R11410 and R11065, 2″ R6041, 2″ R8778 and 1″ R8520.

Akimov *et al.*, 2015; Barrow *et al.*, 2016; Bondar *et al.*, 2017) and R11065 in (Acciarri *et al.*, 2012; Bondar *et al.*, 2015b; Bondar *et al.*, 2017). In particular, R8520 was used in XENON100 experiment (Aprile *et al.*, 2012b), while R11410 and R11065 are currently being

Table 8.1: Properties of cryogenic PMTs with bialkali photocathodes of Hamamatsu production (Hotta, 2014).

PMT type	R8520-406	R8520-506	R8778	R6041-406	R6041-506	R11410	R11065
Size/Shape	1″ square		2″ round	2″ round		3″ round	
Outer size	25.7 mm sq.		57 mm dia.	57 mm dia.		76 mm dia.	
Effective area	20.5 mm sq.		45 mm dia.	45 mm dia.		64 mm dia.	
Tube length	28.3 mm		111.5 mm	32.5 mm		123 mm	
QE	30% at 175 nm	25% at 420 nm	26% at 175 nm	30% at 175 nm	25% at 420 nm	26% at 175 nm	25% at 420 nm
Min. oper. temp.	163 K	87 K	163 K	163 K	87 K	163 K	87 K
Dynode structure	Metal channel		Box & line	Metal channel		Box & line	
Number of dynode stages	10		12	12		12	
Supply voltage	800 V		1500 V	800 V		1500 V	
Gain	1E+06		5E+06	1E+06		5E+06	
Rise time	1.8 ns		5 ns	2.3 ns		5.5 ns	
Pulse linearity at +/−2%	30 mA		13 mA	40 mA		20 mA	
Radioactivity	10 mBq/PMT		70–100 mBq/PMT	150 mBq/PMT		70–100 mBq/PMT	

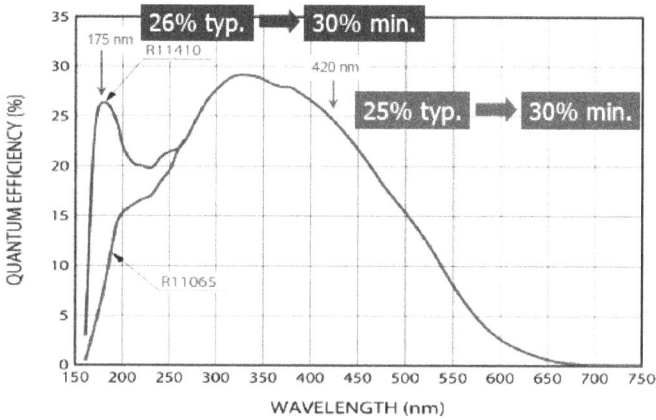

Figure 8.3:   Quantum efficiency spectra of 3″ cryogenic PMTs with advanced bialkali photocathodes of Hamamatsu production: R11410 vs. R11065. Redrawn from Hotta (2014).

used in Xenon1T (Aprile *et al.*, 2017) and Darkside-50 (Agnes *et al.*, 2015) experiments, respectively. The latter two have a special type of bialkali photocathode, characterized by a lower sheet resistance, making unnecessary the metal strips or platinum backing of the photocathode layer. Their quantum efficiency (QE) spectra are shown in Fig. 8.3. One can see that compared to R11065, R11410 has an enhanced sensitivity to Xe scintillation around 175 nm, while in the rest wavelength range their QE are the same, in particular amounting to 25% at emission peak of TPB (420 nm).

The QE of bialkali photocathode of the R11410-type PMTs was reported to increase at lower temperatures (Lyashenko *et al.*, 2014), the effect being different at different wavelengths. It was demonstrated that during the PMT cooldown from room temperature to 165 K the QE increased by a factor of 1.1–1.15 at 175 nm: see Fig. 8.4. The increase of the QE at low temperatures can be accounted for by the reduced photoelectron energy losses in the bulk photocathode material due to decrease of cross-section for optical phonon collision with the photoelectron. The fastest QE growth rate with respect to temperature was found at around 165 nm, while the slowest one was observed at around 200 nm. These results highlight the importance

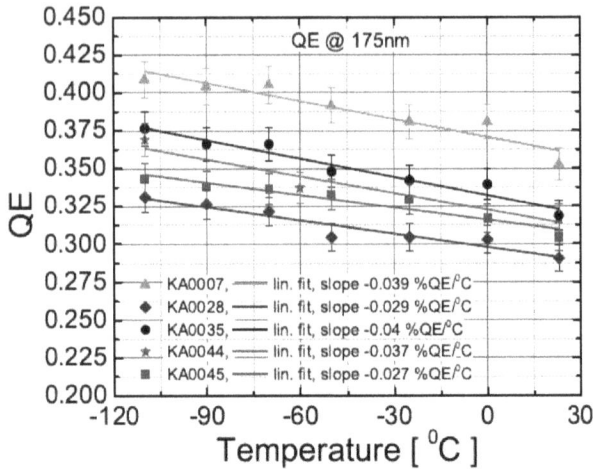

Figure 8.4:   QE as a function of temperature at a wavelength of 175 nm measured for the R11410 PMTs. Redrawn from Lyashenko *et al.* (2014).

of calibration of the photomultiplier tubes under the exact conditions in which they will be used in real experiments.

Radioactivity of the PMT components is a key concern in rare-event experiments. PMTs with glass envelopes should be avoided, since the rate of ^{40}K gamma rays can be quite significant, up to 10 Bq per device (Chepel and Araujo, 2013). Quartz has 10^4 lower content of ^{40}K than normal glass, and so PMTs with quartz windows are preferable from this point of view. A significant effort was made by manufacturers to reduce the background from PMTs by rigorous choice of the raw materials used for their components. As can be seen from Table 8.1, the rather low radioactivity per PMT was reached, namely 10 mBq for 1″ R8520 and 70–100 mBq for 3″ R11410 and R11065.

Good definition of the single photoelectron (PE) signal is also an important requirement for reliable detection of weak signals, implying a well-defined peak in amplitude spectrum separated from the noise. Figure 8.5 shows examples of such single photoelectron spectra for LAr PMTs at 87 K: those for R11065 (Acciarri *et al.*, 2012) and R6041-506MOD (Bondar *et al.*, 2017). One can see

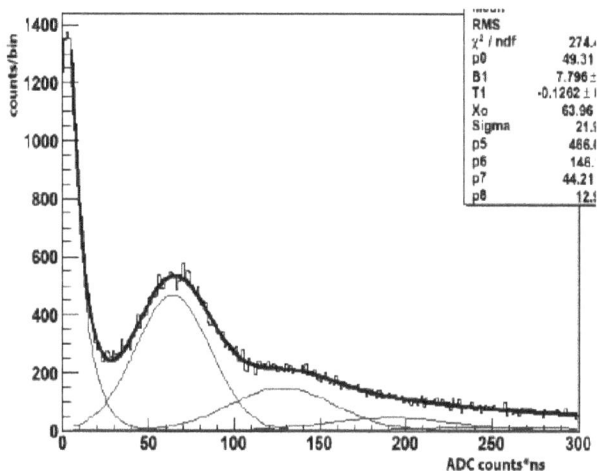

(a)

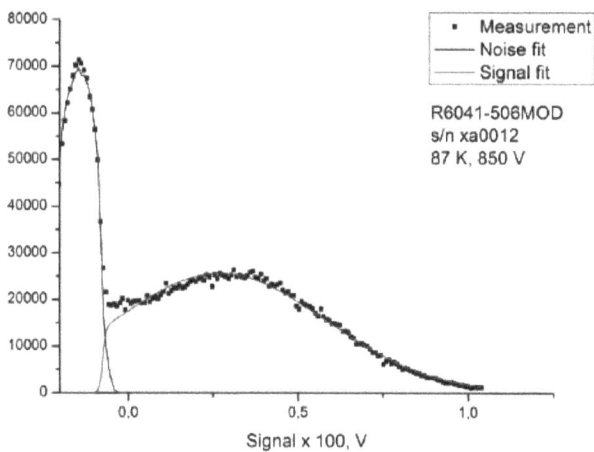

(b)

Figure 8.5: Typical amplitude spectra of single photoelectron PMT response for 3″ R11065 (a) and for 2″ R6041-506MOD (b) at 87 K. Redrawn from Acciarri *et al.* (2012) and Bondar *et al.* (2017), respectively.

that the dynode structure makes sense: the peak-to-valley ratio of the spectrum is better for box-and-line dynode structure used in elongated R11065 PMTs than that of metal-channel used in compact R6041 PMTs. This is the price one has to pay for compactness.

8.3. Cryogenic SiPMs

The current trend in two-phase Ar detectors is to replace PMTs with silicon photomultipliers (SiPMs), because the latter are more radio-pure, more compact, provide better position resolution and have better single-photoelectron spectrum (Aalseth *et al.*, 2018; Aalseth *et al.*, 2020a). The latter property is illustrated in Fig. 8.6 showing the photoelectron spectrum for 1-channel SiPM matrix of a sensitive area of 24 cm^2, i.e., larger than that of 2″ PMT (D'Incecco *et al.*, 2018). One can see superior separation of one-, two- and three-PE signals, compared to that of PMTs: compare Fig. 8.6 to Fig. 8.5. The review on SiPM performance at LAr temperatures by 2012 can be found elsewhere (Buzulutskov, 2012). In this section, the most recent developments of cryogenic SiPMs relevant to the current experiments based on two-phase Ar are presented.

There is an intense R&D of cryogenic VUV-sensitive SiPMs for direct detection of LXe scintillation around 175 nm, intended to be used in single-phase liquid Xe detectors in MEG-II (Baldini *et al.*, 2018) and nEXO (Gallina *et al.*, 2019) experiments. This however is out of scope of the present book; we refer the reader to appropriate references therein.

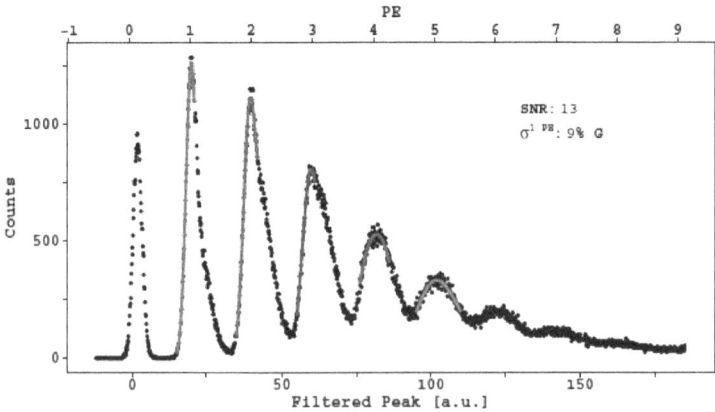

Figure 8.6:   Photoelectron amplitude spectrum at 77 K of a 1-channel 24 cm^2 SiPM matrix composed of 24 SiPMs of NUV-HD-LF (FBK) type of 1×1 cm^2 area each. Redrawn from D'Incecco *et al.* (2018).

Here we focus on the SiPM matrices developed for operation in two-phase Ar detectors and sensitive in the near UV, visible and NIR range, namely sensitive to either TPB emission around 420 nm or non-VUV scintillation of Ar (i.e., to that of neutral bremsstrahlung in the visible range or atomic scintillation in the NIR; see Chapter 4 for details).

It should be remarked that though the SiPM operation at LAr temperatures was repeatedly studied (Lightfoot *et al.*, 2007; Collazuol *et al.*, 2011; Bondar *et al.*, 2011b, 2015a; Aalseth *et al.*, 2017), the understanding of their performance at low temperatures is still incomplete. In particular, it was recognized only recently that the cell quenching resistor (R_q), made of doped polysilicon, and its manufacture parameters are fundamental in terms of the cell recharge time at cryogenic temperatures (Bondar *et al.*, 2014a; Aalseth *et al.*, 2018). It is desirable to have the quenching resistor lower and less dependent on temperature, with its value not exceeding 100 MΩ at 87 K. Otherwise, due to possible dramatic increase of the quenching resistor with the temperature decrease (Fig. 8.7(a)), the SiPM signal amplitude might be significantly reduced at even relatively weak photon fluxes: see Fig. 8.7 illustrating the wrong choice of the quenching resistor for CPTA-made SiPMs (Bondar *et al.*, 2014a).

Two examples of SiPM-matrices successfully developed and operated at LAr temperatures were presented in the literature: these are shown in Fig. 8.8. The first one is a 1-channel 5×5 cm^2 Photodetector Module for DarkSide-20k experiment, which includes a SiPM matrix composed of 24 mounted 12×8 mm^2 SiPMs of NUV-HD (FBK) type (Kochanek *et al.*, 2019). The second one is a 11×11 cm^2 SiPM matrix with 1 cm pitch, composed of 121 mounted 6×6 mm^2 SiPMs of S13360-6050PE (Hamamatsu) type. It was used in a two-phase Ar detector with direct SiPM-matrix readout and with indirect readout via combined THGEM/SiPM-matrix multiplier (Aalseth *et al.*, 2020a). Below, some characteristic properties of such SiPMs are described.

Figures 8.9 and 8.10 presents the spectral sensitivity of the SiPMs used in the matrices, namely the spectra of photon detection efficiency (PDE) which is a product of the quantum efficiency and

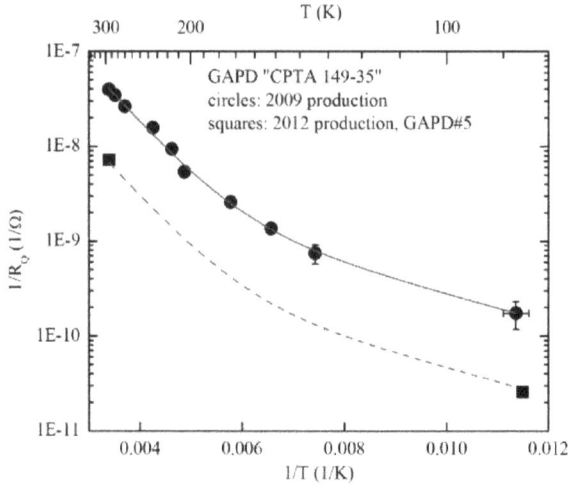

(a)

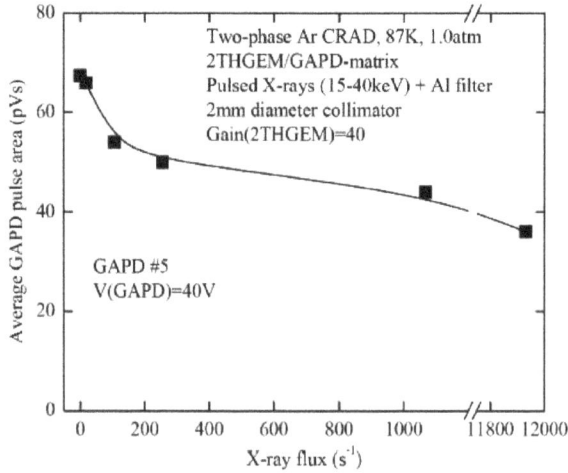

(b)

Figure 8.7: (a) SiPM single-cell quenching-resistor (R_Q) dependence on temperature: the reciprocal value $1/R_Q$ is shown as a function of the reciprocal temperature $1/T$ for two CPTA-made SiPMs production batches. (b) Average amplitude (pulse-area) of the signals of CPTA-made SiPM as a function of the incident X-ray photon flux in the two-phase Ar detector with THGEM/SiPM-matrix readout. One can see a significant reduction of the amplitude at even relatively weak photon fluxes. Redrawn from Bondar *et al.* (2014a).

(a)   (b)

Figure 8.8: (a) 1-channel 5×5 cm^2 Photo Detector Module for DarkSide-20k experiment, which includes a SiPM matrix (tile) composed of 24 mounted 12×8 mm^2 SiPMs of NUV-HD (FBK) type (Kochanek *et al.*, 2019). Right: 11×11 cm^2 SiPM matrix with 1 cm pitch, composed of 121 mounted 6×6 mm^2 SiPMs of S13360-6050PE (Hamamatsu) type. It had 25 central readout channels and was used in a two-phase Ar detector with direct SiPM-matrix readout and with indirect readout via combined THGEM/SiPM-matrix multiplier (Aalseth *et al.*, 2020a).

the probability to produce a discharge in the SiPM cell by the photoelectron. In Fig. 8.9, it is shown in comparison with the PMT quantum efficiency, WLS emission spectrum and transmission spectra of some materials. One can see that at 420 nm (WLS emission peak) the PDE of the SiPM of the S13360-6050PE (Hamamatsu) type exceeds 50%. The same is true for SiPMs of NUV-HD-SF (FBK) type. Figure 8.10 illustrates the fact that the SiPM PDE may substantially increase with the overvoltage: the enhancement factor can reach 1.5.

It should be noticed that the SiPM performance at cryogenic temperatures is superior to that at room temperature: the noise-rate is considerably reduced, while the amplitude resolution and the maximum gain is substantially increased. This is illustrated in Fig. 8.11 showing gain and noise rate characteristics of two types of SiPMs of Hamamatsu production: at LAr temperature, the noise

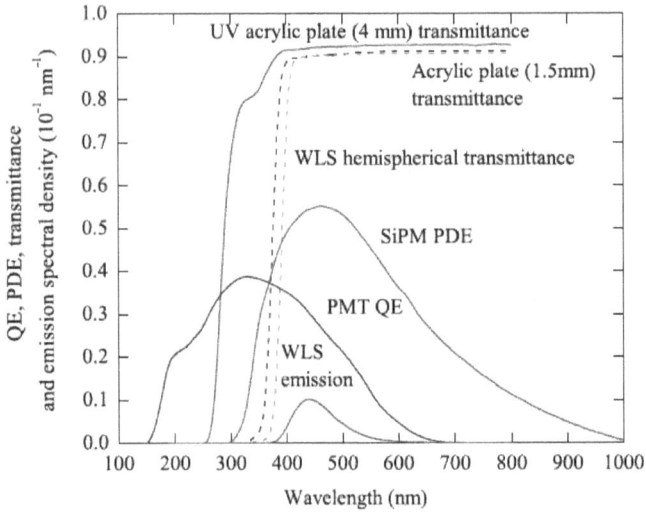

Figure 8.9:   QE of the PMT R11065 and R6041-506MOD at 87 K redrawn from Hotta (2014) and Lyashenko *et al.* (2014) using a temperature dependence derived there, PDE of the SiPM (MPPC 13360-6050PE (Hamamatsu, 2020)) at an overvoltage of 5.6 V obtained from (Otte *et al.*, 2017) using the PDE voltage dependence, transmittance of the ordinary and UV acrylic plate, measured in Buzulutskov *et al.* (2018), and hemispherical transmittance of the WLS (TPB in polystyrene) (Francini *et al.*, 2013). Also shown is the emission spectrum of the WLS (TPB in polystyrene). Redrawn from Gehman *et al.* (2013).

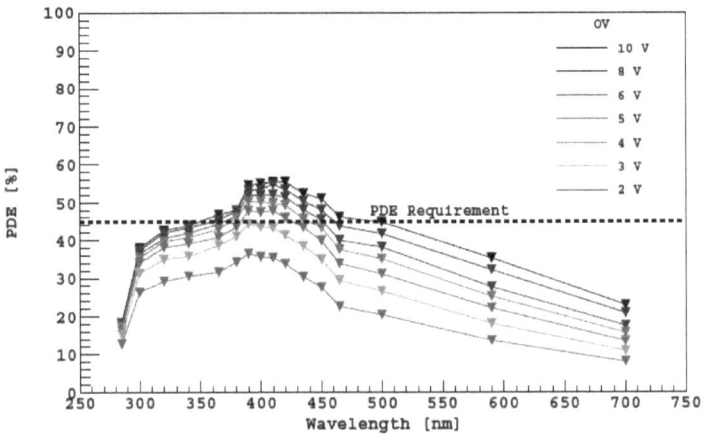

Figure 8.10:   PDE spectra for a SiPM of NUV-HD-SF (FBK) type at 300 K, at different overvoltages. Redrawn from Aalseth *et al.* (2018).

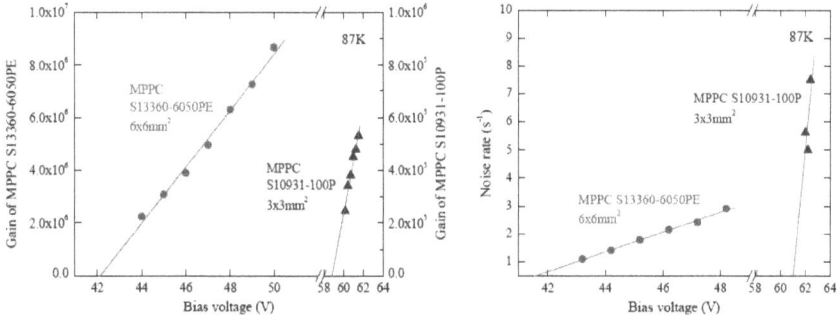

Figure 8.11: Gain and noise rate characteristics of two SiPM types of Hamamatsu production at 87 K: 3×3 mm^2 S10931-100P and 6×6 mm^2 S13360-6050PE. Redrawn from Aalseth *et al.* (2020a).

rate of 6×6 mm^2 SiPMs can be as low as a few Hz at rather high gains, reaching 5×10^6.

The other effects typical for SiPM performance at cryogenic temperatures are those of direct cross-talk, delayed cross-talk and after-pulsing. These effects are reflected in Fig. 8.12, showing the distribution of amplitude vs. delay time for noise signals of NUV-HD SiPM (Aalseth *et al.*, 2018). The main group of events is determined by primary dark count rate (DCR) with the amplitude corresponding to one PE. Direct cross-talk (DiCT) events are practically simultaneous to DCR. These are defined by cross-talk photons producing independent discharges in neighboring cells, characterized by pulses corresponding to two, three, etc. PE. The least populated group, with characteristic delay time of 10 ns, is due to delayed cross-talk (DeCT), caused by cross-talk photons absorbed in the non-depleted region of a neighboring cell. Finally, the group of events with delay times of 1–100 μs and amplitude of one PE or less, are identified as after-pulsing (AP). AP is caused by discharge electrons trapped by some impurity in the silicon lattice and then released with some delay, generating secondary discharge in the same cell.

There are several readout schemes to combine SiPMs into a matrix having one readout channel: these are determined by the amplifier characteristics and specific tasks (D'Incecco *et al.*, 2018; Baldini *et al.*, 2018; Aalseth *et al.*, 2018). As a tutorial example,

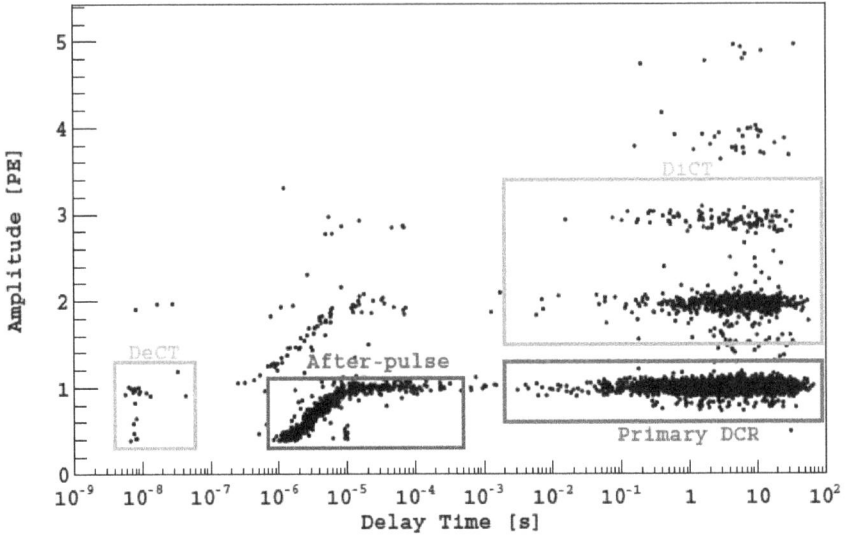

Figure 8.12: Distribution of amplitude vs. delay time for noise signals of SiPM of NUV-HD (FBK) type taken at 77 K. Basic components of the SiPM noise response can be clearly identified: primary DCR, direct cross-talk (DiCT), delayed cross-talk (DeCT), and after-pulsing (AP). Redrawn from Aalseth *et al.* (2018).

let us discuss two readout schemes of 9 SiPMs combined into a 1-channel matrix. In the most trivial scheme, where all the SiPMs are readout in parallel (see Fig. 8.13(a)), the matrix capacitance is just the sum of all SiPM capacities, resulting in a degraded signal-to-noise ratio (SNR) and in reducing the bandwidth of the amplifier response. This scheme forms a slower RC time constant with amplifier input impedance. Although the charge (pulse area) here is preserved, the pulse-height is reduced due to slower rise and longer tail of the pulse. Such a readout scheme is not optimal for timing and high-rate applications.

The alternative is a hybrid readout scheme, with parallel-series combination of SiPMs, known as 3p3s configuration (Baldini *et al.*, 2018; Aalseth *et al.*, 2018): see Fig. 8.13(b). Here, the output signal of a SiPM is reduced by a factor equal to the number of SiPMs put in series, but this disadvantage is compensated by the attenuation of noise gain due to the reduction in the input capacitance. In this

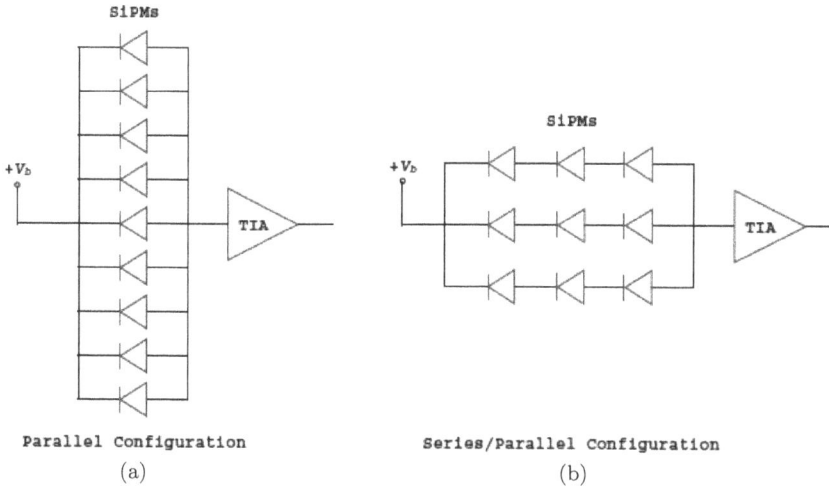

Figure 8.13:   Two readout schemes of a 1-channel SiPM matrix combined from 9 SiPMs: classic scheme with all SiPMs summed in parallel (a) and hybrid scheme with SiPMs summed in series-parallel, known as 3p3s configuration (b). Redrawn from Baldini *et al.* (2018); Aalseth *et al.* (2018).

case, the SNR and the overall matrix capacitance (and accordingly the bandwidth) is the same as that of a single SiPM.

8.4. Light Collection: Reflectors and Wavelength Shifters

There is a substantial difference between two-phase Ar and Xe detectors in terms of light collection. In two-phase Xe detectors, no wavelength shifter (WLS) is used and the TPC uses polytetrafluorethylene (PTFE, Teflon) as reflective/diffusive material to improve the UV light collection by the top and bottom PMT matrices (Aprile *et al.*, 2012b; Aprile *et al.*, 2017). In contrast in two-phase Ar detectors, the WLS is used to convert the VUV light into the visible one (Agnes *et al.*, 2015; Aalseth *et al.*, 2018).

For example, in the DarkSide-50 experiment the top and bottom PMT matrices, submerged in LAr, view the active volume of the two-phase TPC through fused-silica windows, which are coated on both faces with transparent conductive films of indium tin oxide (ITO) of

15 nm thickness. The cylindrical TPC wall is made of PTFE and acts as a reflector. Both the cylindrical wall and the windows are coated with the TPB wavelength shifting films of a thickness of 190–230 $\mu g/cm^2$ on the windows and 160–220 $\mu g/cm^2$ on the walls. Usually the TPB films are deposited by vacuum evaporation. In addition, in some cases to make the WLS film more stable, the TPB is deposited in a polymer (e.g., polystyrene) matrix (Gehman *et al.*, 2013; Francini *et al.*, 2013).

The optical properties of TPB films depend significantly on the film thickness, the deposition process and on the type of substrate: polymer plastic, PTFE, acrylic plastic or the glass of the PMT windows. The effective thickness of the vacuum-evaporated TPB can also depend on the substrate: according to different sources, it varies from 30 $\mu g/cm^2$ (Tanaka *et al.*, 2020) to 60–100 $\mu g/cm^2$ (Lally *et al.*, 1996) and even to 200–600 $\mu g/cm^2$ (Francini *et al.*, 2013).

There are also serious discrepancies in data on the absolute conversion efficiency, in particular on the efficiency of light shifting from 128 to 420 nm (Benson *et al.*, 2018): for evaporated TPB film it can vary from 0.7 to 1.2 (see Fig. 8.14(a)). For TPB in polystyrene matrix the conversion efficiency is typically lower compared to pure TPB, at 128 nm reported to be 0.58 (Gehman *et al.*, 2013) and 0.49 (Buzulutskov *et al.*, 2018; Borisova, 2020): see Fig. 8.14(b) showing conversion efficiency for WLS film composed of 1 part of TPB per three parts of polystyrene.

Another interesting TPB-related issue should be mentioned, namely that of the effect of TPB dissolving in liquid Ar (Asaadi *et al.*, 2019). It may result in TPB molar concentration in liquid Ar reaching 2–10 ppb and appropriate wavelength shifting behavior of TPB in the bulk. This effect applies mostly to evaporated TPB films. It does not work for WLS films composed of non-saturated TPB in a polymer matrix, which seem to be resistant to dissolving in liquid Ar. Rapid dissolution of TPB, deposited on quartz by vacuum evaporation, has been earlier observed in LXe (Sanguino *et al.*, 2016).

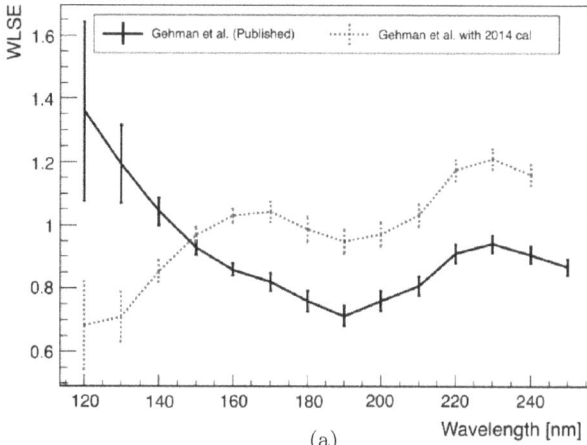

(a)

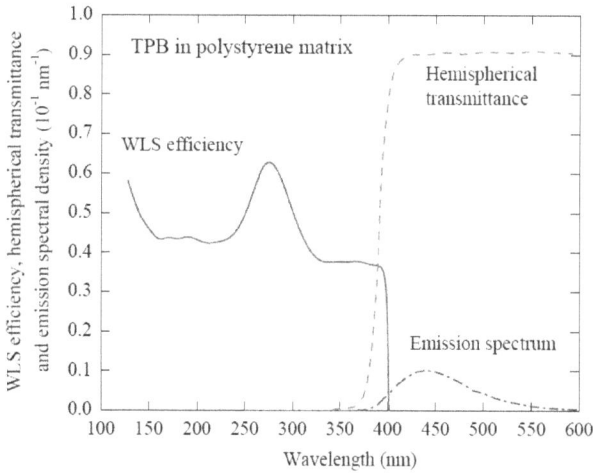

(b)

Figure 8.14: (a) WLS absolute conversion efficiency for evaporated TPB film using two different calibrations (redrawn from Benson *et al.* (2018)). A significant discrepancy can be seen between two data sets. (b) WLS efficiency for TPB in polystyrene matrix (one part of TPB per three parts of polystyrene), drawn from (Borisova, 2020), combining the data from (Gehman *et al.*, 2013; Lally *et al.*, 1996; Francini *et al.*, 2013). Hemispherical transmittance and photon emission spectrum for this WLS film are also shown.

8.5. Purification of Working Media

One of the main technological problems of constructing massive emission detectors based on condensed noble gases is a purification of working media from impurities affecting their operation. Dangerous for the operation of two-phase detectors are electronegative impurities, trapping electrons when they are collecting from the working medium of the detector, molecular impurities that absorb scintillation and electroluminescent photons and cool down drifting electrons that reduces the probability of their emission from the condensed phase, and radioactive impurities, including such relatively short living isotopes captured from the atmosphere as ^{39}Ar, ^{85}Kr, ^{222}Rn.

The technology for purification of the noble gases from chemically active impurities has been under development for a long time (see, for example, Fastovsky *et al.*, 1972; Barabash and Bolozdynya, 1993; Aprile *et al.*, 2006a). The wide variety of developed purification methods can be classified as following:

(1) Chemical methods that provide chemical bonding of active impurities the absorbers (getters) based on chemically active metals (sodium, lithium, copper, titanium), catalytic hydrogenation (oxygen bonding by adding hydrogen admixture).

(2) Physical methods, including distillation and rectification, absorption of impurities by special absorbers (activated carbon, molecular sieves), chromatographic separation, purification by electrical current passing through liquefied gases as a stream of electrons captured by electronegative impurities.

(3) Combined methods such as a spark purification, in which an active titanium getter in the form of nanodispersed dust is generated by electric discharge between titanium electrodes immersed into the noble liquids (Akimov *et al.*, 2019).

Today there are developed industrial methods for purification of noble gases including molecular sieves and hot and cold metal getters that are quite effective for purification of argon to the level of 0.1 ppb of residual concentration of oxygen. This corresponds to an electron lifetime of about 3 ms that is sufficient to collect electrons from a liquid argon layer of about 2.5 m thickness and to ensure operability of detectors with a working mass of thousands of tones (Montanari *et al.*, 2017).

A more serious challenge is purification of xenon, which due to high polarizability of atoms, is an effective absorber of polar molecules and complex molecular compounds in liquid state. The next most difficult problem is the separation of radioactive isotopes from media with similar characteristics in the case of low-background experiments.

The general procedure for ensuring the necessary purity of a liquefied noble gas as a working medium of two-phase emission detectors can be divided into several steps:

(1) Proper selection of materials contacting the working media of the detectors.
(2) Preliminary purification of raw materials used as working media of the detectors in liquid phase.
(3) Cleaning of internal surfaces of detector structure working in contact with the working medium.
(4) Circulation of working medium through purifiers during the whole circle of operation of the detector in order to remove impurities outgassing from the inside structure of the detector and atmospheric gases leaking through joints.

Below we consider the details of the above steps in order to achieve the state of working media that is enough for successful operation of the detectors.

8.5.1. *Hardware selection*

The correct selection of materials for internal infrastructure of the detector is very important in order to provide its successful operation in future. Chemically resistant construction materials such as stainless steel and metal sealing parts should be first considered for manufacturing of the two-phase emission detectors. For vessels of low-background detectors it is preferable to use materials having low content of radioactive isotopes such as copper and titanium.

During the fabrication process, all metal parts should be washed with deionized water and detergent to remove oil and grease then cleaned with ethanol (Adamowski *et al.*, 2014). It is also very useful to treat the surfaces of metal parts (especially electrodes operating at high potentials) using electro polishing.

All joints and valves associated with gas and liquid purification systems should utilize metal seals. Valves operation should be based on bellows or metal diaphragms to prevent the diffusion of oxygen and water contaminations. The exhaust side of relief valves continuously purge with inert gas to prevent diffusion of oxygen and water from ambient air across the O-ring seal. *ConFlat* flanges with copper seals should be used on both cryogenic and room temperature piping. Pipe connections should be made with *VCR* fittings with stainless steel gaskets. In the case of argon working medium Spiral wound graphite gaskets can be also used.

Teflon ribbon cables should be used to connect the circuit boards inside the detectors and outside electronics through ceramic-to-metal feedthroughs such as *Ceramaseal*® products. The *RTD* platinum sensors type K of 100 Ohm are preferable for temperature control.

8.5.2. *Preparation of raw materials to be used as working media*

Before being used to fill the detectors, noble gases should be purified to the level of relative oxygen concentration of ~0.1 ppb or less. The purification system is normally composed of filtration elements and associated equipment required to regenerate them, particulate filters, interconnecting piping, necessary valves and control instrumentation. The Ar purifiers often contain commercially available molecular sieves and copper pellets or *Oxisorb*® in order to remove water and oxygen, respectively. For Xe purification systems the heated metal getters such as *SAES MonoTor*® or *MegaTor*® are preferable; these are also used for Ar, in particular in the DarkSide-50 experiment. During operations, the filters may be switched between active filtration and regenerative modes, with one half of the set either actively filtering or being regenerated, so the purification process is uninterrupted. In case of heavy contaminations, such as happen during xenon isotope separation, the spark purification technology is quite effective in the two-stage implementation (Akimov *et al.*, 2019); it is described below. At the first stage, a massive (up to 300 kg in case of using Mojdodyr spark purifier — Fig. 8.18) liquid Xe sample was exposed to hard UV radiation generated by an electric high-voltage (HV) discharge in a liquid in order to decompose complex high-molecular-weight impurities in photolysis reactions. The resulting solid deposits were removed from the purification chamber mechanically after xenon was removed from the chamber. At the second stage, the liquid-xenon sample was purified by nanosized titanium dust getter, which was produced in the liquid by electric HV discharges between titanium electrodes. At a high contamination of the initial xenon sample, the final phase of purification by nanodispersed titanium was carried out in the absence of UV radiation. In this case, the HV was turned off after the required amount of nanodispersed titanium was produced and the liquid-xenon sample was kept for some time in contact with nanodispersed titanium. This time is needed for the chemical binding of impurities on the surface of nanodispersed titanium at a temperature of about $-100°C$. This technology has made it possible to increase the lifetime of quasi-free electrons in the 205 kg of liquid xenon from ≤ 0.1 to ≥ 400 μs in a drift electric field range of 50–500 V/cm (Akimov *et al.*, 2019).

8.5.3. *Purging detector before liquid filling*

A noble gas purge or recirculation through the detector and gas purifiers should be performed at the beginning of each run period and before cooling down the detector. At the end of the purge of recirculation, the oxygen, water and nitrogen levels can be reduced to 1–10, 1 and 10 ppm, respectively.

The liquid argon purity demonstrator (LAPD) located at Fermilab for the first time has demonstrated that vacuum evacuation of the detector is not necessary and purging is good enough for achieving long electron lifetimes in liquid argon (Adamowsky *et al.*, 2014). After the removal of the ambient air by the argon purge, argon gas was pumped through the molecular sieve and oxygen filter at a rate of a volume exchange every 3.4 h, and then returned to the cryostat. The gas recirculation for the second run period lasted for about 75 volume exchanges. At the end of this phase, the oxygen concentration was reduced to approximately 20 ppb and the water concentration stabilized at about 670 ppb, corresponding to an outgassing rate of 1.03×10^{-6} g/s: see Fig. 8.15 (Abi *et al.*, 2018). The overall results for the gas recirculation step indicate that water outgasses from all inside surfaces of the cryostat and piping, and that the water outgassing rate is matched by the filtration rate after several volume exchanges while the oxygen outgassing rate continues to decline.

The gas recirculation phase provides an important opportunity to look for any final leaks before committing to cryogenic operation. After that, the detector could be cooled down and filled with noble liquids.

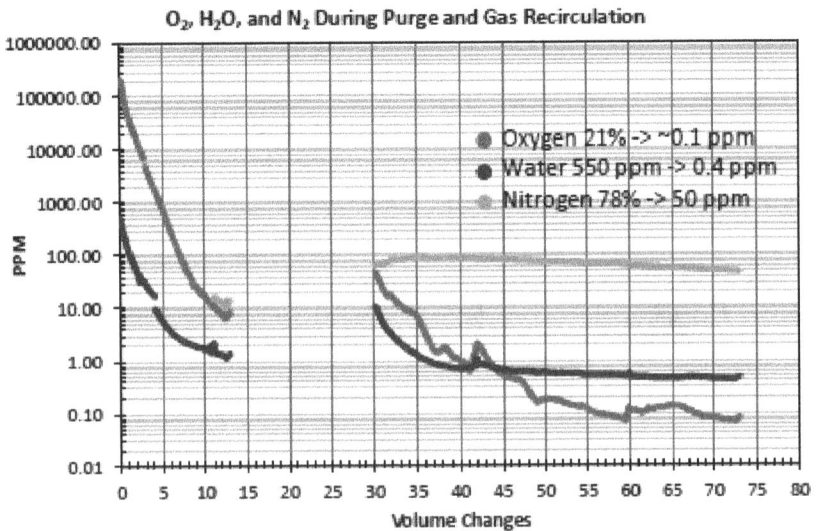

Figure 8.15: Plot of the O_2, H_2O, and N_2 content in liquid argon during the piston purge and gas recirculation stages of the 35 t phase 1 run. Redrawn from Abi *et al.* (2018).

8.5.4. *Purifying circulation in the course of the detector operation*

Purification system of massive detectors is normally performed as two independent circuitries used for circulating liquid and gas phases through suitable filtration elements. The filling of the cryostat of the detector with liquid is carried out by passing the cooled gas through filters that trap electronegative impurities. After filling the cryostat and starting operation of the massive two-phase detector, the circulation of the working substance through the purification filters begins and continues throughout its operation. The most effective method of cleaning the working substance of a two-phase detector is the circulation of both liquid and gaseous phases. In the liquid phase, it is necessary to ensure the maximum possible lifetime of drifting electrons. The gas phase fills the communication lines through which the detector is connected to the signal readout system, and the gas communication lines through which the detector is filled with working gas and evacuated after the detector is finished. Some of the elements of the communication lines (bellows pipelines, cable lines, valves, safety valves, connectors, feedthroughs) are operating at room temperature and therefore they are an inevitable source of atmospheric fumes, which, when absorbed by liquid working medium, pollute it. The purge of communication lines with pure noble gas vaporized from the surface of the liquid working medium is an indispensable condition for achieving high purity of the liquid working medium.

The easiest way to purge is to use a circulation pump, which takes gas in small portions from the detector and pumps it through active filters before returning it to the detector. One of the first examples of such technologies preventing the pollution of a liquid working medium (liquid argon) by outgassing communication lines was demonstrated by the ICARUS collaboration (Vignoli, 2014, 2015): see Fig. 8.16. A similar scheme is used in the RED-100 detector with liquid xenon as a working medium (Akimov *et al.*, 2017b).

The circulation of a liquid working medium can be carried out in two ways. In ICARUS and LAPD detectors (Vignoli, 2014;

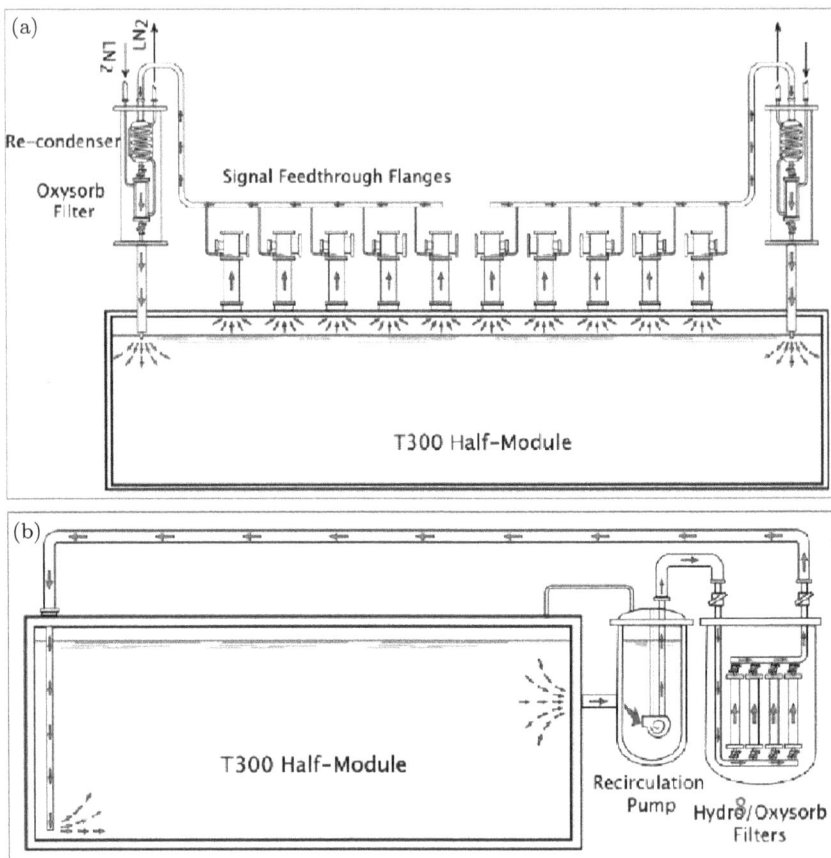

Figure 8.16: Circulating purification of gas (a) and liquid (b) phases in the ICARUS T600 LAr detector. Redrawn from Vignoli (2015).

Adamowski *et al.*, 2014) liquid argon circulated directly through *Oxisorb/Hydrosorb* filters (Fig. 8.16). In the RED-100 detector, liquid xenon was evaporated inside a heat exchanger, pumped through a hot metal getter *SAES MonoTor* and, passing in the opposite direction through the same heat exchanger, condensed back inside the detector (Akimov *et al.*, 2017b).

After several liquid volume exchanges, the effect of filter saturation with adsorbed impurities may be observed in massive liquid noble-gas detectors. Therefore, from time to time they were regenerated when the purification of the working medium of the detector was interrupted. The LAPD team found no correlation between volume exchange rate and measured lifetime values in liquid argon. This fact indicates that the boil off argon vapor intercepts the majority of the outgassing contamination from the ullage before the contamination can diffuse downward into the liquid argon.

As a liquid argon pump can be used the *Barber-Nichols BNCP-32B-000* magnetically driven centrifugal pump which isolates the pump and liquid argon from the electrical motor. The impeller, inducer, and driving section of the magnetic coupling each have their own bearings that are lubricated by the liquid argon at the impeller. The liquid argon flow rate can be measured at the pump discharge by a flow meter, e.g., by that of *Emerson Process Management Micro Motion Coriolis*. In the ICARUS T600, the *ACD Cryo AC-32* submerged motor centrifugal pump was used.

8.5.5. *Purifying from radioactive impurities*

In addition to electronegative impurities, LAr and LXe detectors should be purified from short-living radioactive isotopes captured from the atmosphere: ^{39}Ar, ^{85}Kr, ^{222}Rn. Natural Xe has no long-lived radioactive isotopes and the half-life of the potential double-beta emitter ^{136}Xe is so long that it does not limit the sensitivity of LXe detectors. This is not the case for natural Ar, which is contaminated by ^{39}Ar produced by cosmic rays and interactions with neutrons. Its activity in naturals Ar taken from the atmosphere was measured to be about 1 Bq per kg. In addition, the radioisotope ^{85}Kr contaminant can be present in liquid Xe and Ar. This isotope is released into the environment in nuclear weapon tests and by nuclear reprocessing plants.

In XENON10 experiment, purifying from Kr was done using chromatographic separation (Bolozdynya *et al.*, 2007), purifying 25 kg of Xe to the level of <3 ppt of Kr. In XENON100 and XENON1T experiments, dedicated distillation columns were used where a Kr reduction factor of 10^4–10^5 was achieved for 3.3 tons of Xe (Aprile *et al.*, 2017), reducing the content of natural Kr in Xe from 60 ppb down to 0.4 ppt. With some modifications of hardware and processing, this distillation column was also used for a Rn distillation, demonstrating a Rn reduction factor of >27 (Aprile *et al.*, 2017).

More complicated task is purifying Ar from ^{39}Ar. The method used in the DarkSide-50 and DarkSide-20k experiments involves the production of underground Ar (UAr) extracting it from deep underground CO_2 sources (Aalseth *et al.*, 2018). In UAr, the ^{39}Ar specific activity can be reduced down to about 1 mBq/kg (Aalseth *et al.*, 2018). The Urania project will extract and purify the UAr from the CO_2 wells located in Cortez, Colorado, USA at a production rate of 100 kg/day. Additionally, it would be beneficial to further deplete the UAr of ^{39}Ar, giving extended sensitivity to DarkSide-20k. The Aria project will serve to chemically purify the UAr using cryogenic distillation columns 350 m tall, installed in the coal mine well in Sardinia, Italy. The ultimate goal of the Aria project is to process about 150 kg/day of argon through to achieve an additional depletion factor between 10 and 100, in addition to the reduction of ^{39}Ar already seen in the UAr (Aalseth *et al.*, 2018).

8.5.6. *Purity monitoring*

When purifying massive working media of two-phase detectors, it is important to continuously monitor the purity of the liquid. For this purpose, as a rule, specially designed devices are used.

The purity of liquid Ar can be monitored by a double gridded ionization chamber immersed in the liquid: see, for example (Carugno *et al.*, 1990; Amerio *et al.*, 2004; Adamowski *et al.*, 2014). In this device a Xe flash lamp is used to illuminate the cathode via quartz optical fiber and to extract photoelectrons into the liquid (Fig. 8.17). The fraction of electrons generated at the photo-cathode that arrive at the anode (Q_A/Q_C) after the electron drift time t, is a measure of the electronegative impurity concentration and can also be interpreted as the quasi-free electron lifetime τ such that

$$Q_A/Q_C = \exp(-t/\tau).$$

Another solution for a purity monitor is an X-ray ionization chamber. In particular, a two-electrode X-ray ionization chamber with an

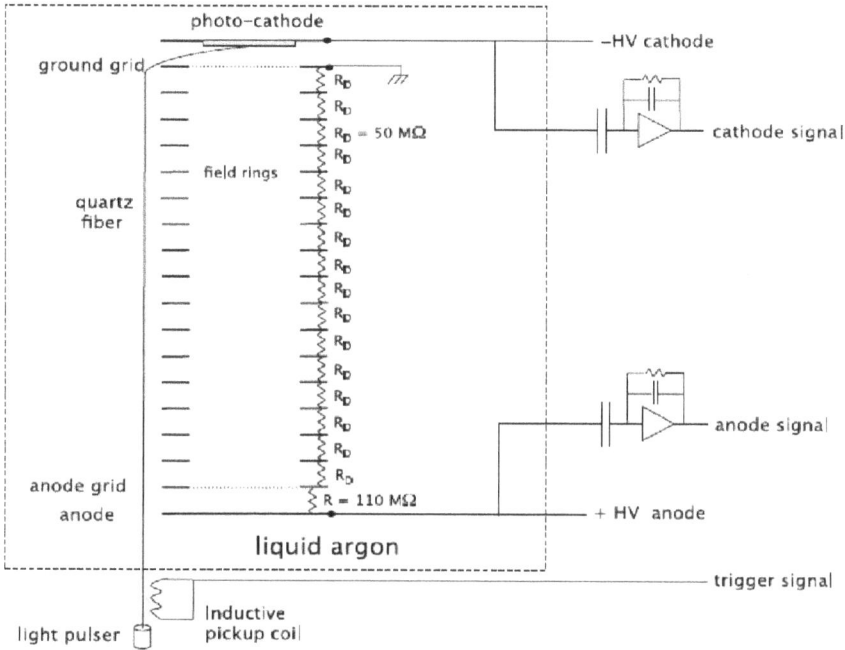

Figure 8.17: Double gridded ionization chamber with cathode illuminated by UV source via quartz optical fiber for measurement of quasi-free electron lifetime before capture by electronegative impurities. Redrawn from Adamowski *et al.* (2014).

18-mm gap was installed at the center of the cold chamber of the *Moidodyr* installation (Fig. 8.18). All the electrodes of the device, in particular, the discharge electrodes *1* and the anode *2* of the ionization chamber are fixed in place on metal rods, through which they are connected to the respective electrical circuits through the cermet terminals *8* and *9* on the upper flange of the cryostat. At the center of the lower flanges of the cryostat are aluminum entrance windows *11* for X-ray radiation. An X-ray pulse with an energy as high as 30 keV and a duration of approximately 0.5 μs is generated by a BSV-7 pulsed X-ray tube *12*, which is installed outside the cryostat and has two beryllium windows for emission of X-ray radiation (upwards to the cryostat and downwards to the scintillation detector *13* that is used to monitor the tube performance). The depth of

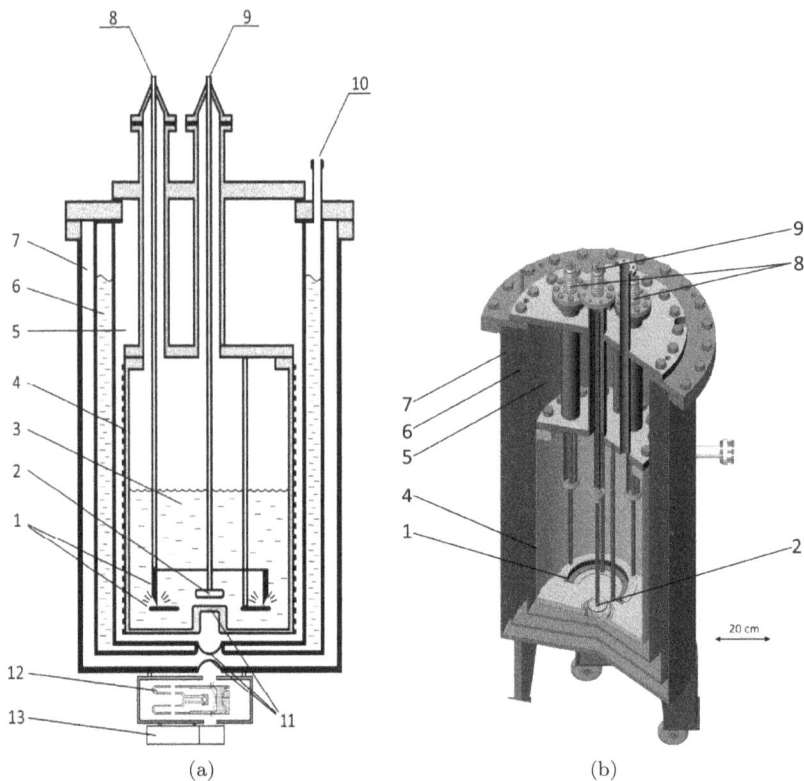

Figure 8.18:   Schematic drawing (a) and general view (b) of Mojdodyr (MDD) installation for spark discharge purification of liquid xenon: 1 — annular Titanium discharge electrodes in the "tip-plane" configuration; 2 — two-electrode ionization chamber for determining the lifetime of electrons before capture by electronegative impurities; 3 — liquefied noble gas of up to 100 liters volume; 4 — cold chamber surrounded by a copper screen, on which heating elements are installed; 5 — gas nitrogen jacket; 6 — liquid nitrogen jacket; 7 — vacuum jacket; 8 — high voltage feed-through; 9 — feed-through for the ionization signal; 10 — nozzle for liquid nitrogen filling; 11 — aluminum windows for input of X-ray radiation into the inter-electrode gap of the ionization chamber, 12 — X-ray gated tube BSV-7; 13 — scintillation detector for monitoring the operation of the X-ray tube. Redrawn from Akimov *et al.*, 2019.

X-ray absorption in liquid xenon is approximately 1 mm. In the case of an ideally pure liquid, the drift of a cloud of electrons produced by X-ray pulses in liquid xenon provides an almost constant charging current in the charge-sensitive preamplifier connected to the anode of

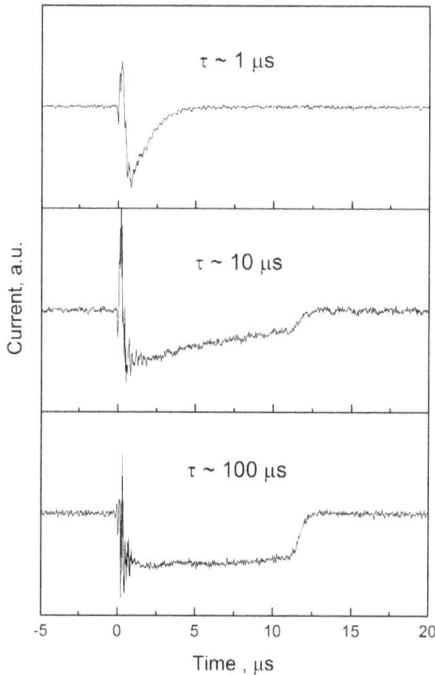

Figure 8.19: The averaged waveforms of the current pulses in the X-ray ionization chamber of the MDD installation for various degrees of purity of liquid Xenon, corresponding to the lifetimes of quasi-free electrons prior to capture by electronegative impurities $\tau \sim 1\ \mu s$, $\tau \sim 10\ \mu s$ and $\tau \sim 100\ \mu s$, at anode potential of $+\ 500$ V and drift gap of 17 mm. Redrawn from Akimov *et al.* (2019).

the ionization chamber. An exponential deviation of the current from a constant value indicates the capture of electrons by electronegative impurities during their drift between the electrodes of the ionization chamber (Fig. 8.19).

The lifetime of electrons can be also determined from the waveform of signals from cosmic muons crossing the sensitive volume of the detector filled with liquid Xe, such as the RED-100 detector. Figure 8.20 shows a typical waveform from a single muon that has crossed the entire sensitive volume of the RED-100 detector (Akimov *et al.*, 2020). The waveform consists of two components: fast scintillation of liquid Xe at the moment of interaction with cosmic muon (a narrow peak near 0 μs), and the subsequent decaying

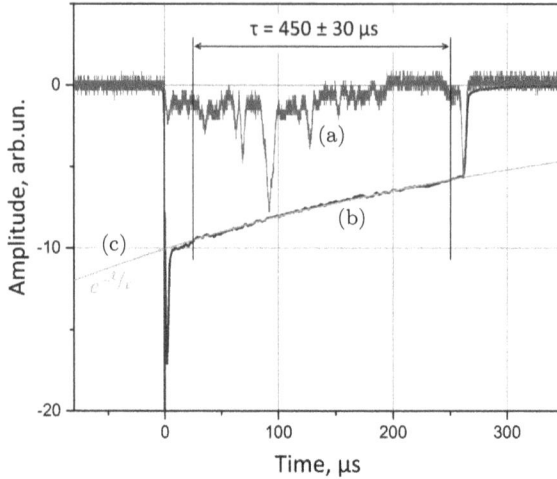

Figure 8.20:   Determination of the lifetime of drifting electrons in liquid Xe before capture by electronegative impurities: (a) — the waveform from a single muon that has crossed the entire sensitive volume of the RED-100 detector filled with liquid Xe; (b) — the averaged waveform of 10,000 muon signals; (c) — the fitting exponent used to define the lifetime of the electrons which is ∼450 μs in given case. Redrawn from Akimov *et al.* (2020).

electroluminescence caused by electrons extracted from the muon track into liquid Xe, drifted through the liquid, extracted into the gas phase and drifting through the gas at high electric field (the falling step from ∼10 μs to ∼260 μs). The averaged electroluminescent waveform was fitted with an exponent, to determine the lifetime τ of the drifting electrons prior to capture by electronegative impurities in liquid Xenon.

8.6. Cryogenics for Two-phase Xenon Emission Detectors

The cost of the advantages of condensed argon and xenon as a working medium for two-phase emission detectors is the need to work in cryogenic conditions. When using the most popular xenon, a complicating circumstance is the fact that the temperature range from freezing to boiling point of xenon is very small: from $-112°C$ to $-108°C$ at ambient pressure. One can moderately alleviate the problem by moving away from the triple point, i.e., choosing a higher operating pressure. The boiling temperature increases much faster than the freezing one. Typical operating temperatures are therefore around $-95°$C, and the pressure is around 1.5 bar. Still one needs a tight regulation of the cooling power especially for two-phase detectors whose proportional electroluminescence gain is pressure sensitive and varies substantially with changes in the liquid level.

8.6.1. *Cooling bath system*

Cooling baths for liquid xenon detectors based on dry ice immersed in liquid ethanol have been used in the past for relatively small detectors working at high pressure, in particular at 5 bar. This imposed a serious restriction on the possibility of using high-performance photodetectors such as cryogenic vacuum photomultipliers immersed in the working medium of the detector. In addition, at elevated working pressures, the light output of electroluminescence is reduced, which makes it difficult to detect weak ionization signals, and a relatively massive thick-walled cryostat has to be used, in order to conduct experiments in low-background conditions.

Detector cooling by immersion in a refrigerant bath is currently used in the EXO experiment. (Auger *et al.*, 2012). The detector EXO-200 is installed in a double-wall cryostat and surrounded by HFE-7000 Engineered Fluid used as a heat transfer fluid capable of reaching $-120°$C and providing additional radiation shielding. HFE-7000 is a dense, radiopure fluid that is liquid both at room temperature and at the LXe operating temperature of 165 K. Circulating the HFE-7000 fluid to external heat exchangers can provide the required cooling power while keeping the heat exchangers outside of the low background region.

However, for detectors with a working medium mass in scale of many tons, such a solution seems economically impractical. In addition, as was first formulated in Bolozdynya *et al.* (1995), the best protective bath for the sensitive volume of the detector is liquid xenon itself if the detector has three-dimensional positional sensitivity and the search for useful events is carried out in a central (fiducial) volume working environment of the detector. In this case, the parietal layer of liquid xenon plays the role of an excellent active radiation protection (see Fig. 1.1).

8.6.2. *LN$_2$ purging system*

At the early stages of development of liquid noble gas detector technology, industrial Freon refrigeration units were used (see, for example, a description of the bubble chamber DIANA by Barmin *et al.*, 1984). However, not enough low temperature of this popular refrigerant allowed the detectors to work with liquefied noble gases only at elevated pressures (60 bar for xenon).

To cool massive detectors, which are a kind of high-capacity heat energy accumulators, one can use a liquid-nitrogen tubular cooling system. Technically, such a system looks like a copper heat shield and a tight-fitting detector placed in a heat-insulating jacket. Copper tubes are soldered to the heat shield, through which liquid nitrogen is supplied from the Dewar vessel under pressure. A similar system was used, for example, to cool the Nadezhda emission streamer chamber (Fig. 6.6, Bolozdynya, 2010), but it required a large flow of liquid nitrogen, because nitrogen from the cooling pipe was released into the atmosphere.

In an improved form, such a system is also used at present, for example, for cooling elements of the gas system of the LZ detector: liquid nitrogen under pressure circulates in a closed circuit as a coolant, in turn, cooled by a powerful cryocooler (see Fig. 8.21).

8.6.3. *LN$_2$ cooling with a "cold finger"*

Early R&D projects often used the method of cooling the detectors using free-boiling liquid nitrogen, which draws heat from the detector using a copper heat conductor in the form of a ferrule surrounding the detector (see Fig. 1.2), or a rod (so called "cold finger" as shown in Fig. 8.22), which transfers a heat from the detector into a vessel with liquid nitrogen. Sometimes, a nitrogen gas jacket was used as a heat conductor, separating the device filled with liquid noble gas from the liquid nitrogen jacket (Akimov *et al.*, 2017b). In the latter

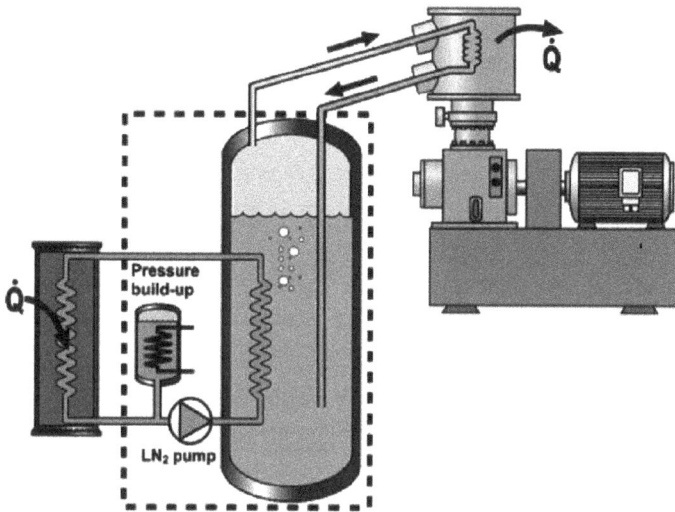

Figure 8.21: Cryocooler based on the Stirling cycle system removes heat, Q, from a closed liquid nitrogen storage reservoir that is used as a heat pump for several secondary cooling loops and thermosyphons thermally connected to that reservoir. Redrawn from Mount *et al.* (2017).

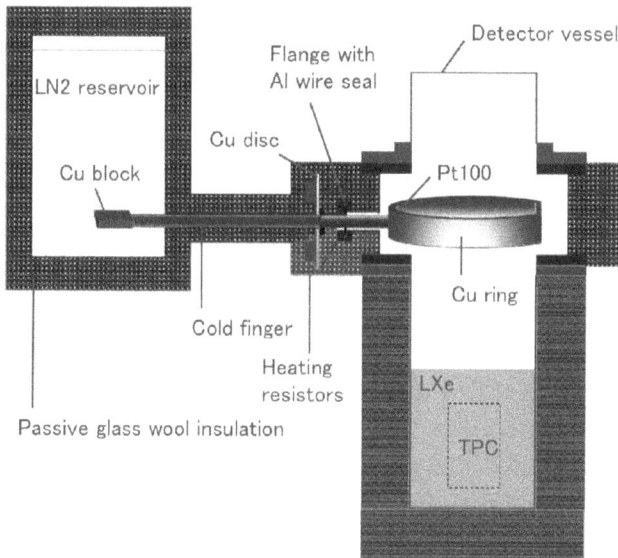

Figure 8.22: Schematic view of the cold finger cooling system: the end of the cold finger is a $1/4''$ thick copper ring; the center of the working volume is free for electrical connections from the top flange. Redrawn from Giboni *et al.* (2019).

case, the heat transfer between the liquid nitrogen jacket and the detector could be controlled by the nitrogen pressure in the gaseous jacket. Nevertheless, in any of the above methods, the heat transfer was finely tuned by electric heaters, which inevitably increased the flow of liquid nitrogen and significantly limited the cost-effectiveness of thermostating of massive devices using this method.

A LN_2 free boiling bath with a cold finger is commonly used to improve the performance of a high resolution germanium (Ge) detector, by keeping the crystal at temperature close to 77.8 K. In like these systems the copper cold finger is used as a heat conductor connecting a LN_2 reservoir and the Ge crystal.

Thus any kind of "cold finger" systems shifts the temperature. If during operation less cooling power is required, or if the operating temperature is higher, the excess cooling power has to be dissipated by the heaters. The "cold finger" heat transfer system acts like a resistor in an electrical circuit reducing the current.

8.6.4. *Pulse tube refrigerators*

With the development of the experimental base for fundamental experiments on the search for dark matter, DM detectors operated by the XMASS (Abe *et al.*, 2013), XENON (Aprile *et al.*, 2012b, 2017) and PandaX (Cao *et al.*, 2014) collaborations were cooled by Pulse Tube Refrigerators (PTRs) (Haruyama *et al.*, 2004) (with the exception of the LUX detector). The principle of PTR based system operation is shown in Fig. 8.23. The cooling power is regulated by a heater installed on the cold head of the PTR.

The XENON10 detector used the first version of the PTR developed by the Iwatani Company, i.e., a P90 with 100 W cooling power. XENON100 DM detector has used Iwatani PC150 PTR with 150 W cooling power (Aprile *et al.*, 2012b). XENON1T experiment using 3.2 tons LXe two-phase emission detector also used PC150 PTR, which demonstrated 250 W cooling power with a modified 7 kW Cryomech CP2870 helium compressor. During operation the cooling power is regulated by a heater installed on the cold head as shown in Fig. 8.23. The heater is powered by a proportional-integral-derivative (PID) controller which delivers resistive heating based on

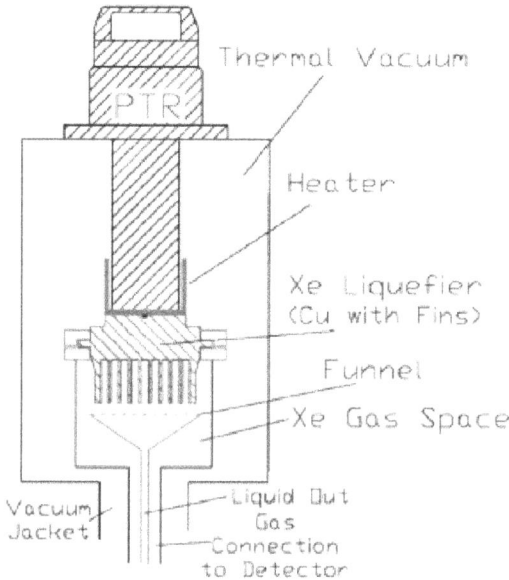

Figure 8.23: Schematic drawing of the cooling module for LXe two-phase emission detector based on Pulse Tube Refrigerator. Redrawn from Giboni *et al.* (2019).

the actual and set temperatures. The cold head of the PTR connects to a copper heat conductor penetrating the vessel wall. This deviates from the initial use of PTRs in the MEG experiment (Adam *et al.*, 2010).

The cooling power of commercially available PTRs is limited to about 200 W. The next generation DM detectors of up to 50 tons mass of LXe will require about 1 kW cooling power. The power can be provided by several PTRs working in parallel, but that may be impractical because of the high costs of the units and the high power consumption of the associated helium compressors.

8.6.5. *Two-phase closed tubular thermosyphon*

A two-phase closed tubular thermosyphon is a recently developed high performance heat transfer device that can be used to transfer a large amount of heat at a high rate with fine regulation of the cooling power (Lock, 1992). The thermosyphon or gravity assisted heat pipe

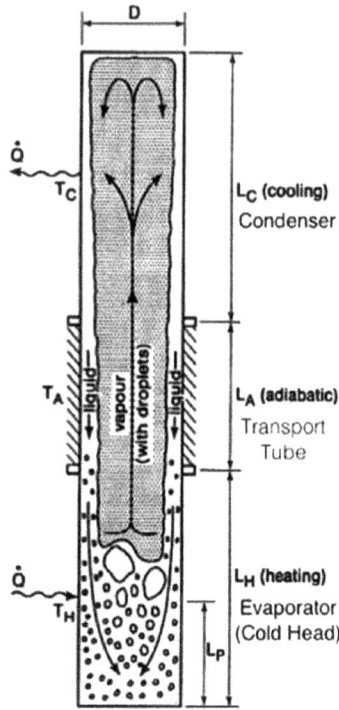

Figure 8.24: The operating principle of a two-phase closed tubular thermosyphon. Redrawn from Lock (1992).

consists of three basic sections as shown in Fig. 8.24: a cooling section (condenser) located above a heating section (evaporator) and a passive adiabatic section connecting the two active sections. A condenser is a part of the thermosyphon of length L_c used to deposit the heat energy Q into a cooling machine operating in good thermal contact with this section. The condensate generated inside the condenser falls down due to gravity through the adiabatic section into the evaporator, where the liquid is boiling and absorbing the heat Q. The generated here vapor rises into the condenser, returning the heat to the cooling machine, condensing to the liquid, then, the heat transfer cycle repeats. Since the operation of the thermosyphon relies upon the gravitational force, the evaporator must be located below the condenser. The adiabatic section is performed as a tubular heat transfer line placed in a vacuum jacket and wrapped with multilayer aluminized Mylar foil thermal-insulation.

There are possibly a few working fluids that can be used in cryogenic thermosyphons (see Table 1 in Bolozdynya *et al.*, 2015). However for liquid argon and xenon detectors nitrogen is an optimal cooling agent, which is available in large amounts and can provide safe operations in the temperature range between 63.15 and 126.2 K and pump out the heat from the evaporator with the latent heat of vaporization LHV = 199 kJ/kg.

Considering the thermosyphon as a heat pipe pumping the heat energy out of the cold head, with cooling power dQ/dt, one can define the thermal conductivity of the system, as

$$k = dQ/dt \cdot L/A \cdot \Delta T,$$

where A is a cross-section of the thermosyphon tube, ΔT is a temperature gradient along the tube of L length. Knowing the thermal conductivity we can calculate the total thermal resistance of the thermosyphon:

$$R = L/A \cdot k.$$

The system exhibited extremely high thermal conductivity exceeding that value for copper by 200–500 times and close to that for nanotubes (Berber *et al.*, 2000). A nitrogen filled thermosyphon system using 1 cm diameter tube provides about 100 W cooling power; a nitrogen filled thermosyphon system using 3.5 cm diameter tube provides about 900 W cooling power (Bolozdynya *et al.*, 2016).

The thermosyphon cryogenic system has the advantage of low cost, compactness, high cooling power and can be used to support operations of large cryogenic installations such as liquid noble gas detectors, oxygen storage tanks, etc. The simple construction and cooling ability of relatively flexible copper tube of about 5 m length has been tested as important for arrangement of low background experiments in fundamental research (Bolozdynya *et al.*, 2009; Akimov *et al.*, 2013a), in which cooled devices must operate remotely inside a massive shielding and mass of construction materials surrounding the installation must be minimized. Thermosyphon cooling technology based on nitrogen cooling agent has been introduced in experimental practice by LUX DM experiment (Akerib *et al.*, 2013a), in which an arrangement of three thermosyphons, one on the bottom of the LXe detector and two on the top, connected to a thermal shield, have been used.

As an example, Fig. 8.25 presents a schematic drawing of the cooling system for the RED-100 detector (Bolozdynya *et al.*, 2015, 2016). The detector is cooled down by four tubular thermosyphons made of copper tubes with a diameter of 12 mm and a 0.5-mm-thick

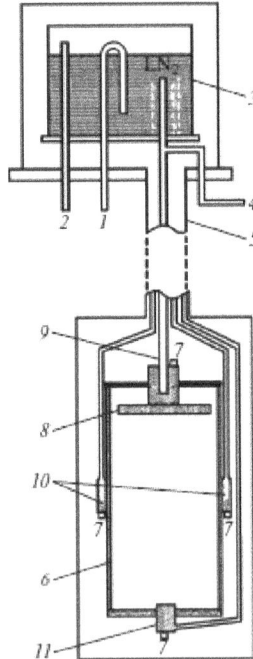

Figure 8.25: Schematic diagram of the cryogenic system of RED-100 LXe detector based on four tubular thermosyphons: 1 — inlet pipe for supplying liquid nitrogen to the liquid nitrogen flask, 2 — ventilation branch pipe, 3 — liquid nitrogen flask, 4 — nitrogen gas supply, 5 — vacuum vessel, 6 — cold vessel wrapped in copper heat shield, 7 — temperature sensors, 8 — upper copper cold head for condensing xenon, 9 — upper thermosyphon tube, 10 — two side thermosyphon cold heads, 11 — cold head for lower thermosyphon used for regulation of temperature gradient along the detector cold chamber height. Redrawn from Bolozdynya *et al.* (2015, 2016).

wall. The heat exchangers are massive copper elements (with a mass of $1-10$ kg) being in thermal contact with the heat shield surrounding the cold chamber, or with refrigerators inside the chamber, which are intended for xenon condensation and control of the temperature gradient inside the cold chamber to avoid a convection flow of LXe affecting UV light collection.

Cooling of the Xe relies upon the temperature difference between the LN_2 and Xe vaporization temperatures as well as the adjustable cooling power of the thermosyphon. Thermosyphon cooling is completely passive, with a fixed amount of nitrogen in the secondary cooling loop. Therefore no pumps are required and there is no direct path for large quantities of LN_2 (typically stored in Dewars). As the thermosyphon transport tubing may be 9.5 or 12.7 mm OD, a minimal amount of nitrogen can be placed into the tubes, of the order of 15 g to reach 10 bar. This mass gradually increases as the nitrogen condenses but the mass remains modest. Over-pressure protection is by relief valves and burst disks. The thermosyphons are charged with nitrogen via high-pressure cylinders. Control of the gas mass (and cooling power) is accomplished with control valves, mass flow controllers, sensors, and processing devices connected to slow control via Ethernet.

The thermosyphon technology is considered in a project of nEXO detector in order to remove 500 W heat power (Al Kharusi *et al.*, 2018). The detector with about 5 tons of liquid xenon is submerged in the HFE-7000 cooling liquid jacket which is cooled with five thermosyphons connected to five cold plates. This fluid provides ultra-low background shielding, thermal uniformity to the TPC and the ability of transferring pressure loads from the TPC vessel to the cryostat. The fluid also transfers the pressure load to the 25 mm thick inner cryostat copper vessel that is designed to tolerate absolute implosive (explosive) loads >100 kPa (>300 kPa).

Conclusion

The detector technology described in this book has been invented at MEPhI 50 years ago and since that time has demonstrated a great potential to be used in a variety of fundamental research programs. To date, this technology has found the most successful application for the search of hypothetical cold dark matter in the form of weakly interacting massive particles (WIMPs). After about twenty years of impressive progress, this area of basic research is inevitably approaching its natural conclusion. The reason is that a gradual decrease in the lower limit of the possible cross-section for the interaction of WIMPs with ordinary baryonic matter due to an increase in the mass of the detector working medium is steadily approaching the limit of registration of elastic coherent scattering of solar neutrinos. In the WIMP mass region of 40–50 GeV/c^2, the limit is the cross-section 10^{-49} cm^2; in the mass region <10 GeV/c^2, the limit is the cross-section of the order of 5–10^{-45} cm^2. These limits will be reached already at the stage of operation of the next G3 generation of WIMP detectors. A further increase in the masses of detectors will contribute to an increase in solar neutrino data, the registration of which will become an irreducible background for the search of WIMPs. In this case, the detectors of the G3 generation will be the source of new information on neutrinos of relatively low energies, including the neutrino pp cycle from the Sun, diffuse supernova neutrinos, atmospheric neutrinos etc. In addition, multi-ton active mass WIMP detectors of the G3 generation shall become, even with naturally occurring isotope abundances, sensitive to double-beta decay at the today achieved level of sensitivity.

In a few years, the Deep Underground Neutrino Experiment (DUNE) will begin data taking with hundreds-ton liquid argon radiation detectors including two-phase emission detectors to study long basic neutrino oscillations, as well as to get new data for neutrino astrophysics and to receive new limits in search for nucleon decay.

The next promising area of research and practical applications will be open in result of the RED-100 detector exposition and Kalinin NPP for studying the effect of coherent scattering of reactor neutrinos off xenon nuclei. The detector is designed to be installed practically on the Earth's surface in the vicinity of nuclear reactors or accelerators. In case of successful testing of the RED-100 detector at the Kalinin NPP over the next couple of years, a way for effective remote monitoring of the active zone of nuclear reactors by means of relatively compact and mobile detectors will be open.

Thus, the further development of the technology of two-phase emission detectors will increasingly focus on neutrino physics and astrophysics, nucleon decay as well as its practical applications in the field of improving the safety of nuclear energy production and supporting international programs on the non-proliferation of nuclear weapons. It should be noted that radioisotope tomography in nuclear medicine remains another attractive application of the technology described in this book. The authors will be delighted if this monograph appears to be useful to support the development of instrumentation based on two-phase emission detector technology for these exciting areas.

Bibliography

Aalbers, J., Agostini, F., Alfonsi, M. *et al.* (2016). DARWIN: towards the ultimate dark matter detector, *JCAP* 1611: 11017.

Aalseth, C. E., Acerbi, F., Agnes, P. *et al.* (2017). Cryogenic characterization of FBK RGB-HD SiPms, *J. Instrum.* 12: P09030.

Aalseth, C. E., Acerbi, F., Agnes, P. *et al.* (2018). DarkSide-20k: a 20 tonne two-phase LAr TPC for direct dark matter detection at LNGS, *Eur. Phys. J. Plus* 133: 131.

Aalseth, C. E., Abdelhakim, S., Agnes, P. *et al.* (2020a). SiPM-matrix readout of two-phase argon detectors using electroluminescence in the visible and near infrared range, e-print arXiv: 2004.02024, 4 April 2020, *Eur. Phys. J. C* 81: 153 (2021).

Aalseth, C. E., Abdelhakim, S., Acerbi, F. *et al.* (2020b). Design and construction of a new detector to measure ultra-low radioactive-isotope contamination of argon, *J. Instrum.* 15:P02024.

Abe, K., Heida, K., Hiraide, K. *et al.* (2013). XMASS detector, *Nucl. Instrum. Meth. A* 716: 78–85.

Abe, K., Hosaka, J., Iida, T. *et al.* (2009). Distillation of Liquid Xenon to Remove Krypton, *Astroparticle Phys.* 31: 290–296.

Abi, B., Acciarri, R., Acero, M. A. *et al.* (2018). The DUNE Far Detector Interim Design Report, Volume 3: Dual-Phase Module, e-print arXiv:1807.10340, 26 July 2018.

Abi, B., Acciarri, R., Acero, M. A. *et al.* (2020). Underground Neutrino Experiment (DUNE) Far Detector Technical Design Report Volume I: Introduction to DUNE, e-print ArXiv:2002.02967.

Abramov, A. V., Dolgoshein, B. A., Kruglov, A. A., and Rodionov, B. U. (1975). Electrostatic emission of free electrons from solid xenon, *JETP Lett.* 21: 82–85 (in Russian).

Acciarri, R., Antonello, M., Baibussinov, B. *et al.* (2010). Effects of nitrogen contamination in liquid argon, *J. Instrum.* 5: P06003.

Acciarri, R., Antonello, M., Baibussinov, B. *et al.* (2011). The WArP Experiment, *J. Phys.: Conf. Ser.* 308: 012005

Acciarri, R., Antonello, M., Boffelli, F. *et al.* (2012). Demonstration and comparison of photomultiplier tubes at liquid Argon temperature, *J. Instrum.* 7: P01016.

Acciarri, R., Carls, B., James, C. *et al.* (2014). Liquid argon dielectric breakdown studies with the MicroBooNE purification system, *J. Instrum.* 9: P11001.

Acosta-Kane, D., Acciarri, R., Amaize, O. *et al.* (2008). Discovery of underground argon with low level of radioactive ^{39}Ar and possible applications to WIMP dark matter detectors, *Nucl. Instrum. Meth. A* 587: 46–51.

Adam, J., Bai, X., Baldini, A. *et al.* (2010). A limit for the decay from the MEG experiment, *Nucl. Phys. B* 834: 1–12.

Adamowski, M., Carls, B., Dvorak, E. *et al.* (2014). The liquid argon purity demonstrator, *J. Instrum.*, 9: P07005.

Adams, D. L., Baird, M., Barr, G. *et al.* (2019). Design and performance of a 35-ton liquid argon time projection chamber as a prototype for future very large detectors, e-print arXiv: 1912.08739v1, 18 December 2019.

Agnes, P., Alexander, T., Alton, A. *et al.* (2015). First results from the DarkSide-50 Dark Matter Experiment at Laboratori Nazionali del Gran Sasso, *Phys. Lett. B* 743: 456–466.

Agnes, P., Agostino, L., Albuquerque, I. F. M. *et al.* (2016). Results From the First Use of Low Radioactivity Argon in a Dark Matter Search. *Phys. Rev. D* 93: 081101; *Phys. Rev. D* (2017) 95: 069901 (addendum).

Agnes, P., Dawson, J., De Cecco, S., Fan, A., Fiorillo, G. *et al.* (2018a). Measurement of the liquid argon energy response to nuclear and electronic recoils, *Phys. Rev. D* 97: 112005.

Agnes, P., Albuquerque, I. F. M., Alexander, T. *et al.*(2018b). DarkSide-50 532-day dark matter search with low-radioactivity argon, *Phys. Rev. D* 98:102006.

Agnes, P. on behalf of DarkSide Collaboration (2020). Simulation of the argon response and light detection in a dual-phase TPC. *J. Instrum.* 15: C01044.

Agostinelli, S., Allison, J., Amako, K., Apostolakis, J., Araújo, H., Arce, P. *et al.* (2003). Geant4 — A simulation toolkit, *Nucl. Instrum. Meth. Phys. Res. A* 506: 250–303.

Ahlen, S. P. (1980). Theoretical and experimental aspects of the energy loss of relativistic heavy ionizing particles, *Rev. Mod. Phys.* 52: 121–173.

Aimard, B., Alt, Ch., Asaadi, J., Auger, M. *et al.* (2018). A 4 tonne demonstrator for large-scale dual-phase liquid argon time projection chambers, *J. Instrum.* 13: P11003.

Akerib, D. S., Bai, X., Bernard, E. *et al.* (2013a). An ultra-low background PMT for liquid xenon detectors, *Nucl. Instrum. Meth. A* 703: 1–6.

Akerib, D. S., Bai, X., Bedikian, S. *et al.* (2013b). The large underground xenon (LUX) experiment, *Nucl. Instrum. Meth. A* 704: 111–126.

Akerib, D. S., Araújo, H. M., Bai, X. *et al.* (2014). First results from the LUX dark matter experiment at the Sanford underground research facility, *Phys. Rev. Lett.* 112: 091303.

Akerib, D. S., Araújo, H. M., Bai, X. *et al.* (2015). Radiogenic and muon-induced backgrounds in the LUX dark matter detector, *Astropart. Phys.* 62: 33.

Akerib, D. S., Alsum, S., Araújo, H.M. *et al.* (2016b). Low-energy (0.7–74 keV) nuclear recoil calibration of the LUX dark matter experiment using D-D neutron scattering kinematics, e-print arXiv:1608.05381v2, 26 October 2016.

Akerib, D. S., Araújo, H. M., Bai, X. *et al.* (2016a). Tritium calibration of the LUX dark matter experiment, *Phys. Rev. D* 93: 072009.

Akerib, D. S., Alsum, S., Araújo, H. M., Bai, X., Bailey, A. J. *et al.* (2017a). Ultralow energy calibration of LUX detector using ^{127}Xe electron capture, *Phys. Rev. D* 96: 112011.

Akerib, D. S., Alsum, S., Araújo, H. M. *et al.* (2017b). ^{83m}Kr calibration of the 2013 LUX dark matter search, *Phys. Rev. D* 96: 112009.

Akerib, D. S., Alsum, S., Araújo, H. M., Bai, X., Bailey, A. J. *et al.* (2017c). Signal yields, energy resolution, and recombination fluctuations in liquid xenon, *Phys. Rev. D* 95: 012008.

Akerib, D. S., Alsum, S., Araújo, H. M. *et al.* (2017d). Results from a search for dark matter in the complete LUX exposure, *Phys. Rev. Lett.* 118: 021303.

Akerib, D. S., Akerlof, C. W., Akimov, D. Yu. *et al.* (2020a). The LUX-ZEPLIN (LZ) Experiment. *Nucl. Instrum. Meth. A* 953: 163047.

Akerib, D. S., Alsum, S., Araújo, H. M. *et al.* (2020b). Investigation of background electron emission in the LUX detector, arXiv:2004.07791.

Akimov, D., Burenkov, A., Danilov, M. *et al.* (1998). Scintillation two-phase xenon detector with gamma and electron-background rejection for dark matter search, in D. B. Cline (ed.), *Proc. of 3rd International Symposium on Sources and Detection of Dark Matter in the Universe*, 17–20 February 1998, Marina del Rey, California, USA, Elsevier Science B.V., p. 461.

Akimov, D., Bewick, A., Davidge, D. *et al.* (2002). Measurements of scintillation efficiency and pulse shape for low-energy recoils in liquid xenon, *Phys. Lett. B* 524: 245–251.

Akimov, D. Yu., Batyaev, V. F., Borovlev, S. P. *et al.* (2003). Liquid xenon for WIMP searches: measurement with a two-phase prototype, in N.J.C. Spooner and V. Kudryavtsev (ed.), *Proc. 4th International Workshop on The Identification of Dark Matter*, 2–6 September 2002, York, UK, World Scientific Publishing Co. Pte. Ltd., pp. 371–376.

Akimov, D. Yu., Alner, G. J., Araújo, H. M. *et al.* (2007). The ZEPLIN-III dark matter detector: instrument design, manufacture and commissioning, *Astropart. Phys.* 27: 46–60.

Akimov, D., Bondar, A., Burenkov, A., and Buzulutskov, A. (2009). Detection of reactor antineutrino coherent scattering off nuclei with a two-phase noble gas detector, *J. Instrum.* 4: P06010.

Akimov, D. (2011). Techniques and results for the direct detection of dark matter (review), *Nucl. Instrum. Meth. A* 628:50–58.

Akimov, D. Y., Araújo, H. M., Barnes, E. J. *et al.* (2012a). WIMP-nucleon cross-section results from the second science run of ZEPLIN-III, *Phys. Lett. B* 709: 14–20.

Akimov, D. Yu., Aleksandrov, I. S., Belov, V. A. *et al.* (2012b). Measurement of single-electron noise in a liquid-xenon emission detector, *Instrum. Exp. Tech.* 55: 423–428.

Akimov, D. Yu., Alexandrov, I. S., Aleshin, V. I. *et al.* (2013a). Prospects for observation of neutrino-nuclear neutral current coherent scattering with two-phase Xenon emission detector. *J. Instrum.* 8: P10023.

Akimov, D. Akindinov, A.V., Alexandrov, I. S. *et al.* (2013b). Two-phase xenon emission detector with electron multiplier and optical readout by multipixel avalanche Geiger photodiodes, *J. Instrum.* 8: P05017.

Akimov, D. Yu., Bolozdynya, A. I., Efremenko, Yu. V. *et al.* (2014). A controllable voltage divider for Hamamatsu R11410-20 photomultipliers for use in the RED 100 emission detector, *Instrum. Exp. Tech.* 57: 615–619.

Akimov, D. Yu., Bolozdynya, A. I., Efremenko, Yu. V. *et al.* (2015). Observation of light emission from Hamamatsu R11410-20 photomultiplier tubes, *Nucl. Instrum. Meth. A* 794: 1–2.

Akimov, D. Yu., Belov, V. A., Bolozdynya, A. I. *et al.* (2016). Observation of delayed electron emission in a two-phase liquid xenon detector, *J. Instrum.* 11: C03007.

Akimov, D. Y., Albert, J. B., An, P. *et al.* (2017a). Observation of coherent elastic neutrino-nucleus scattering, *Science* 357: 1123–1126.

Akimov, D. Yu., Alexandrov, I. S., Belov, V. A. *et al.* (2017b). The RED-100 two-phase emission detector, *Instrum. Exp. Tech.* 60: 175–181.

Akimov, D. Yu., Belov, V. A., Berdnikova, A. K. *et al.* (2017c). Purification of liquid xenon with the spark discharge technique for use in two-phase emission detectors, *Instrum. Exp. Tech.* 60: 782–788.

Akimov, D. Yu., Bolozdynya, A.I., Konovalov, A.M. *et al.* (2018). Two-phase emission low-background detector (in Russian), *Utility model patent* RU 184222 U1.

Akimov, D. Yu., Belov, V. A., Bolozdynya, A. I. *et al.* (2019). An integral method for processing xenon used as a working medium in the RED-100 two-phase emission detector, *Instrum. Exp. Tech.* 62: 457–463.

Akimov, D. Yu., Belov, V. A., Bolozdynya, A. I. *et al.* (2020). First ground-level laboratory test of the two-phase xenon emission detector RED-100, *J. Instrum.* 15: P02020.

Al Kharusi, S., Alamre, A., Albert, J. B. *et al.* (2018). nEXO Pre-Conceptual Design Report, e-Print arXiv:1805.11142v2, 13 August 2018.

Al Samarai, I., Berat, C., Deligny, O. *et al.* (2016). Molecular bremsstrahlung radiation at GHz frequencies in air, *Phys. Rev. D* 93: 052004.

Albert, J., Barbeau, P., Beck, D. *et al.* (2017). Measurement of the drift velocity and transverse diffusion of electrons in liquid xenon with the EXO-200 detector, *Phys. Rev.C* 95: 025502.

Alexander, T., Alton, D., Arisaka, K. *et al.* (2013a). Light yield in DarkSide-10: a prototype two-phase argon TPC for dark matter searches, *Astropart. Phys.* 49: 44–51.

Alexander, T., Back, H. O., Cao, H. *et al.* (2013b). Observation of the dependence on drift field of scintillation from nuclear recoils in liquid argon, *Phys. Rev. D* 88: 092006.

Allen, M. P. and Tildesley, D. J. (2017). *Computer Simulation of Liquids*. Oxford University Press, Oxford.

Alner, G. J., Araújo, H., Arnison, G. J. *et al.* (2005). First limits on nuclear recoil events from the ZEPLIN I galactic dark matter detector. *Astropart. Phys.* 23: 444–462.

Alner, G. J., Araújo, H. M., Bewick, A. *et al.* (2007). First limits on WIMP nuclear recoil signals in ZEPLIN-II: a two-phase xenon detector for dark matter detection, *Astropart. Phys.* 28: 287–302.

Amerio, S., Amoruso, S., Antonello, M. *et al.* (2004). Design, construction and tests of the ICARUS T600 detector, *Nucl. Instrum. Meth. A* 527: 329–410.

Amey, R. L. and Cole, R. H. (1964). Dielectric constants of liquefied noble gases and methane, *J. Chem. Phys.* 40, 146–148.

Amoruso, S., Antonello, M., Aprili, P., Arneodo, F., Badertscher, A. *et al.* (2004). Study of electron recombination in liquid argon with the ICARUS TPC, *Nucl. Instr. Meth. Phys. Res. A* 523: 275–286.

Anderson, D. F., Charpak, G., Holroyd, R. A., and Lamb, D. C. (1987). Liquid ionization chambers with electron extraction and multiplication in the gaseous phase, *Nucl. Instrum. Meth. A* 261: 445–448.

Anger, H. (1958). Scintillation camera, *Rev. Sci. Instrum.* 29: 27–33.

Angle, J., Aprile, E., Arneodo, F. *et al.* (2008). First results from the XENON10 dark matter experiment at the Gran Sasso National Laboratory, *Phys. Rev. Lett.* 100: 021303.

Angle, J., Aprile, E., Arneodo, F., Baudis, L., Bernstein, A. *et al.* (2011). Search for light dark matter in XENON10 data, *Phys. Rev. Lett.* 107: 051301.

Anisimov, S. N., Bolozdynya, A. I., and Stekhanov, V. N. (1984). Electron localization and drift under the surface of condensed krypton, *JETP Lett.* 40: 829–832.

Anisimov, S. N., Barabash, A. S., Bolozdynya, A. I., and Stekhanov, V. N. (1989a). Control of electro-negative impurities contents in liquid krypton with emission detector, *Pribory i Tehnika Eksperimenta* 1: 79–82 (in Russian).

Anisimov, S. N., Barabash, A. S., Bolozdynya, A. I., and Stekhanov, V. N. (1989b). Measuring the ^{85}Kr content in krypton using a liquid ionization chamber, *Atomic Energy* 66: 415–417 (in Russian).

Ankowski, A., Antonello, M., Aprili, P. *et al.* (2006). Characterization of ETL 9357FLA photomultiplier tubes for cryogenic temperature applications, *Nucl. Instrum. Meth. A*556: 146–157.

Aprile, E., Mukherjee, R., and Suzuki, M. (1991). Performance of a liquid xenon ionization chamber irradiated with electrons and gamma-rays, *Nucl. Instr. and Meth. in Phys. Res. A* 302: 177–185.

Aprile, E., Bolotnikov, A., Chen, D. *et al.* (1994). Performance of CsI photocathodes in liquid Xe, Kr, and Ar, *Nucl. Instrum. Meth. A* 338: 328–335.

Aprile, E., Curioni, A., Egorov, V. *et al.* (2000). Spectroscopy and imaging performance of the liquid xenon gamma-ray imaging telescope (LXeGRIT). *Proc. SPIE* 4140:333-343; e-Print astro-ph/0012297.

Aprile, E., Baltz, E.A., Curioni, A. *et al.* (2002). XENON: a 1 tonne liquid xenon experiment for a sensitive dark matter search, in *Proc. International Workshop on Technique and Application of Xenon Detectors*, 3–4 December 2001, pp. 165–178. e-Print astro-ph/0207670.

Aprile, E., Giboni, K., Majevski, P., Ni, K., and Yamashita, M. (2004). Proportional light in a dual-phase xenon chamber, *IEEE Trans. Nucl. Sci.* 51: 1986–1990.

Aprile, E., Giboni, K. L., Majewski, P. *et al.* (2005a). Scintillation response of liquid xenon to low energy nuclear recoils, *Phys. Rev. D* 72: 072006.

Aprile, E., Giboni, K. L., Majewski, P. *et al.* (2005b). The XENON dark matter search experiment, *New Astron. Rev.* 49: 289–295.

Aprile, E., Bolotnikov, A., Bolozdynya, A., and Doke, T. (2006a). *Noble Gas Detectors*, Weinheim, Germany: Wiley-VCH Verlag GmbH & Co. KGaA, 362 p.

Aprile, E., Dahl, C.E., DeViveiros, L. *et al.* (2006b). Simultaneous measurement of ionization and scintillation from nuclear recoils in liquid xenon as target for a dark matter experiment, *Phys. Rev. Lett.* 97: 081302.

Aprile, E., Giboni, K., Majewski, P., Ni, K., and Yamashita, M. (2007). Observation of anticorrelation between *scintilçlation and ionization for MeV gamma rays in liquid xenon, Phys Rev. B* 76: 014115.

Aprile, E., Baudis, L., Choi, B., Giboni, K. L., Lim, K., Manalaysay, A. *et al.* (2009). New measurement of the relative scintillation efficiency of xenon nuclear recoils below 10 keV, *Phys. Rev. C* 79: 045807.

Aprile, E. and Doke, T. (2010). Liquid xenon detectors for particle physics and astrophysics, *Rev. Mod. Phys.* 82: 2053–2097.

Aprile, E., Angle, J., Arneodo, F. *et al.* (2011). Design and performance of the XENON10 dark matter experiment, *Astropart. Phys.* 34: 679–698.

Aprile, E., Budnik, R., Choi, B., Contreras, H.A., Giboni, K.-L. *et al.* (2012a). Measurement *of the scintillation yield of low-energy electrons in liquid xenon, Phys. Rev. D* 86: 112004.

Aprile, E., Arisaka, K., Arneodo, F. *et al.* (2012b). The XENON100 dark matter experiment, *Astropart. Phys.* 35: 573–590.

Aprile, E., Alfonsi, M., Arisaka, K., Arneodo, F., Balan, C. *et al.* (2013). Response of the XENON100 dark matter detector to nuclear recoils, *Phys. Rev. D* 88: 012006.

Aprile, E., Alfonsi, M., Arisaka, K., Arneodo, F., Balan, C. *et al.* (The XENON100 Collaboration) (2014a). Observation and applications of single-electron charge signals in the XENON100 experiment, *J. Phys. G: Nucl. Part. Phys.* 41: 035201.

Aprile, E., Contreras, H., Goetzke, L. W. *et al.* (2014b). Measurements of proportional scintillation and electron multiplication in liquid xenon using thin wires, *J. Instrum.* 9: P11012.

Aprile, E., Aalbers, J., Agostini, F. *et al.* (2016). XENON100 dark matter results from a combination of 477 live days, *Phys. Rev. D* 94: 122001.

Aprile, E., Aalbers, J., Agostini, F. *et al.* (2017). The XENON1T dark matter experiment, *Europ. Phys. J. C* 77: 881.

Aprile, E., Aalberts, J., Agostini, F. *et al.* (2018a). Dark matter search results from a one tonne×year exposure of XENON1T, *Phys. Rev. Lett.* 121: 111302.

Aprile, E., Anthony, M., Lin, Q., Greene, Z., de Perio, P. *et al.* (2018b). Simultaneous measurement of the light and charge response of liquid xenon to low-energy nuclear recoils at multiple electric fields, *Phys. Rev. D.* 98: 112003.

Aprile, E., Aalbers, J., Agostini, F., Alfonsi, M., Althueser, L. *et al.* (2019a). XENON1T dark matter data analysis: Signal and background models and statistical inference, *Phys. Rev. D* 99: 112009.

Aprile, E., Aalbers, J., Agostini, F. *et al.* (2019b). Observation of two-neutrino double electron capture in ^{124}Xe with XENON1T, *Nature* 568: 532–535.

Araújo, H. M., Bewick, A., Davidge, D. *et al.* (2004). Low-temperature study of 35 photo-multiplier tubes for the ZEPLIN-III experiment, *Nucl. Instrum. Meth. A* 521: 407–415.

Araújo, H. M., Akimov, D. Y., Barnes, E. J., Belov, V. A., Bewick, A., Burenkov, A. A. *et al.* (2012). Radioactivity backgrounds in ZEPLIN-III, *Astropart. Phys.* 35: 495–502.

Arazi, L., Coimbra, A. E. C., Itay, R. *et al.* (2013). First observation of liquid-xenon proportional electroluminescence in THGEM holes, *J. Instrum.* 8: C12004.

Arazi, L., Erdal, E., Coimbra, A. E. C. *et al.* (2015). Liquid hole multipliers: bubble-assisted electroluminescence in liquid xenon, *J. Instrum.* 10: P08015.

Arneodo, F., Baiboussinov, B., Badertscher, A. *et al.* (2000). Scintillation efficiency of nuclear recoil in liquid xenon, *Nucl. Instrum. Meth. A* 449: 147–157.

Asaadi, J., Jones, B. J. P., Tripathi, A. *et al.* (2019). Emanation and bulk fluorescence in liquid argon from tetraphenyl butadiene wavelength shifting coatings, *J. Instrum.* 14: P02021.

Atrazhev, V. M., Berezhnov, A. V., Dunikov, D. O. *et al.* (2005). Electron transport coefficients in liquid xenon, in *Proc. 2005 IEEE Int. Conf. on Diel. Liquids ICDL2005*, Coimbra June 26–July 1, 2005, pp. 329–332.

Auger, M., Auty, D. J., Barbeau, P. S., Beauchamp, E., Belov, V., Benitez-Medina, C. *et al.* (2012). Search for neutrinoless double-beta decay in 136Xe with EXO-200, *Phys. Rev. Lett.* 109: 032505.

Auger, M., Auty, D. J., Barbeau, P. S. *et al.* (2012). The EXO-200 detector, part I: detector design and construction, *J. Instrum.* 7: P05010.

Auger, M., Blatter, A., Ereditato, A. *et al.* (2016). On the electric breakdown in liquid argon at centimeter scale, *J. Instrum.*, 11: P03017.

Badertscher, A., Curioni, A., Knecht, L. *et al.* (2011). First operation of a double phase LAr large electron multiplier time projection chamber with a 2D projective readout anode, *Nucl. Instrum. Meth. A* 641: 48–57.

Badertscher, A., Curioni, A., Degunda, U. *et al.* (2013). First operation and performance of a 200 lt double phase LAr LEM-TPC with a 40 × 76 cm^2 readout, *J. Instrum.* 8: P04012.

Bakale, G., Sowada, U., and Schmidt, W.F. (1976). Effect of an electric field on electron attachment to SF_6, N_2O, and O_2 in liquid argon and xenon, *J. Phys. Chem.* 80: 2556–2559.

Balashov, V. V. (1997). *Interaction of Particles and Radiation with Matter.* Springer Verlag.

Balau, F., Solovov, V., Chepel, V. *et al.* (2009). GEM operation in double-phase xenon, *Nucl. Instrum. Meth. A* 598: 126–129.

Baldini, A. M., Chippini, M., Gerone, M. *et al.* (2018). The design of the MEG II experiment, *Eur. Phys. J. C* 78: 380.

Barabash, A. S. and Bolozdynya, A. I. (1989). How to detect "dark matter" of the Galaxy if it is composed of neutral weakly interacting particles with 1-10 GeV/c^2 masses?, *Lett. JETP* 49(6): 314–317 (in Russian).

Barabash, A. S. and Bolozdynya, A. I. (1993). *Liquid Ionization Detectors.* Moscow: Energoatomizdat, 240 p. (in Russian).

Barmin, V. V., Borissov, V. N., Golubchikov, V. M. *et al.* (1984). The 700-liters Xenon bubble chamber DIANA, *Instrum. Exp. Tech.* 27: 850–853.

Barocchi, F., Chieux, P., Magli, R., Reatto, L., and Tau, M. (1993). Static structure of dense krypton and interatomic interaction. *J. Phys. C* 5: 4299–4314.

Barrow, P., Baudis, L., Cichon, D. *et al.* (2016). Qualification tests of the R11410-21 photomultiplier tubes for the XENON1T detector, *J. Instrum.* 12: P01024.

Baudis, L., Dujmovic, H., Geis, C., James, A., Kish, A. *et al.* (2013). Response of liquid xenon to Compton electrons down to 1.5 keV, *Phys. Rev. D* 87: 115015.

Belli, P., Bernabei, R., D'Angelo, S. *et al.* (1991). Liquid Xenon Scintillators, *Nucl. Instrum. Meth. A* 310: 150–153.

Belogurov, S., Bressi, G., Carugno, G., Conti, E., Iannuzzi, D., and Meneguzzo, A. T. (2000). Measurement of the light yield of infrared scintillation in xenon gas, *Nucl. Instrum. Meth. A* 452: 167–169.

Belonoshko, A. B., LeBacq, O., Ahuja, R., and Johansson, B. (2002). Molecular dynamics study of phase transitions in Xe. *J. Chem. Phys.* 117: 7233–7244.

Benetti, P., Calligarich, E., Dolfini, R. *et al.* (1993). Detection of energy deposition down to the keV region using liquid xenon scintillation, *Nucl. Instrum. Meth. A* 327: 203–206.

Benetti, P., Calaprice, F., Calligarich, E. *et al.* (2007). Measurement of the specific activity of ^{39}Ar in natural argon, *Nucl. Instrum. Meth. A* 574: 83–88.

Benetti, P., Acciarri, R., Adamo, F. *et al.* (2008). First results from a dark matter search with liquid argon at 87 K in the Gran Sasso Underground Laboratory, *Astropart. Phys.* 28: 495–507.

Benson, C., Gann, G. D. O., and Gehman, V. (2018). Measurements of the intrinsic quantum efficiency and absorption length of tetraphenyl butadiene thin films in the vacuum ultraviolet regime, *Eur. Phys. J. C* 78: 329.

Berber, S., Kwon, Y.-K., and Tomnek, D. (2000). Unusually high thermal conductivity of carbon nanotubes, *Phys. Rev. Lett.* 84: 4613–4616.

Bernabei, R., Belli, P., Cerulli, R. *et al.* (2001). Light response of a pure liquid Xenon scintillator irradiated with 2.5 MeV neutrons, *Eur. Phys. J.* 3: 1–8.

Bernardes, N. and Primakoff, H. (1959). Molecule formation in the inert gases, *J. Chem. Phys.* 30: 691–694.

Bezrukov, F., Kahlhoefer, F., and Lindner, M. (2011). Interplay between scintillation and ionization in liquid xenon dark matter searches, *Astropart. Phys.* 35: 119–127.

Billard, J., Figueroa, E., and Strigari, L. (2014). Implication of neutrino backgrounds on the reach of next generation dark matter direct detection experiments, *Phys. Rev. D* 89: 023524.

Birks, J. B. (1951). Scintillations from organic crystals: Specific fluorescence and relative response to different radiations, *Proc. Phys. Soc. A* 64: 874–877.

Blatter, A., Ereditato, A., Hsu, C.-C. *et al.* (2014). Experimental study of electric breakdowns in liquid argon at centimeter scale, *J. Instrum.* 9: P04006.

Bloch, F. and Bradbury, N. E. (1935). On the mechanism of unimolecular electron capture, *Phys. Rev.* 48: 689–695.

Bolozdynya, A., Egorov, V., Korshunov, A. A., Sokolov, L. I., Miroshnichenko, V. P., and Rodionov, B. U. (1977a). The first observation of particle tracks in condensed matter obtained by the emission method, *Lett. JETP* 25: 401–404 (in Russian).

Bolozdynya, A. I., Miroshnichenko, V. P., Rodionov, B. U. (1977b). Electrostatic emission of free electrons from liquid and solid argon, *Lett. JTP* 2: 64–67 (in Russian).

Bolozdynya, A. I., Lebedenko, V. N., Rodionov, B. U., Balakin, A. A., Boriev, I. A., and Yakovlev, B. S. (1978). Electrostatic emission of electrons into the gas phase from alpha-tracks in the liquid isooctane, *J. Tech. Phys.* 48: 1514–1519 (in Russian).

Bolozdynya, A. I., Egorov, O. K., Miroshnichenko, V. P., Rodionov, B. U., and Shuvalova, E. N. (1980). A new possibility to search for low-ionizing particles, *Elementary Particles and Cosmic Rays* 5: 65–72. Moscow: Atomizdat (in Russian).

Bolozdynya, A. I., Egorov, V. V., Kalashnikov, S. D., Krivoshein, L., Miroshnichenko, V. P., and Rodionov, B. U. (1981). Detector of coordinates for low energy particles based on condensed krypton, preprint ITEP-113, Moscow: Institute for Theoretical and Experimental Physics, 8 p. (in Russian).

Bolozdynya, A. I., Egorov, V. V., Kalashnikov, S. D., Krivoshein L., Miroshnichenko, V. P., and Rodionov, B. U. (1982). Emission electroluminescence gamma camera based on condensed xenon, preprint ITEP-37, Moscow: Institute for Theoretical and Experimental Physics. 33 p. (in Russian).

Bolozdynya, A. I. and Stekhanov, V. N. (1984). Capture of quasi-free electrons by oxygen in condensed krypton, preprint ITEP-27, Moscow: Institute for Theoretical and Experimental Physics. 20 p. (in Russian).

Bolozdynya, A. I. (1985). Registration of ionizing radiation in condensed krypton by the emission method, PhD thesis, Moscow: ITEP. 136 p. (in Russian).

Bolozdynya, A. I., Egorov, V. V., Kalashnikov, S. D., Krivoshein L., Miroshnichenko, V. P., and Rodionov, B. U. (1985). Emission electroluminescence chamber with condensed Xenon working medium, *Instrum. Exp. Tech.* 4: 43–45 (in Russian).

Bolozdynya, A. I. (1986a). To electron emission from liquid isooctane, preprint ITEP, 86-103, 8 p. (in Russian).

Bolozdynya, A. I. (1986b). Excess electron emission from condensed krypton and other nonpolar dielectrics, preprint ITEP 172-86, Moscow: CNIIatominform (in Russian).

Bolozdynya, A. I. (1991). Transport of excess electrons through and along a condensed krypton interface, in *Proc. 3rd Int. Conf. Porp. And Appl. Diel. Materials*, July 8–12, 1991, Tokyo, Japan, pp. 841–844.

Bolozdynya, A., Egorov, V., Rodionov, B., Miroshnichenko, V. (1995). Emission detectors, *IEEE Trans. Nucl. Sci.* 42: 565–569.

Bolozdynya, A., Egorov, V., Koutchenokov, A., Safronov, G., Smirnov, G., Medved, S., and Morgunov, V. (1997a). A high pressure xenon self-triggered scintillation drift chamber with 3D sensitivity in the range of 20–140 keV deposited energy, *Nucl. Instrum. Meth.* A 385: 225–238.

Bolozdynya, A., Egorov, V., Koutchenokov, A., Safronov, G., Smirnov, G., Medved, S., and Morgunov, V. (1997b). An electroluminescence emission detector to search for double beta positron decays of ^{134}Xe and ^{78}Kr, *IEEE Trans. Nucl. Sci.* 44: 1046–1051.

Bolozdynya A. (1999) Two-phase emission detectors and their applications, *Nucl. Instrum. Meth. Phys. Res.* A 422: 314–320.

Bolozdynya, A. I., Brusov, P. P., Shutt, T. *et al.* (2007). A chromatographic system for removal of radioactive ^{85}Kr from xenon, *Nucl. Instrum. Meth.* A 579: 50–53.

Bolozdynya, A., Bradley, A., Bryan, S. *et al.* (2009). Cryogenics for the LUX detector, *IEEE Trans. Nucl. Sci.* 56: 2309.

Bolozdynya, A. (2010). *Emission Detectors*. Singapore: World Scientific Publishing Co. 209 p.

Bolozdynya, A. I., Dmitrenko, V. V., Efremenko, Yu. V. *et al.* (2015). The two-phase closed tubular cryogenic thermosyphon, *Int. J. Heat Mass Transf.* 80: 159–162.

Bolozdynya, A. I., Efremenko, Yu. V., Sidorenko, A. V. *et al.* (2016). Thermostatting of the RED-100 liquid-xenon emission detector, *Instrum. Exp. Tech.* 59: 483–486.

Bondar, A., Buzulutskov, A., and Shekhtman, L. (2002a). High pressure operation of the triple-GEM detector in pure Ne, Ar and Xe, *Nucl. Istrum. Meth. A* 481: 200–203.

Bondar, A., Buzulutskov, A., Snopkov, L., and Vasiliev, A. (2002b). Triple GEM operation in compressed He and Kr, *Nucl. Istrum. Meth. A* 493: 8–15.

Bondar A., Buzulutskov A., Shekhtman L. *et al.* (2004). Cryogenic avalanche detectors based on gas electron multipliers, *Nucl. Istrum. Meth. A* 524: 130–141.

Bondar, A., Buzulutskov, A., Grebennuk, A. *et al.* (2006). Two-phase argon and xenon avalanche detectors based on gas electron multipliers, *Nucl. Instr. Meth. A* 556: 273–280.

Bondar, A., Buzulutskov, A., Grebennuk, A. *et al.* (2007). A two-phase argon avalanche detector operated in a single electron counting mode, *Nucl Istrum Meth A* 574: 493–499.

Bondar, A., Buzulutskov, A., Grebennuk, A. *et al.* (2008) Thick GEM versus thin GEM in two-phase argon avalanche detectors, *J. Instrum.* 3: P07001.

Bondar, A., Buzulutskov, A., Grebennuk, A. *et al.* (2009a). Recent results on the properties of two-phase argon avalanche detectors, *Nucl. Istrum. Meth. A* 598:121–125.

Bondar, A., Buzulutskov, A., Grebennuk, A. *et al.* (2009b). Electron emission properties of two-phase argon and argon-nitrogen avalanche detectors, *J. Instrum.* 4: P09013.

Bondar, A., Buzulutskov, A., Grebennuk, A. *et al.* (2010). Direct observation of avalanche scintillations in a THGEM-based two-phase Ar avalanche detector using Geiger-mode APD, *J. Instrum.* 5: P08002.

Bondar, A., Buzulutskov, A., Grebennuk, A. *et al.* (2011a). On the low-temperature performances of THGEM and THGEM/G-APD multipliers in gaseous and two-phase Xe, *J. Instrum.* 6: P07008.

Bondar, A., Buzulutskov, A., Grebennuk, A. *et al.* (2011b). Geiger Mode APD performance in a cryogenic two-phase Ar avalanche detector based on THGEMs, *Nucl. Instrum. Meth. A* 628: 364–368.

Bondar, A., Buzulutskov, A., Dolgov, A. *et al.* (2012). Study of infrared scintillations in gaseous and liquid argon. Part II: light yield and possible applications, *J. Instrum.* 7: P06014.

Bondar, A., Buzulutskov, A., Dolgov, A. *et al.* (2013a). First demonstration of THGEM/GAPD-matrix optical readout in a two-phase cryogenic avalanche detector in Ar, *Nucl. Instrum. Meth A* 732: 213–216.

Bondar, A., Buzulutskov, A., Dolgov, A. *et al.* (2013b). Two-phase cryogenic avalanche detectors with THGEM and hybrid THGEM/GEM multipliers operated in Ar and Ar+N_2, *J. Instrum.* 8: P02008.

Bondar, A., Buzulutskov, A., Dolgov, A. *et al.* (2014a). Performance degradation of Geiger-mode APDs at cryogenic temperatures, *J. Instrum.* 9: P08006.

Bondar, A., Buzulutskov, A., Dolgov, A. *et al.* (2014b). Measurement of the ionization yield of nuclear recoils in liquid argon at 80 and 233 keV, *Europhys. Lett.* 108: 12001.

Bondar, A., Buzulutskov, A., Dolgov, A. *et al.* (2015a). MPPC versus MRS APD in two-phase Cryogenic Avalanche Detectors, *J. Instrum.* 10: P04013.

Bondar, A., Buzulutskov, A., Dolgov, A. *et al.* (2015b). Characterization of photo-multiplier tubes for the cryogenic avalanche detector, *J. Instrum.* 10: P10010.

Bondar, A., Buzulutskov, A., Dolgov, A. *et al.* (2016). X-ray ionization yields and energy spectra in liquid argon, *Nucl. Instrum. Meth. A* 816: 119–124.

Bondar, A., Buzulutskov, A., Dolgov, A. *et al.* (2017). Study of cryogenic photomultiplier tubes for the future two-phase cryogenic avalanche detector, *J. Instrum.* 12: C05002.

Bondar, A., Buzulutskov, A., Frolov, E. *et al.* (2019). Electron transport and electric field simulations in two-phase detectors with THGEM electrodes, *Nucl. Instrum. Meth. A* 943: 162431.

Bondar, A., Buzulutskov, A., Dolgov, A. *et al.* (2020a). Neutral bremsstrahlung in two-phase argon electroluminescence: Further studies and possible applications, *Nucl. Instrum. Meth. A* 958: 162432.

Bondar, A., Borisova, E., Buzulutskov, A. *et al.* (2020b). Observation of unusual slow components in electroluminescence signal of two-phase argon detector, *J. Instrum.* 15: C06064.

Bondar, A., Borisova, E., Buzulutskov, A. *et al.* (2020c). Observation of primary scintillations in the visible range in liquid argon doped with methane, *J. Instrum.* 15: C06053.

Boness, M. J. W. and Schulz, G. J. (1980). Structure of O_2, *Phys. Rev. A* 2: 2182–2186.

Borghesani, A. F., Carugno, G., Cavenago, M., and Conti, E. (1990). Electron transmission through the Ar liquid-vapor interface, *Phys. Lett. A* 149: 481–484.

Borghesani, A. F., Carugno, G., and Santini, M. (1991). Experimental determination of the conduction band of excess electrons in liquid Ar, *IEEE Trans. Elec. Insulation.* 26: 615–622.

Boriev, I. A., Balakin, A. A., and Yakovlev, B. S. (1978). Electron emission from non-polar liquids, *High Energy Chem.* 12: 20–25 (in Russian).

Borisova, E. (2020). Study of proportional electroluminescence effect argon for two-phase dark matter detectors, PHD thesis, Budker INP, Novosibirsk.

Boyle, G. J., McEachran, R. P., Cocks, D. G., and White, R. D. (2015). Electron scattering and transport in liquid argon, *J. Chem. Phys.* 142: 154507.

Boyle, G. J., Cocks, G.J., McEachran, R. P. *et al.* (2016). *Ab initio* electron scattering cross sections and transport in liquid xenon, *J. Phys. D* 49: 355201.

Brandt, W. and Kitigawa, M. (1982). Effective stopping-power charges of swift ions in condensed matter, *Phys. Rev. B* 25: 5631–5637.

Breskin, A. (2013). Liquid hole-multipliers: a potential concept for large single-phase noble-liquid TPCs of rare events, *J. Phys. Conf. Ser.* 460: 012020.

Breskin, A., Alon, R., Cortesi, M. *et al.* (2009). A concise review on THGEM detectors, *Nucl. Instrum. Meth. A* 598:107–111.

Bressan, A., Ropelewski, L., Sauli, F., Buzulutskov, A., and Shekhtman, L. (1999). High gain operation of GEM in pure argon, *Nucl. Instrum. Meth. A* 423: 119–124.

Bressi, G., Carugno, G., Conti, E., Iannuzzi, D., and Meneguzzo, A. T. (2000). Infrared scintillation in liquid Ar and Xe, *Nucl. Instrum. Meth. A* 440: 254–257.

Bressi, G., Carugno, G., Conti, E., Del Noce, C., and Iannuzz,i D. (2001a). Infrared scintillation: a comparison between gaseous and liquid xenon, *Nucl. Instrum. Meth. A* 461: 378–380.

Bressi, G., Carugno, G., Conti, E., Iannuzzi, D., and Meneguzzo, A. T. (2001b). A first study of the infrared emission in argon excited by ionizing particles, *Phys. Lett. A* 278: 280–285.

Burenkov, A. A., Akimov, D. Yu., Grishkin, Yu. L. *et al.* (2009). Detection of a single electron in xenon-based electroluminescent detectors, *Phys. Atom. Nucl.* 72: 653–661, *Yad. Fiz.*, 72: 693–701.

Butikov, Yu. A., Dolgoshein, B. A., Lebedenko, V. N., Rogozhin, A. M., and Rodionov, B. U. (1970). Electroluminescence of the noble gases, *Sov. Phys. JETP* 30: 24–28.

Buzulutskov, A., Breskin, A., Chechik, R., Garty, G., Sauli, F., Shekhtman, L. (2000). The GEM photomultiplier operated with noble gas mixtures, *Nucl. Instrum. Meth. A* 443: 164–180.

Buzulutskov, A., Bondar, A., Shekhtman, L., Snopkov, R., and Tikhonov, Y. (2003). First results from cryogenic avalanche detectors based on gas electron multipliers, *IEEE Trans. Nucl. Sci.* 50: 2491–2493.

Buzulutskov, A., Dodd, J., Galea, R. *et al.* (2005). GEM operation in helium and neon at low temperatures, *Nucl. Istrum. Meth. A* 548: 487–498.

Buzulutskov, A. and Bondar, A. (2006). Electric and Photoelectric Gates for ion backflow suppression in multi-GEM structures, *J. Instrum.* 1: P08006.

Buzulutskov, A. F. (2007). Radiation detectors based on gas electron multipliers (review), *Instrum. Exp. Tech.* 50: 287–310.

Buzulutskov, A. F. (2008). Gaseous photodetectors with solid photocathodes, *Phys. Part. Nucl.* 39: 424–453.

Buzulutskov, A., Bondar, A., and Grebenuk, A. (2011). Infrared scintillation yield in gaseous and liquid argon, *Europhys. Lett.* 94: 520011.

Buzulutskov, A. (2012). Advances in cryogenic avalanche detectors, *J. Instrum.*, 7: C02025.

Buzulutskov, A. (2017). Photon emission and atomic collision processes in two-phase argon doped with xenon and nitrogen, *Europhys. Lett.* 117: 39002.

Buzulutskov, A., Shemyakina, E., Bondar, A. *et al.* (2018). Revealing neutral bremsstrahlung in two-phase argon electroluminescence, *Astroparticle Phys.* 103: 29–40.

Buzulutskov, A. (2020). Electroluminescence and electron avalanching in two-phase detectors, *Instruments* 4: 16.

Calvo, J., Cantini, C., Crivelli, P. *et al.* (2017). Commissioning of the ArDM experiment at the Canfranc underground laboratory: first steps towards a tonne-scale liquid argon time projection chamber for Dark Matter searches, *J. Cosmol. Astropart. Phys.* 03: 003.

Cantini, C., Eppercht, L., Gendotti A. *et al.* (2015). Performance study of the effective gain of the double phase liquid Argon LEV Time Projection Chamber, *J. Instrum.* 10: P03017.

Cao, X., Chen, X., Chen, Y. *et al.* (2014). PandaX: a liquid xenon dark matter experiment at CJPL, *Sci. China Phys., Mech. & Astron.* 57: 1476–1494.

Cao, H., Alexander, T., Aprahamian, A. *et al.* (2015). Measurement of scintillation and ionization yield and scintillation pulse shape from nuclear recoils in liquid argon, *Phys. Rev. D* 91: 092007.

Carugno, G., Dainese, B., Pietropaolo, F. *et al.* (1990). Electron lifetime detector for liquid argon, *Nucl. Instrum. Meth. A* 292: 580–584.

Cennini, P., Cittolin, S., Revol, J.-P., Rubbia, C., Tian, W. H., Picchi, P. *et al.* (1994). Performance of a three-ton liquid argon time projection chamber, *Nucl. Instr. Meth. Phys. Res. A* 345: 230–242.

Charpak, G., Bouclier, R., Bressani, T., Favier, J., and Zupančič, C. (1968). The use of multiwire proportional counters to select and localize charged particles, *Nucl. Instrum. Meth.* 62: 262–268.

Chechik, R., Breskin, A., Shalem, C., and Mörmann, D. (2004). Thick GEM-like hole multipliers: properties and possible applications, *Nucl. Instrum. Meth. A* 535: 303–308.

Chen, S., McEachran, R. P., and Stauffer, A. D. (2008). *Ab initio* optical potentials for elastic electron and positron scattering from the heavy noble gases, *J. Phys. B: At. Mol. Opt. Phys.* 41: 025201.

Chepel, V., Solovov, V., Neves, F., Pereira, A., Mendes, P. J., Silva, C. P. *et al.* (2006), Scintillation efficiency of liquid xenon for nuclear recoils with the energy down to 5 keV, *Astropart. Phys.* 26, 58–63.

Chepel, V. and Araújo, H. (2013). Liquid noble gas detectors for low energy particle physics, *J. Instrum.* 8: R04001.

Christophorou, L. G. (1978a). Interactions of O_2 with slow electrons, *Rad. Phys. Chem.* 12: 19-34.

Christophorou, L. G. (1978b). The lifetimes of metastable negative ions, In: Marton L. (ed.), *Advances in Electronics and Electron Physics*, vol. 46. Elsevier Academic Press, pp. 55–129.

Cline, D., Curioni, A., Lamarina, A. *et al.* (2000). A WIMP detector with a two-phase xenon, *Astroparticle Phys.* 12: 373–377.

Cohen, M. H. and Lekner, J. (1967). Theory of hot electrons in gases, liquids and solids, *Phys. Rev.* 158: 305–309.

Collazuol, G., Giuseppina, M.B., Marcatili, S. *et al.* (2011), Study of silicon photomultipliers at cryogenic temperatures, *Nucl. Instrum. Meth. A* 628: 389–392.

Conde, C. A. N., Requicha, Ferreira L., and Ferreira, M.F.A. (1977). The secondary scintillation output of xenon in a uniform field gas proportional scintillation counter, *IEEE Trans. Nuc. Sci.* 24: 221–224.

Conti, E., DeVoe, R., Gratta, G. *et al.* (2003). Correlated fluctuations between luminescence and ionization in liquid xenon, *Phys. Rev. B* 68: 054201.

Cuesta, C. on behalf of DUNE collaboration (2019). Status of ProtoDUNE Dual Phase, *European Physical Society Conference on High Energy Physics — EPS-HEP*2019 — 10-17 July 2019; e-print arXiv:1910.10115 v1 22 Oct 2019.

Cui, X., Abdukerim, A., Chen, W. *et al.* (2017). Dark matter results from 54-Ton-day exposure of PandaX-II experiment, *Phys. Rev. Lett.* 119: 181302.

D'Incecco, M., Galbiati, C., Giovanetti, G. K. *et al.* (2018). Development of a novel single-channel, 24 cm^2, SiPM-based, cryogenic photodetector, *IEEE Trans. Nucl. Sci.* 65: 591–596.

Dahl, C. E. (2009), The physics of background discrimination in liquid xenon and first results from XENON10 in the hunt for WIMP dark matter. PhD thesis, Princeton University, 2009.

Dajon, M. I., Dolgoshein, B. A., Efremenko, V. I., Leksin, G. A., and Lyubimov, V. A. (1967). *Spark Chamber.* Moscow: Atomizdat, 319 p.

Davies, G. J., Davies, J. D., Lewin, J. D. *et al.* (1994). Liquid xenon as a dark matter detector. Prospects for nuclear recoil discrimination by photon timing, *Phys. Lett. B* 320: 395–399.

Derenzo, S. E. (1974), Lawrence Berkeley Laboratory, Group A Physics Note No. 786.

Derenzo, S. E., Mast, T. S., Zaklad, H., and Muller R.A. (1974). Electron avalanche in liquid xenon, *Phys. Rev. A* 9: 2582–2591.

Dodelet, J.-P., Shinsaka, K., Kortsch, U. and Freeman, G.R. (1973). Electron ranges in liquid alkenes, dienes, and alkynes: range distribution function in hydrocarbons, *J. Chem. Phys.* 56: 2376–2386.

Doke, T. (1981). Fundamental properties of liquid argon, krypton and xenon as radiation detector media, *Portug. Phys.* 12: 9–48.

Doke, T. (1982). Recent developments of liquid xenon detectors, *Nucl. Instr. Meth.* 196: 87–96.

Doke, T., Hitachi, A., Kikuchi, J., Masuda, K., Tamada, S. *et al.* (1985). Estimation of the fraction of electrons escaping from recombination in the ionization of liquid argon with relativistic electrons and heavy ions. *Chem. Phys. Lett.* 115: 164–166.

Doke, T., Crawford H. J., Hitachi A., Kikuchi J., Lindstrom P. J., Masuda K. *et al.* (1988). LET dependence of scintillation yields in liquid argon, *Nucl. Instrum. Meth. Phys. Res. A* 269: 291–296.

Doke, T., Masuda, K., and Shibamura, E. (1990). Estimation of absolute photon yields in liquid argon and xenon for relativistic (1 MeV) electrons, *Nucl. Instrum. Meth. Phys. Res. A* 291: 617–620.

Doke, T. and Masuda, K. (1999). Present status of liquid rare gas scintillation detectors and their new application to gamma-ray calorimeters, *Nucl. Instrum. Meth. Phys. Res. A*420: 62–80.

Doke, T., Hitachi, A., Kikuchi, J. *et al.* (2002). Absolute scintillation yields in liquid argon and xenon for various particles, *Jap. J. Appl. Phys.* Part 1, 41: 1538–1545.

Doke, T. (2005). Ionization and excitation by high energy radiation, in W. F. Schmidt and E. Illenberger (eds.), *Electronic Excitations in Liquefied Rare Gases.* American Scientific Publishers, 2005, pp. 71–93.

Dolgoshein, B. A., Kruglov, A. A., Lebedenko, V. N., Miroshnichenko, V. P., and Rodionov, B. U. (1973). Electronic particle detection method for two-phase systems: liquid-gas, *Sov. J. Part. Nucl.* 4: 70–77.

Dolgoshein, B. A., Lebedenko, V. N., and Rodionov, B. U. (1970). New method of registration of ionizing particle tracks in condensed matter, *JETP Lett.* 11(11): 351–353.

Dolgoshein, B. A., Lebedenko, V. N., and Rodionov, B. U. (1973). Some electron methods of detection of particle tracks in liquids, *Elementary Particles Cosmic Rays* 3: 86–91 (in Russian).

Druger, S. D. and Knox, R. S. (1969). Theory of trapped-hole centers in rare-gas solids, *J. Chem. Phys.* 50: 3143–3153.

Drukier, A. and Stodolsky, L. (1984). Principles and applications of a neutral current detector for neutrino physics and astronomy, *Phys. Rev. D* 30: 2295–2309.

Duval, S., Breskin, A., Carduner, H. *et al.* (2009). MPGDs in Compton imaging with liquid-xenon, *J. Instrum.* 4: P12008.

Duval, S., Breskin, A., Budnik, R. *et al.* (2011). On the operation of a micropattern gaseous UV-photomultiplier in liquid-Xenon, *J. Instrum.* 6: P04007.

Dyachkov, L. G., Kobzev, G. A., and Norman, G. E. (1974). Effect of resonances in elastic scattering on the bremsstrahlung of electrons in an atomic field, *Sov. Phys. JETP* 38: 697–700.

Edwards, B. N. V., Bernard, E., Boulton, E. M. *et al.* (2018). Extraction efficiency of drifting electrons in a two-phase xenon time projection chamber, *J. Instrum.* 13: P01005.

Edwards, B., Araújo, H. M., Chepel, V. *et al.* (2008). Measurement of single electron emission in two-phase xenon, *Astropart. Phys.* 30: 54–57; e-Print arXiv:0708.0768.

Egorov, V. V., Ermilova, V. K., and Rodionov, B. U. (1982). Preprint FIAN N°166 (in Russian).

Egorov, V. V., Miroshnichenko, V. P., Rodionov, B. U., Bolozdynya, A. I., Kalashnikov, S. D., and Krivoshein, V. L. (1983). Electroluminescence emission gamma-camera, *Nucl. Instrum. Meth.* 205: 373–374.

Erdal, E., Arazi, L., Breskin, A. *et al.* (2020). Bubble-assisted liquid hole multipliers in LXe and LAr: towards "local dual-phase TPCs", *J. Instrum.* 15: C04002.

Erdal, E., Arazi, L., Chepel, V. *et al.* (2015). Direct observation of bubble-assisted electro-luminescence in liquid xenon, *J. Instrum.* 10: P11002.

ESTAR — Stopping power and range tables for electrons, National Institute of Standards and Technology, http://physics.nist.gov/PhysRefData/ Star/Text/ESTAR.html

Ewig, C. S. and Tellinghuisen, J. (1991). *Ab initio* study of the electronic states of O_2^- in vacuo and in simulated ionic solids, *J. Chem. Phys.* 95: 1097–1106.

Fastovsky, V. G., Rovinsky, A. E., and Petrovsky, Yu. V. (1972). *Inert Gases.* Moscow, Atomizdat, 352 p. (in Russian).

Feldman, G. J. and Cousins, R. D. (1998). Unified approach to the classical statistical analysis of small signals, *Phys. Rev. D* 57: 3873–3889.

Fonseca, A. C., Meleiro, R., Chepel, V., Pereira, A., Solovov, V., and Lopes, M. I. (2004). Study of secondary scintillation in xenon vapour, *IEEE Nuc. Sci. Symp. 2004 Conf. Record* 1: 572–576.

Fraga, M. M., Fetal, S. T. G., Fraga, F. A. F. *et al.* (2000). Study of scintillation light from microstructure based detectors, *IEEE Trans. Nucl. Sci.* 47: 933–938.

Francini, R., Monteriali, R. M., Nichelatti, E. *et al.* (2013). Tetraphenyl-butadiene films: VUV-Vis optical characterization of Tetraphenyl-butadiene films on glass and specular reflector substrates from room to liquid Argon temperature, *J. Instrum.* 8: P09006.

Freedman, D. Z. (1974). Coherent effects of a weak neutral current. *Phys. Rev. D* 9: 1389–1392.

Gai, M., Alon, R., Breskin, A. *et al.* (2007). Toward application of a thick gas electron multiplier (THGEM) readout for a dark matter detector, e-print arxiv: 0706.1106 10 June 2007.

Galea, R., Dodd, J., Ju, Y. *et al.* (2006). Gas purity effect on GEM performance in He and Ne at low temperatures, *IEEE Trans. Nucl. Sci.* 53: 2260–2263.

Gallina, G., Giampa, P., Retiere, F. *et al.* (2019). Characterization of the Hamamatsu VUV4 MPPCs for nEXO, *Nucl. Instrum. Meth. A* 940: 371–379.

Gastler, D., Kearns, E., Hime, A., Stonehill, L. C., Seibert, S. *et al.* (2012). Measurement of scintillation efficiency for nuclear recoils in liquid argon, *Phys. Rev. C* 85: 065811.

Gehman, V. M., Ito, T. M., Griffith, W. C. *et al.* (2013). Characterization of protonated and deuterated tetraphenyl butadiene film in a polystyrene matrix, *J. Instrum.* 8: P04024.

Gerhold, J., Hubmann, M., Telser, E. (1994). Gap size effect on liquid helium breakdown, *Cryogenics* 34: 579–586.

Ghelfenstein, M., Szwarc, H., and López-Delgado, R. (1977). On xenon molecule excited state lifetimes, *Chem. Phys. Lett.* 52: 236–238.

Giboni, K. L., Juyal, P., Aprile, E. *et al.* (2019). A LN2 based cooling system for a next generation liquid xenon dark matter detector, e-Print arXiv:1909.09698.

Goetzke, L. W., Aprile, E., Anthony, M., Plante, G., and Weber, M. (2017). Measurement of light and charge yield of low-energy electronic recoils in liquid xenon, *Phys. Rev. D* 96: 103007.

Gonzalez-Diaz, D., Monrabal, F., and Murphy, S. (2018). Gaseous and dual-phase time projection chambers for imaging rare processes, *Nucl. Instrum. Meth. A* 878: 200–255.

Gorodkov, Yu. B., Lyubimov, V. A., Sidorov, I. V., and Soloschenko, V. A. (1974). Cryogenic tracking spark chamber, *Instr. Exp. Techn.* 6: 46–47 (in Russian).

Grebinnik, V. G., Dodokhov, V. Kh., Zhukov, V. A. *et al.* (1978). High-pressure proportional counter at low temperatures, *Instr. Exp. Tech.* 21: 1225–1227.

Gushchin, E. M., Kruglov, A. A., and Obodovski, I. M. (1982a). Electron dynamics in condensed argon and xenon, *Sov. Phys. JETP* 55: 650–655.

Gushchin, E. M., Kruglov, A. A., and Obodovski, I. M. (1982b). Emission of "hot" electrons from liquid and solid argon and xenon, *Sov. Phys. JETP* 55: 860–862.

Guschin, E. M., Kruglov, A. A., Obodovsky, I. M., Pokachalov, S. G. and Shilov, V. A. (1982c). Liquid xenon position-sensitive detector of gamma radiation, *Instr. Exp. Tech.* 3: 49–52 (in Russian).

Hagmann, C. and Bernstein, A. (2004). Two-phase emission detector for measuring coherent neutrino-nucleus scattering, *IEEE Trans. Nucl. Sci.* 51: 2151–2155.

Hamamatsu Photonics K.K (2020). https://www.hamamatsu.com.

Haruyama, T., Kasami, K., Inoue, H. *et al.* (2004). Development of a high-power coaxial pulse tube refrigerator for a liquid xenon calorimeter, *AIP Conf. Proc.* 710: 1459.

Heindl, T., Dandl, T., Hofmann, M. *et al.* (2010). The scintillation of liquid argon, *Europhys. Lett.* 91: 62002.

Henshow, D. G. (1957). Atomic distribution in liquid argon by neutron diffraction and the cross sections of A^{36} and A^{40}, *Phys. Rev.* 105: 976–981.

Herzenberg, A. (1969). Attachment of slow electrons to oxygen molecules, *J. Chem. Phys.* 54: 4942–4950.

Hilt, O. and Schmidt, W.F. (1994a). Positive hole mobility in liquid xenon, *Chem. Phys.* 183: 147–153.

Hilt, O., Schmidt, W. F., and Khrapak, A. G. (1994b). Ionic mobilities in liquid xenon, *IEEE Trans. Dielect. Elect. Insul.* 1: 648–656.

Hitachi, A., Takahashi, T., Funayama, N. *et al.* (1983). Effect of ionization density on the time dependence of luminescence from liquid argon and xenon, *Phys. Rev. B* 27: 5279–11470.

Hitachi, A., Yunoki, A., Doke, T., and Takahashi, T. (1987). Scintillation and ionization yield for (alpha-symbol)-particles and fission fragments in liquid argon, *Phys. Rev. A* 35: 3956–3958.

Hitachi, A., Doke, T., and Mozumder, A. (1992b). Luminescence quenching in liquid argon under charged particle impact: Relative scintillation yield at different linear energy transfers, *Phys. Rev. B* 46: 11463–11470.

Hitachi, A. (2005a). Properties of liquid xenon scintillation for dark matter searches, *Astropart. Phys.* 24: 247–256.

Hitachi, A. (2005b). Properties of liquid rare gas scintillation for WIMP search, in N. J. C. Spooner and V. Kudryavtsev (eds.), *Proc. of the 5th Int. Workshop on the identification of Dark matter (IDM2004)*, World Scientific, 2005, pp. 396–401.

Hogenbirk, E., Decowski, M. P., McEwan, K., and Colijn, A. P. (2018). Field dependence of electronic recoil signals in a dual-phase liquid xenon time projection chamber, *J. Instrum.* 13: P10031.

Hollywood, D., Majumdar, K., Mavrokoridis, K. *et al.* (2020). ARIADNE — A novel optical LArTPC: technical design report and initial characterisation using a secondary beam from the CERN PS and cosmic muons, *J. Instrum.* 15: P03003.

Horn, M., Belov, V. A., Akimov, D. Y., Araújo, H. M., Barnes, E. J., Burenkov, A. A. *et al.* (2011). Nuclear recoil scintillation and ionisation yields in liquid xenon from ZEPLIN-III data, *Phys. Lett. B* 705: 471–476.

Hotta, Y. (Hamamatsu Photonics K.K.) (2014). Latest developments in PMTs for low temperature operation, *UCLA's 11th Symposium on Sources and Detection of Dark Matter and Dark Energy in the Universe*, 26–28 February 2014, UCLA, USA.

Hübner, H. (1998). Diatomic molecules, in *Landolt-Börnstein Numerical Data and Functional Relationships in Science and Technology*, vol. 24. Molecular Constants. Ed. Hüttner W. Springer-Verlag, pp. 7–156.

Hutchinson, G. W. (1948). Ionization in liquid and solid argon, *Nature* 162: 610–611.

Huxley, L. G. H. and Crompton, R. W. (1974). *The Diffusion and Drift of Electrons in Gases*, Wiley. 669 p.

Jackson, J. D. (1999). *Classical Electrodynamics*. John Wiley & Sons, 3rd edition.

Jaffé, G. (1913). Zur Theorie der Ionisation in Kolonnen, *Ann. Phys*, 347: 303–344.

Jakse, N., Bomont, J. M., and. Bretonnet, J. L. (2002). Effects of three-body interactions on the structure and thermodynamics of liquid krypton, *J. Chem. Phys.* 116: 8504–8508.

Jeng, S-C, Fairbank, W. M., and Miyajima, M. (2009). Measurements of the mobility of alkaline earth ions in liquid xenon, *J. Phys. D: Appl. Phys.* 42: 035302.

Jensen, F. (2017). *Introduction to Computational Chemistry*, New York: John Wiley & Sons Ltd..

Jortner, J., Meyer, L., Rice, S. A., and Wilson, E.G. (1965). Localized excitations in condensed Ne, Ar, Kr, and Xe, *J. Chem. Phys.* 42: 4250–4253.

Ju, Y. L., Dodd, J., Galea, R. *et al.* (2007). Cryogenic design and operation of liquid helium in an electron bubble chamber towards low energy solar neutrino detectors, *Cryogenics* 47: 81–88.

Kalashnikov, S. D. (1985). *Basic Physics of Scintillation Gamma Camera Design*. Moscow: Energoatomizdat, 120 p. (in Russian).

Karnbach, R., Joppien, M., Stapelfeldt, J., Wörmer, J., and Möller, T. (1993). CLULU: An experimental setup for luminescence measurements on van der Waals clusters with synchrotron radiation. *Rev. Sci. Instrum.* 64: 2838–2849.

Kastens, L. W., Cahn, S. B., Manzur, A., and McKinsey, D. N. (2009). Calibration of a Liquid Xenon Detector with Kr-83m, *Phys. Rev. C* 80: 045809.

Kastens, L. W., Bedikian, S., Cahn, S. B. *et al.* (2010). A ^{83}Kr^m source for use in low-background liquid xenon time projection chambers, *J. Instrum.* 5: P05006.

Khrapak, A. G., Schmidt, W. F., and Illenberger, E. (2005). Localized electrons, holes and ions, in W. F. Schmidt and E. Illenberger (eds.), *Electronic Excitations in Liquefied Rare Gases*. American Scientific Publishers, pp. 317–330.

Kim, J. G., Dardin, S. M., Kadel, R. W. *et al.* (2004). Electron avalanches in liquid argon mixtures, *Nucl. Instrum. Meth. A* 534: 376–396.

Kimura, M., Aoyama, K., Tanaka, M., and Yorita, K. (2020). Liquid argon scintillation response to electronic recoils between 2.8–1275 keV in a high light yield single-phase detector, arXiv:2003.14248.

Knoll, G. F. (2010). *Radiation Detection and Measurements*. John Wiley & Sons, 4th edition.

Kochanek, I. for the DarkSide Collaboration (2019). SiPMs for cryogenic temperature, *Nuovo Cim.* 42 C 62: 1–4.

Kramers, H. A. (1952). On a modification of Jaffe's theory of column ionization, *Physica* 18: 665–675.

Kubota, S., Nakamoto, A., Takahashi T. *et al.* (1976). Evidence of the existence of the exciton states in liquid argon and exciton-enhanced ionization from xenon doping, *Phys. Rev. B* 13: 1649–1653.

Kubota, S., Hishida, M, and Raun(Gen), J. (1978). Evidence for a triplet state of the self-trapped exciton states in liquid argon, krypton and xenon, *J. Phys. C: Solid State Phys.* 11: 2645–2651.

Kubota, S., Hishida, M., Suzuki, M., and Ruan(Gen), J.-Z. (1979). Dynamical behavior of free electrons in the recombination process in liquid argon, krypton and xenon, *Phys. Rev. B* 20: 3486–3496.

Kubota, S., Suzuki, M., and Ruan(Gen), J.-Z. (1980). Specific-ionization-density effect on the time dependence of luminescence in liquid xenon, *Phys. Rev. B* 21: 2632.

Kubota, S., Hishida, M., Suzuki M., and Ruan(Gen) J.-Z. (1982a). Liquid and solid argon, krypton and xenon scintillators, *Nucl. Instrum. Meth.* 196: 101–105.

Kubota, S., Takahashi, T. and Ruan (Gen), J.-Z. (1982b). Hot electron relaxation in solid and liquid argon, krypton and xenon. *J. Phys. Soc. Japan* 51, 3274–3277.

Kudryavtsev, V. A. (2005). Dark matter experiments at Boulby mine, *Springer Proc. Phys.* 98: 139–143.

Lally, C. H., Davies, G. J., Jones, W. G., Smith, N. J. T. *et al.* (1996). UV quantum efficiencies of organic fluors, *Nucl. Instrum. Meth. B* 117: 421–427.

Lansiart, A., Seigneur, A., Moretti, J.-L., and Morucci, J. P. (1976). Development research on a highly luminous condensed xenon scintillator, *Nucl. Instrum. Meth.* 135: 47–52.

Laschuk, E. F., Martins, M. M., and Evangelisti, S. (2003). *Ab initio* potentials for weakly interacting systems: Homonuclear rare gas dimers, *Quant. Chem.* 95: 303–312.

Le Comber, P. G., Loveland, R. J., and Spear, W.E. (1975). Hole transport in the rare-gas solids Ne, Ar, Kr and Xe, *Phys. Rev. B* 11: 3124–3130.

Lebedenko, V. N., Araújo, H. M., Barnes, E. J. *et al.* (2009). Results from the first science run of the ZEPLIN-III dark matter search experiment, *Phys. Rev. D* 80: 052010.

Lekner, J. (1967). Motion of electrons in liquid argon, *Phys. Rev.* 158: 130–137.

Lemmon, E. W., McLinden, M. O., and Friend, D. G. (2018). Thermophysical properties of fluid systems. in P. J. Linstrom and W. G. Mallard (eds.), *NIST Chemistry WebBook, NIST Standard Reference*

Database Number 69. National Institute of Standards and Technology, Gaithersburg MD, 20899, https://doi.org/10.18434/T4D303 (retrieved November 28, 2019).

Lenardo, B., Kazkaz, K., Manalaysay, A., Mock, J., Szydagis, M., and Tripathi, M. (2015). A global analysis of light and charge yields in liquid xenon, *IEEE Trans. Nucl. Sci.* 62: 3387–3396.

Lenardo, B. G., Xu, J., Pereverzev, S. *et al.* (2019). Measurement of the ionization yield from nuclear recoils in liquid xenon between 0.3–6 keV with single-ionization-electron sensitivity, e-print arXiv:1908.00518, 1 August 2019.

Leo, W. R. (1994). *Techniques for Nuclear and Particle Physics Experiments: A How-to Approach*. Springer-Verlag,.

Lewin, J. D. and Smith, P. F. (1996). Review of mathematics, numerical factors, and corrections for dark matter experiments based on elastic nuclear recoil, *Astropart. Phys.* 6: 87–112.

Li, Y., Tsang, T., Thorn, C., Qian, X., Diwan, M. *et al.* (2016). Measurement of longitudinal electron diffusion in liquid argon, *Nucl. Instr. Meth. Phys. Res. A* 816, 160–170.

Lightfoot, P. K., Hollingworth, R., Spooner, N. J. C., Tovey, D. (2005). Development of a double-phase xenon cell using Micromegas charge readout for applications in dark matter physics, *Nucl. Instrum. Meth. A* 554: 266–285.

Lightfoot, P. K., Barker, G. J., Mavrokoridis, K. *et al.* (2007). Characterisation of a silicon photomultiplier device for applications in liquid argon based neutrino physics and dark matter searches, *J. Instrum.* 3: P10001.

Lightfoot, P. K., Barker, G. J., Mavrokoridis, K., Ramachers, Y. A., and Spooner, N. J. C. (2009). Optical readout tracking detector concept using secondary scintillation from liquid argon generated by a thick gas electron multiplier, *J. Instrum.* 4: P04002.

Lindblom, P. and Solin, O. (1988). Atomic infrared noble gas scintillations I: Optical spectra, *Nucl. Instrum. Meth. A* 268: 204–208.

Lindhard, J., Nielsen, V., Sharff, M., and Thomsen, P.V. (1963a). Integral equations governing radiation effects (Notes on atomic collisions, II), *Mat. Fys. Medd. Dan. Vid. Selsk.* 33(10): 1–41.

Lindhard, J., Scharff, M., and Schiøtt, H. E. (1963b). Range concepts and heavy ion ranges (Notes on atomic collisions, II), *Mat. Fys. Medd. Dan. Vid. Selsk.* 33, n°14, pp. 1–42.

Lindhard, J., Nielsen, V., and Sharff, M. (1968). Approximation method in classical scattering by screened Coulomb fields, *Mat. Fys. Medd. Dan. Vid. Selsk.* 36(10).

Lindote, A., Araújo, H., Pinto da Cunha, J. *et al.* (2007). Preliminary results on position reconstruction for ZEPLIN III, *Nucl. Instrum. Meth. A* 573: 200–203.

Lippincott, W. H., Coakley, K. J., Gastler, D. *et al.* (2008). Scintillation time dependence and pulse shape discrimination in liquid argon, *Phys. Rev. C* 78: 035801.

Lippincott, W. H., Coakley, K. J., Gastler, D., Kearns, E., McKinsey, D. N., and Nikkel, J. A. (2012). Scintillation yield and time dependence from electronic and nuclear recoils in liquid neon, *Phys. Rev. C* 86: 015807.

Lock, G. S. H. (1992). *The Tubular Thermosyphon. Variation on a Theme.* Oxford University Press.

Lopes, M. I. and Chepel, V. (2005). Rare gas liquid detectors, in W. F. Schmidt and E. Illenberger (eds.), *Electronic Excitations in Liquefied Rare Gases.* American Scientific Publishers, pp. 331–388.

Lorents, D. C. (1976). The physics of electron beam excited rare gases at high densities, *Physica* 82C: 19–26.

Loveland, R. J., Le Comber, P. G., and Spear, W. E. (1972). Experimental evidence for electronic bubble states in liquid Ne, *Phys. Lett. A* 39: 225–226.

Lyashenko, A., Nguyen, T., Snyder, A. *et al.* (2014). Measurement of the absolute Quantum Efficiency of Hamamatsu model R11410-10 photomultiplier tubes at low temperatures down to liquid xenon boiling point, *J. Instrum.* 9: P11021.

Machulin, I. N., Miroshnichenko, V. P., Rodionov, B. U., Chepel, V. Yu., Rostovtsev, A. A. *et al.* (1983). Feasibility of precision spectrometry of 1 MeV electrons in liquid xenon, *Sov. Tech. Phys. Lett.* 9: 484–486.

Majewski, P., Solovov, V. N., Akimov, D. Y. *et al.* (2012). Performance data from the ZEPLIN-III second science run, *J. Instrum.* 7: C03044.

Mangiarotti, A., Lopes, M. I., Benabderrahmane, M. L., Chepel, V., Lindote, A. *et al.* (2007). A survey of energy loss calculations for heavy ions between 1 and 100 keV, *Nucl. Instrum. Meth. Phys. Res. A* 580: 114–117.

Mano, R., Monteiro, C., and Freitas, E. (2019). A gas proportional scintillation counter with krypton filling, *LIDINE 2019: Light Detection In Noble Elements.* Manchester, UK, 28–30 August 2019.

Manzur, A., Curioni, A., Kastens, L., McKinsey, D. N., Ni, K., and Wongjirad, T. (2010). Scintillation efficiency and ionization yield of liquid xenon for monoenergetic nuclear recoils down to 4 keV, *Phys. Rev. C* 81: 025808.

Marchionni, A., Amsler, C., Badertscher, A. *et al.* (2011). ArDM: a ton-scale LAr detector for direct Dark Matter searches, *J. Phys. Conf. Ser.* 308: 012006.

Marcoux, J. (1970). Dielectric constants and indices of refraction of Xe, Kr, and Ar, *Can. J. Phys.* 48: 244–245.

Masaoka, S., Katano, S., Kishimoto, S., and Isozumi, Y. (2000). A model for the operation of a helium-filled proportional counter at low temperatures near 4.2 K, *Nucl. Instrum. Meth.* B 171: 360–372.

Masuda, K., Takasu, S., Doke, T. *et al.* (1979). A liquid xenon proportional scintillation counter, *Nucl. Instrum. Meth.* 160: 247–253.

Mavrokoridis, K., Ball, F., Carroll, J. *et al.* (2014). Optical readout of a two phase liquid argon TPC using CCD camera and THGEMs, *J. Instrum.*, 9: P02006.

McConkey, N., Barker, G.J., Bennieston, A. J. *et al.* (2010). Optical readout technology for large volume liquid argon detectors, *Nucl. Phys. B (Proc. Suppl.)* 215: 255–257.

McKinsey, D. N., Brome, C. R., Butterworth, J. S., Dzhosyuk, S. N., Huffman, P. R., Mattoni, C. E. H. *et al.* (1999). Radiative decay of the metastable $He_2(a^3\Sigma_u^+)$ molecule in liquid helium, *Phys. Rev. A* 59: 200–204.

Mei, D.-M., Yin, Z.-B., Stonehill, L. C., and Hime, A. (2008). A model of nuclear recoil scintillation efficiency in noble liquids, *Astropart. Phys.* 30: 12–17.

Miller, L. S., Howe, S., and Spear, W. E. (1968). Charge transport in solid and liquid Ar, Kr, and Xe, *Phys. Rev.* 166: 871–878.

Miller, T. M. (2019). Electron affinities, in *CRC Handbook of Chemistry and Physics.* Editor-in-Chief: John R. Rumble, 100th edition.

Miroshnichenko, V. P., Nevski, P. L., and Rodionov, B. U. (1982). In "Elementary Particles and Cosmic Rays," p. 60. Energoizdat, Moscow (in Russian).

Miyajima, M., Sasaki, S., and Tawara, H. (1994). Search for double beta-decay products of ^{136}Xe in liquid xenon, *IEEE Trans. Nucl. Sci.* NS-41: 835–839.

Montanari, D., Adamowski, M., Bremer, J. *et al.* (2017). Status of the LBNF cryogenic system, *IOP Conf. Series: Mater. Sci. Eng.* 278: 012117.

Monteiro, C. M. B., Fernandes, L. M. P., Lopes, J. A. M. *et al.* (2007). Secondary scintillation yield in pure xenon, *J. Instrum.* 2: P05001.

Monteiro, C. M. B., Lopes, J. A. M., Veloso, J. F. C. A. *et al.* (2008). Secondary scintillation yield in pure argon, *Phys. Lett. B* 668: 167–170.

Morikawa, E., Reininger, R., Gürtler, P., Saile, V., and Laporte, P. (1989). Argon, krypton and xenon excimer luminescence: From the dilute gas to the condensed phase, *J. Chem. Phys.* 91: 1469–1477.

Moriyama, S. on behalf of the XENON collaboration (2019). Direct Dark Matter Search with XENONnT, *International symposium on Revealing the history of the Universe with underground particle and nuclear research*, 7–9 March 2019, Tohoku University, Sendai, Japan.

Morozov, A., Heindl, T., Krucken, R. *et al.* (2008). Conversion efficiencies of electron beam energy to vacuum ultraviolet light for Ne, Ar, Kr, and Xe excited with continuous electron beams, *J. Appl. Phys.* 103(10): 103301.

Morozov, A., Solovov, V., Martins, R. *et al.* (2016). ANTS2 package: simulation and experimental data processing for Anger camera type detectors, *J. Inst.* 11(04): P04022.

Mount, B.J., Hans, S., Rosero, R. *et al.* (2017). The LUX-ZEPLIN (LZ) Technical Design Report, BN-1007256; e-Print arXiv: 1703.09144.

Mozumder, A. (1995a). Free-ion yield in liquid argon at low-LET, *Chem. Phys. Lett.* 238: 143–148.

Mozumder, A. (1995b). Free-ion yield and electron-ion recombination rate in liquid xenon, *Chem. Phys. Lett.* 245: 359–363.

Mulliken, R. S. (1970). Potential curves of diatomic rare-gas molecules and their ions, with particular reference to Xe_2^*, *J. Chem. Phys.* 52: 5170–5180.

Neumeier, A., Dandl, T., Heindl, T., Himpsl, A., Oberauer, L., Potzel, W., Roth, S., Schönert, S., Wieser, J. and Ulrich, A. (2015). Intense vacuum ultraviolet and infrared scintillation of liquid Ar-Xe mixtures, *Europhys Lett* 109: 12001.

Neves, F., Solovov, V., Chepel, V. *et al.* (2007). Position reconstruction in a liquid xenon scintillation chamber for low-energy nuclear recoils and γ-rays, *Nucl. Instrum. Meth. A* 573: 48–52.

nEXO Collaboration (2019). Imaging individual barium atoms in solid xenon for barium tagging in nEXO, *Nature* 569: 203–207.

Nikkel, J. A., Hasty, R., Lippincott, W. H., and McKinsey, D. N. (2008). Scintillation of liquid neon from electronic and nuclear recoils, *Astropart. Phys.* 29: 161–166.

Nikolaev, V. S. and Dmitriev, I. S. (1968). On the equilibrium charge distribution in heavy element ion beams, *Phys. Lett. A* 28: 277–278.

NIST Chemistry WebBook (2018). NIST Chemistry WebBook, NIST Standard Reference Database Number 69, P.J. Linstrom and W.G. Mallard (Eds.), National Institute of Standards and Technology, Gaithersburg MD, 20899, https://doi.org/10.18434/T4D303 (retrieved January 17, 2020).

Njoya, O., Tsang, T., Tarka, M. *et al.* (2020). Measurements of electron transport in liquid and gas xenon using a laser-driven photocathode, *Nucl. Instrum. and Meth. in Phys. Res. A*972: 163965; arXiv:1911.11580, 25 November 2019.

Obodovski, I. (2005) Saturation curves and energy resolution of LRG ionization spectrometers, in *Proc. 2005 IEEE Int. Conf. on Diel. Liquids ICDL2005*, Coimbra June 26–July 1, pp. 321–324.

Ogilvie, J. F. and Wang, F. Y. H. (1992). Potential-energy functions of diatomic molecules of the noble gases. I. Like nuclear species, *J. Mol. Struct.* 273: 277–290.

Oliveira, C. A. B., Schindler, H., Veenhof, R. J. *et al.* (2011). A simulation toolkit for electroluminescence assessment in rare event experiments, *Phys. Lett. B* 703: 217–222.

Oliveira, C. A. B., Correia, P., Ferreira, A. L. *et al.* (2013). Simulation of gaseous Ar and Xe electroluminescence in the near infrared range, *Nucl. Instrum. Meth. A* 722: 1–4.

Otte, A. N., Garcia, D., Nguyen, T., and Purushotham, D. (2017). Characterization of three high efficiency and blue sensitive silicon photomultipliers, *Nucl. Instrum. Meth. A* 846: 106–125.

Pagani, L. (2017). Direct dark matter detection with the DarkSide-50 experiment, PhD thesis. Department of Physics, University of Genoa.

Pansky, A., Breskin, A., and Chechik, R. (1997). Fano factor and the mean energy per ion pair in counting gases, at low x-ray energies, *J. Appl. Phys.* 82: 871–877.

Periale, L., Peskov, V., Iacobaeus, C. *et al.* (2005). The successful operation of hole-type gaseous detectors at cryogenic temperatures, *IEEE Trans. Nucl. Sci.* 52: 927–931.

Plante, G., Aprile, E., Budnik, R., Choi, B., Giboni, K.-L., Goetzke, L.W. *et al.* (2011). New measurement of the scintillation efficiency of low-energy nuclear recoils in liquid xenon, *Phys. Rev. C*84: 045805.

Plante, I. and Cucinotta, F. A. (2009). Cross sections for the interactions of 1 eV–100 MeV electrons in liquid water and application to Monte-Carlo simulation of HZE radiation tracks, *New J. Phys.* 11: 063047.

Platzman, R. L. (1961). Total ionization in gases by high-energy particles: An appraisal of our understanding, *Int. J. Appl. Rad. Isot.* 10: 116–127.

Policarpo, A. J. P. L. (1981). Light production and gaseous detectors, *Physica Scripta* 23: 539–549.

Policarpo, A. P. L., Chepel, V., Lopes, M. I. *et al.* (1995). Observation of electron multiplication in liquid xenon with a Microstrip plate, *Nucl. Instrum. Meth. A* 365: 568–571.

Radzig, A. A. and Smirnov, B. M. (1985). *Reference Data of Atoms, Molecules and Ions, Springer Series in Chemical Physics*, Vol. 31. Springer-Verlag, Berlin, p. 88.

Raju, G. G. (2006). *Gaseous Electronics: Theory and Practice*. CRC Press, Taylor and Francis Group.

Raju, G. G. (2011). *Gaseous Electronics: Tables, Atoms, and Molecules*. CRC Press, Taylor and Francis Group.

Regenfus, C., Allkofer, Y., Amsler, C., Creus, W., Ferella, A., Rochet, J., and Walte, M. (2012). Study of nuclear recoils in liquid argon with monoenergetic neutrons, *J. Phys.: Conf. Series* 375: 012019.

Rodionov, B. U. (1969). Study of processes on tracks of ionizing particles in noble gases and liquids and the possibility of developing a controlled track detector based on liquefied noble gases, PhD thesis. Moscow: MEPhI. 137 p. (in Russian).

Rubbia, A. (2006). Experiments For CP-Violation: A Giant Liquid Argon Scintillation, Cerenkov And Charge Imaging Experiment, *J. Phys. Conf. Ser.* 39: 129–350.

Rutkai, G., Thol, M., Span, R., and Vrabec, J. (2016) How well does the Lennard-Jones potential represent the thermodynamic properties of noble gases?, *Mol. Phys.* 115: 1104–1121.

Sakai, Y. (2005). Hot electrons, in W. F. Schmidt and E. Illenberger (eds.), *Electronic Excitations in Liquefied Rare Gases*. American Scientific Publishers, pp. 51–69.

Sakai, Y. (2007). Quasifree electron transport under electric field in nonpolar simple-structured condensed matters, *J. Phys. D: Appl. Phys.* 40: R441–R452.

Sakai, Y., Nakamura, S. and Tagashira, H. (1985). Drift velocity of hot electrons in liquid Ar, Kr, and Xe, *IEEE Trans. Electr. Insul* EI-20: 133–137.

Salvat, F., Fernández-Varea, J. M., and Sempau, J. (2011). PENELOPE-2011. A code system for Monte Carlo simulation of electron and photon transport, Workshop Proceedings, Barcelona, Spain, 4–7 July 2011, Data Bank NEA/NSC/DOC(2011)5.

Samson, J. A. R. and Haddad, G. N. (1976). Average energy loss per ion pair formation by photon and electron impact on xenon between threshold and 90 eV, *Radiat. Res.* 66: 1–10.

Sangiorgio, S., Bernstein, A., Coleman, J. *et al.* (2012). R&D for the observation of coherent neutrino-nucleus scatter at a nuclear reactor with a dual-phase argon ionization detector, *Phys. Proc.* 37: 1266–1272.

Sangiorgio, S, Joshi, T. H., Bernstein, A. *et al.* (2013). First demonstration of a sub-keV electron recoil energy threshold in a liquid argon ionization chamber, *Nucl. Instrum. Meth. A* 728: 69–72.

Sanguino, P., Balau, F., Botelho do Rego, A. M., Pereira A., and Chepel, V. (2016). Stability of tetraphenyl butadiene thin films in liquid xenon, *Thin Solid Films* 600: 65–70.

Santos, E., Edwards, B., Chepel, V. *et al.* (2011). Single electron emission in two-phase xenon with application to the detection of coherent neutrino-nucleus scattering, *JHEP* 2011(12): 115.

Sauli, F. (2014). Gaseous Radiation Detectors, Fundamentals and Applications. Cambridge University Press, Cambridge. 460 pp.

Sauli, F. (2016). The gas electron multiplier (GEM): Operating principles and applications, *Nucl. Instr. Meth. A* 805: 2–24.

Scalettar, R. T., Doe, P. J., Mahler, H.-J., and Chen, H. H. (1981). Critical test of geminate recombination in liquid argon, *Phys. Rev. A* 25: 2419–2422.

Schmidt, P. W. and Tompson, C. W. (1968). X-ray scattering in simple fluids, in H. L. Frish and Z. W. Salsburg (eds.), *Simple Dense Fluids*. Academic Press, New York and London.

Schmidt,W. F (1997). *Liquid State Electronics of Insulating Liquids.* CRC Press.

Schmidt, W. F. and Yoshino, K. (2005). Electric discharges, in W. F. Schmidt and E. Illenberger (eds.), *Electronic Excitations in Liquefied Rare Gases*. American Scientific Publishers, pp. 295–315.

Schussler, A. S., Burghorn, J., Wyder, P., and Lembrikov, B. I., and Baptist, R. (2000). Observation of excimer luminescence from electron-excited liquid xenon, *Appl. Phys. Lett.*, 77: 2786–2788.

Schwentner, N., Koch, E. E., and Jortner, J. (1985). *Electronic Excitations in Condensed Rare Gases*, Springer Tracts in Modern Physics, Springer-Verlag, Berlin.

Searles, D. J. and Huber, H. (2005). Ground State Properties of Rare Gas Fluids and their Microscopic Foundations, in W. F. Schmidt and E. Illenberger (eds.), *Electronic Excitations in Liquefied Rare Gases*. American Scientific Publishers, pp. 51–69.

Séguinot, J., Passardi, G., Tischhauser, J., and Ypsilantis, T. (1992). Liquid xenon ionization and scintillation studies for a totally active-vector electromagnetic calorimeter, *Nucl. Instrum. Meth. in Phys. Res. A* 323: 583–600.

Shibamura, E. *et al.* (1975). Drift velocities of electrons, saturation characteristics of ionization and W-values for conversion electrons in liquid argon, liquid argon-gas mixtures and liquid xenon, *Nucl. Instrum. Meth.* 24: 249.

Shibamura, E., Takahashi, T., Kubota, S., and Doke, T. (1979). Ratio of diffusion coefficient to mobility for electrons in liquid argon, *Phys. Rev. A* 20: 5247–2554.

Shibamura, E., Masuda, K., and Doke, T. (1984). Measurement of the ratio D_L/D_T for electrons in liquid xenon, in *8th Workshop on Electron Swarms*, Institute of Technology, Tokyo.

Shutt, T., Dahl, C. E., Kwong, J., Bolozdynya, A., and Brusov, P. (2007). Performance and fundamental processes at low energy in a two-phase liquid xenon dark matter detector, *Nucl. Instrum. Meth. Phys. Res. A* 579: 451–453.

Sidorov, I. V. (1975). Cryogenic streamer chamber, *3d ITEP School*, issue IV, Moscow: Atomizdat, pp. 52–60 (in Russian).

Sigmund, P. (2004). *Stopping of Heavy Ions: A Theoretical Approach*, Springer Tracts in Modern Physics, vol. 204, Springer, 164 p.

Sigmund, P. (2006). *Particle Penetration and Radiation Effects: General Aspects and Stopping of Swift Point Charges, Springer Tracts of Solid-State Sciences*, vol. 151, Springer, 437 p.

Simonović, I., Garland, N. A., Bošsnjaković, D. *et al.* (2019). Electron transport and negative streamers in liquid xenon, *Plasma Sources Sci. Technol.* 28: 015006.

Slaviček, P., Kalus, R., Paška, P., Odvárková, I., Hobza, P. *et al.* (2003). State-of-the-art correlated *ab initio* potential energy curves for heavy rare gas dimers: Ar_2, Kr_2, and Xe_2, *J Chem. Phys.* 119: 2102–2119.

Solovov, V., Balau, F., Neves, F., Chepel, V., Pereira, A., and Lopes, M.I. (2007). Operation of gas electron multipliers in pure xenon at low temperatures, *Nucl. Instrum. Meth. A* 580: 331–334.

Solovov, V. N., Belov, V. A., Akimov, D. Yu. *et al.* (2012). Position reconstruction in a dual phase xenon scintillation detector, *IEEE Trans. Nucl. Sci.* 59: 3286–3293.

Sorensen, P., Manzur, A., Dahl, C. E. *et al.* (2009). The scintillation and ionization yield of liquid xenon for nuclear recoils, *Nucl. Instrum. Meth. A* 601: 339–346.

Sorensen, P. (2010). A coherent understanding of low-energy nuclear recoils in liquid xenon, *J. Cosm. Astropart. Phys.* 09: 033.

Sorensen, P. and Dahl, C. E. (2011). Nuclear recoil energy scale in liquid xenon with application to the direct detection of dark matter, *Phys. Rev. D* 83: 063501.

Sorensen, P. (2017). Electron train backgrounds in liquid xenon dark matter search detectors are indeed due to thermalization and trapping, e-print arXiv:1702.04805.

Sorensen, P. and Kadmin, K. (2018). Two distinct components of the delayed single electron noise in liquid xenon emission detectors, *J. Instrum.* 13: P02032.

Sowada, U., Warman, J. M., and de Haas, M. P. (1982). Hot-electron thermalization in solid and liquid argon, krypton and xenon, *Phys. Rev. B* 25, 3434–3437.

Steinberger, I. T. (2005). Band structure parameters of classical rare gas liquids, in W. F. Schmidt and E. Illenberger (eds.), *Electronic Excitations in Liquefied Rare Gases*. American Scientific Publishers, pp. 51–69.

Stewart, D. Y., Barker, G. J., Benninston, A. J. *et al.* (2010). Modelling electroluminescence in liquid argon, *J. Instrum.* 5: P10005.

Suzuki, S. and Hitachi, A. (2011). Applications of Rare Gas Liquids to Radiation Detectors, in Y. Hatano, Y. Katsumura, and A. Mozumder (eds.), *Charged Particle and Photon Interactions with Matter: Recent Advances, Applications and Interfaces*. CRC Press, pp. 879–922.

Szydagis, M., Barry, N., Kazkaz, K. *et al.* (2011). NEST: a comprehensive model for scintillation yield in liquid xenon, *J. Instrum.* 6: P10002.

Szydagis, M., Fyhrie, A., Thorngren, D., and Tripathi, M. (2013). Enhancement of NEST capabilities for simulating low-energy recoils in liquid xenon, *J. Instrum.* 8: C10003.

Tachiya, M. (1988). Breakdown of the Onsager theory of geminate recombination, *J. Chem. Phys.* 89: 6929–6935.

Takahashi, T., Konno, S., Hamada, T., Miyajima, M., Kubota, S., Nakamoto, A. *et al.* (1975). Average energy expended per ion pair in liquid xenon, *Phys. Rev. A* 12: 1771–1775.

Tan, A., Xiao, X., Cui, X. *et al.* (2016a). Dark matter search results from the commissioning run of PandaX-II, *Phys. Rev. D* 93: 122009.

Tan, A., Xiao, X., Cui, X. *et al.* (2016b). Dark matter results from first 98.7 days of data from the PandaX-II experiment, *Phys. Rev. Lett.* 117: 121303.

Tanaka, M., Doke, T., Hitachi, A., Kato, T., Kikuchi, J., Masuda, K. *et al.* (2001). LET dependence of scintillation yields in liquid xenon, *Nucl. Instrum. Meth. Phys. Res. A* 457: 454–464.

Tanaka, N., Aoyama, K., Kimura, M. *et al.* (2020). Studies on liquid argon S1 and S2 properties for low mass WIMP search experiments, *J. Phys. Conf. Series*, 1468, 012052.

Thomas, J. and Imel, D. A. (1987). Recombination of electron-ion pairs in liquid argon and liquid xenon, *Phys. Rev. A* 36: 614–616.

Thomas, J., Imel, D.A., and Biller, S. (1988), Statistics of charge collection in liquid argon and liquid xenon, *Phys. Rev. A* 38: 5793–5800.

Tilinin, I.S. (1995), Quasiclassical expression for inelastic energy losses in atomic particle collisions below the Bohr velocity, *Phys. Rev. A* 51: 3058–3065.

Tvrznikova, L., Bernard, E.P., Kravitz, S. *et al.* (2019). Direct comparison of high voltage breakdown measurements in liquid argon and liquid xenon, *J. Instrum.* 14: P12018.

Vignoli, C. (2014). The ICARUS T600 liquid argon purification system, presented at the *25th International Cryogenic Engineering Conference and the International Cryogenic Materials Conference in 2014* (ICEC 25 — ICMC 2014), July 7–11, Enschede, The Netherlands, 2014.

Vignoli, C. on behalf of the ICARUS Collaboration (2015). The ICARUS T600 liquid argon purification system, *Physics Procedia* 67: 796–801.

Voronova, T. Ya., Kirsanov, M. A., Kruglov, A. A., Obodovski, I. M., Pokachalov, S. G., Shilov, V. A., and Kristich, E. B. (1989). Ionization yield from electron tracks in liquid xenon, *Sov. Phys. Tech. Phys.* 34: 825–827.

Walters, A. J. and Mitchell, L. W. (2003). Mobility and lifetime of ^{208}Tl ions in liquid xenon, *J. Phys. D: Appl. Phys.*, 36, 1323–1331.

Wamba, K., Hall, C., Breidenbach, M. *et al.* (2005). Mobility of thorium ions in liquid xenon, *Nucl. Instr. Meth. Phys. Res. A* 555: 205–210.

Wang, H. (1998). Xenon as a detector for dark matter search, *Phys. Rep.* 307: 263–267.

Xiao, M., Xiao, X., Zhao, L. *et al.* (2014). First dark matter search results from the PandaX-I experiment, *Sci. China Phys. Mech. Astron.* 57: 2024.

Yamashita, M. (2005). R&D results and status of the XENON dark matter experiment. In *Applications of Rare Gas Xenon to Science And Technology* (XeSAT2005), Waseda University, Tokyo, Japan, 8–10 March 2005.

Yarnell, J. L., Katz, M. J., Wenzel, R. G., and Koenig, S. H. (1973). Structure factor and radial distribution function for liquid argon at 85 K, *Phys. Rev. A* 7: 2130–2144.

Ye, T., Giboni, K. L., and Ji, X. (2014). Initial evaluation of proportional scintillation in liquid Xenon for direct dark matter detection, *J. Instrum* 9: P12007.

Zhang, H., Abdukerim, A., Chen, X. *et al.* (2019). Dark matter direct search sensitivity of the PandaX-4T experiment, *Sci. China Phys. Mech. Astron* 62(3): 31011.

Ziegler, J. F. (1980). *Handbook of Stopping Cross Sections for Energetic Ions in All Elements*, vol. 2 of Series "Stopping and Ranges of Ions in Matter". Pergamon Press, New York.

Ziegler, J. F., Biersack, J. P., and Littmark, U. (1985). *The Stopping and Range of Ions in Solids*, vol. 1. Pergamon Press, New York.

Ziegler, J. F. (1999). Stopping of energetic light ions in elemental matter, *J. Appl. Phys.* 85: 1249–1272.

Ziegler, J. F., Biersack, J. P., and Ziegler, M. D. (2008). SRIM — The Stopping Power and Range of Ions in Matter, SRIM Co, 2008, available at http://www.srim.org.

Ziegler, J. F., Ziegler, M. D., and Biersac, J., P. (2010). SRIM — The stopping and range of ions in matter (2010). *Nucl. Instrum. Meth. Phys. Res. B* 268: 1818–1823.

Zworykin, V. K. and Ramberg, E. G. (1949). *Photoelectricity and Its Application*, John Wiley & Sons, New York, 494 p.

Index

www.ingramcontent.com/pod-product-compliance
Lightning Source LLC
Chambersburg PA
CBHW050538190326
41458CB00007B/1829